CONVEXITY AND GRAPH THEORY

annals of discrete mathematics

General Editor

Peter L. HAMMER, University of Waterloo, Ontario, Canada

Advisory Editors

NORTH-HOLLAND — AMSTERDAM • NEW YORK • OXFORD

NORTH-HOLLAND
MATHEMATICS STUDIES 87

Annals of Discrete Mathematics (20)

General Editor: Peter L. Hammer

University of Waterloo, Ontario, Canada

CONVEXITY AND GRAPH THEORY

Proceedings of the Conference
on Convexity and Graph Theory
Israel
March 1981

Edited by:
M. ROSENFELD, Ben Gurion University, Israel
J. ZAKS, University of Haifa, Israel

1984

NORTH-HOLLAND — AMSTERDAM ● NEW YORK ● OXFORD

ISBN: *0 444 86571 3*

Publishers
ELSEVIER SCIENCE PUBLISHERS B.V.
P.O. BOX 1991
1000 BZ AMSTERDAM
THE NETHERLANDS

Sole distributors for the U.S.A. and Canada
ELSEVIER SCIENCE PUBLISHING COMPANY, INC.
52 VANDERBILT AVENUE,
NEW YORK, N.Y. 10017

Library of Congress Cataloging in Publication Data

Conference on Convexity and Graph Theory (1981 : Israel)
Convexity and graph theory.

(North-Holland mathematics studies ; 87) (Annals of discrete mathematics ; 20)
Includes index.
1. Graph theory--Congresses. 2. Convex domains--Congresses. I. Rosenfeld, M. II. Zaks, J. III. Title. IV. Series. V. Series: Annals of discrete mathematics ; 20.
QA166.C62 1981 511'.5 83-23666
ISBN 0-444-86571-3 (U.S.)

PRINTED IN THE NETHERLANDS

PREFACE

These are the proceedings of the conference on "Convexity and Graph Theory" which was held in Israel on 16–20 March 1981. In all, 69 participants from 13 different countries took part in the conference.

The conference started well above the clouds on the top of a tower in the northern part of Israel and ended up enjoying a variety of weather conditions in the Negev in the south.

Fifty-nine talks were given by the 69 participants. We regret that two of those who registered and sent us their abstracts, and who intended to participate, could not do so; they are A. Kelmans and M. Lomonosov, both from Moscow, U.S.S.R. A letter of sympathy, expressing the hope to meet them in future meetings, was signed by many participants and sent to them. They have received the letter and wish to thank all of those who signed it.

The papers appearing in this volume cover a spectrum of topics in Graph Theory, Geometry, Convexity and Combinatorics. All the papers have been carefully refereed. We express our sincere thanks to all authors for their cooperation, to the referees for their support and to the editors of the Annals of Discrete Mathematics for their help, patience and willingness to include this volume in the series. Our special thanks go to Professor Yair Censor at Haifa who volunteered to take carc of most of the correspondence and the registration before the conference, and who was the link connecting the authors of the proceedings to its editors. His help was essential for the success of the conference and for the proceedings.

On behalf of all the participants we wish to express our thanks to Professor Gadi Moran for making the many trips during the conference so enjoyable and educational for all of us; these include his guidance during the trip south via the Big Crater, guiding the tour through the ancient Nabatian city of Avdat, leading the descent of the cliffs of Ein Avdat and, finally, guiding the trip back to Jerusalem, via Ben Gurion's hut at Kibbutz Sde Boker and the Flour Tunnel in Peratzim Kenyon near the Dead Sea. We wish to thank Professor David Wolf, the Rector of Ben Gurion University in the Negev, for his encouragement and help, to Professor Amos Richmond, the head of the Desert Research Institute, who hosted the second half of the conference, to Dr. Shabtai Dover of the Desert Research Institute, and Dubit Ater of the Public Relations Office, B.G.U. for their most valuable organizational help. Our thanks to Mr. Dov Rosentzweig and Ms. L. Harman for their organizational help in Haifa and to all the others who made valuable contributions to the success of the conference.

M. Rosenfeld
J. Zaks

CONTENTS

LIST OF PARTICIPANTS

R.M. Adin, Technion, Haifa, Israel
J. Akiyama, Nippon Ika University, Kawasaki, Japan
N. Alon, Tel Aviv University, Israel
B. Alspach, Ben Gurion University, Beer Sheva, Israel and Simon Fraser University
D. Amar, Université de Paris-Sud, Orsay, France
I. Beck, Technion, Haifa, Israel
L.J. Berezina, University of Haifa, Israel
C. Berge, C.N.R.S., Paris, France
A. Berman, Technion, Haifa, Israel
U. Betke, University of Siegen, Fed. Rep. Germany
A. Blokhuis, Eindhoven University of Technology, The Netherlands
J. Bokowski, Ruhr University, Bochum, Fed. Rep. Germany
B. Bollobás, University of Cambridge, U.K.
T.C. Braun, Bgaziçi Üniv., Istanbul, Turkey, and S.F.U., Canada
Y. Censor, University of Haifa, Israel (Co-organizer)
D.I.A. Cohen, Rockefeller University, New York, U.S.A.
S. Cohen, N.Y., U.S.A.
H.S.M. Coxeter, University of Toronto, Canada
L. Danzer, University of Dortmund, Fed. Rep. Germany
E. Degreef, Vrije University, Brussels, Belgium
P. Duchet, University of Le Mans, France
P. Erdös, Hungarian Academy of Sciences, Budapest, Hungary and Technion, Israel
C. Fisher, University of Regina, Canada
I. Fournier, Université de Paris-Sud, Orsay, France
A.S. Fraenkel, Weizmann Institute, Rehovot, Israel
H. Fried, Bar-Ilan University, Ramat Gan, Israel
A. Gewirtz, CUNY Brooklyn C, New York, U.S.A.
C. Godsil, Montanuniversität, Leoben, Austria
M. Goldberg, Technion, Haifa, Israel
M.C. Golumbic, Bell Laboratories, U.S.A.
P.R. Goodey, Royal Holloway College, Egham, Surrey, U.K.
P. Gritzman, University of Siegen, Fed. Rep. Germany
R. Häggkvist, Inst. Mittag-Leffler, Sweden
F. Harary, University of Michigan, Ann. Arbor, U.S.A.
K. Heinrich, Ben Gurion University, Beer Sheva, Israel, and Simon Fraser University, Canada

P. Hell, Rutgers University and Simon Fraser University, Canada
F. Hering, University of Dortmund, Fed. Rep. Germany
W. Imrich, Montanuniversität, Leoben, Austria
A. Israeli, Hebrew University, Jerusalem, Israel
G. Kalai, Hebrew University, Jerusalem, Israel
M. Kallay, Hebrew University, Jerusalem, Israel
A. Kelmans, Profsouznaya Str. 130, K. 3, KV. 33, 117321 Moscow, U.S.S.R.
A. Kotzig, University of Montreal, Quebec, Canada
Y. Kupitz, Hebrew University, Jerusalem, Israel
D.G. Larman, University College London, U.K.
C.W. Lee, IBM Research Center, Yorktown Heights, and University of Kentucky, U.S.A.
M.V. Lomonosov, Chertanovskaya 34–1–286, 113525 Moscow, U.S.S.R.
E. Mendelsohn, University of Toronto, Canada
G. Moran, University of Haifa, Israel
P.J. Owens, University of Surrey, U.K.
Y. Perl, Weizmann Institute, Rehovot, Israel
M.D. Plummer, University of Bonn and Vanderbilt University, U.S.A.
M. Rosenfeld, Ben Gurion University, Beer Sheva, Israel (organizer)
U.G. Rothblum, Yale University, U.S.A.
H. Schneider, University of Wisconsin, Madison, U.S.A.
S. Schreiber, Bar-Ilan University, Ramat Gan, Israel
C. Schulz, Fernuniversität Hagen, Fed. Rep. Germany
E. Shamir, Hebrew University, Jerusalem, Israel
A. Shapiro, Ben Gurion University, Beer Sheva, Israel
I. Shemer, Hebrew University, Jerusalem, Israel
G. Sierksma, University of Groningen, The Netherlands
D. Sotteau, Orsay, France
J. Srivastava, Colorado St. University and University of Columbia at Berkeley, U.S.A.
J.M. Turgeon, University of Montreal, Quebec, Canada
J.M. Wills, University of Siegen, Fed. Rep. Germany
Y. Yomdin, Ben Gurion University, Beer Sheva, Israel
J. Zaks, Texas A&M University and Haifa, Israel (Organizer).
C.M. Zamfirescu, CUNY Hunter College, New York, U.S.A.
T. Zamfirescu, University of Dortmund, Fed. Rep. Germany

Annals of Discrete Mathematics 20 (1984) 1–11
North-Holland

ON EXTREME POSITIVE OPERATORS BETWEEN POLYHEDRAL CONES

Ron M. ADIN*

Department of Mathematics, Technion — Israel Institute of Technology, Haifa, Israel

Fiedler and Pták called a cone *minimal* if it is n-dimensional and has $n+1$ extreme rays. We call a cone *almost-minimal* if it is n-dimensional and has $n+2$ extreme rays. Duality properties stemming from the use of Gale pairs lead to a general technique for identifying the extreme cone-preserving (positive) operators between polyhedral cones. This technique is most effective for cones with dimension not much smaller than the number of their extreme rays. In particular, the Fiedler–Pták characterization of extreme positive operators between minimal cones is extended to the following cases: (i) operators from a minimal cone to an arbitrary polyhedral cone, and (ii) operators from an almost-minimal cone to a minimal cone.

1. Terminology and notations

All scalars, vectors, and matrices mentioned below are real. The field of real numbers, the vector-space of real n-tuples, and the vector space of $m \times n$ real matriccs will be denoted by $\mathbb{R}$, $\mathbb{R}^n$, and $M_{m\times n}$, respectively.

Occasional use of *empty matrices* (elements of $M_{m\times 0}$ or $M_{0\times n}$) follows the conventions of [10]. The terms *vector* and *n-tuple*, as well as *linear operator* and *matrix*, are used synonymously.

The transpose of the matrix A is denoted A^{T}.

Throughout the paper, standard terminology concerning convex cones is used (cf., for instance, [3]).

All cones considered are full, pointed, polyhedral (and thus closed) convex cones. The interior of a cone K is denoted int K, while ext K is the union of the extreme rays of K (elements of ext K are *extremals* of K). The cone of positive (cone-preserving) maps from a cone K_1 to cone K_2 is denoted $\Pi(K_1, K_2)$.

2. Gale pairs and bipositive operators

Motivated by the concepts of "Gale diagram" and "Gale transformation" [5, 10, 2] we introduce

* Due to space limitations this paper gives the main definitions and results of the work. For proofs and further details the reader is referred to [1].

Research supported in part by the Technion's Vice President for Research Grant no. 100–473.

Definition 1. Let X and Z be two matrices, both having n columns. The pair (X, Z) will be called a *Gale pair* if:

(i) $XZ^{\mathrm{T}} = 0$, and

(ii) $\operatorname{rank} X + \operatorname{rank} Z = n$.

(Equivalently, if the row-spaces of X and of Z are orthogonal complements.)

The roles of X and of Z in this definition are symmetric — but this is no longer so in the use made of Gale pairs in the special context of

Definition 2. Let (X, Z) be a Gale pair, where $X \in M_{k\times n}$ and $Z \in M_{l\times n}$. Let K be a (full and pointed) polyhedral cone in $\mathbb{R}^k$. The pair (X, Z) *represents* K if:

(i) the columns of X are representatives of the extreme rays of K (i.e. they are nonzero extremals of K, and each extreme ray contains exactly one of them), and

(ii) $\operatorname{rank} Z = l$.

Note that $\operatorname{rank} X = k$ follows from the fullness of K.

Definition 3. Let (X_1, Z_1) and (X_2, Z_2) be Gale pairs. Assume that X_i and Z_i have n_i columns $(i = 1, 2)$. A *bipositive operator from* (X_1, Z_1) *to* (X_2, Z_2) is an $n_2 \times n_1$ matrix B, satisfying:

(i) $B \geqslant 0$ (elementwise), and

(ii) $X_2 B Z_1^{\mathrm{T}} = 0$.

Notation. Let $\mathcal{BP}[(X_1, Z_1), (X_2, Z_2)]$ denote the set of bipositive operators from (X_1, Z_1) to (X_2, Z_2). If no ambiguity results, this will be abbreviated to $\mathcal{BP}$.

Note that $\mathcal{BP}$ is a pointed — though not necessarily full in $M_{n_2\times n_1}$ — polyhedral cone. Also, $\mathcal{BP}$ is representation-invariant in the sense that if (X_i, Z_i) represent cones K_i $(i = 1, 2)$, then they lead to the same cone $\mathcal{BP}$ as any other pair of Gale-pair representations of K_1 and K_2.

Definition 4. Let (X_1, Z_1) and (X_2, Z_2) be Gale pairs, representing cones K_1 and K_2, and let $\mathcal{BP}$ be as above. Denote

$$V = \{B \in M_{n_2\times n_1} \mid X_2 B Z_1^{\mathrm{T}} = 0\},$$

$$V_x = M_{k_2\times k_1},$$

$$V_z = M_{l_1\times l_2}.$$

Define now:

$$\pi_x : V \to V_x \text{ by: } \pi_x(B) X_1 = X_2 B,$$

and

$$\pi_z : V \to V_z \text{ by: } \pi_z(B)Z_2 = Z_1 B^T.$$

$\pi_x(\mathcal{BP})$ is called the *cone of X-positive operators* (from X_1 to X_2), while $\pi_z(\mathcal{BP})$ is the *cone of Z-positive operators* (from Z_2 to Z_1).

Notice that the properties of cone-representing Gale pairs imply that π_x and π_z are well-defined.

Lemma 5. *Let K_1 and K_2 be polyhedral cones represented by the Gale pairs (X_1, Z_1) and (X_2, Z_2), respectively. Then an operator is X-positive from X_1 to X_2 if and only if it is a positive operator (in the cone-preserving sense) from K_1 to K_2.*

In other words: π_x maps $\mathcal{BP}$ onto the cone of positive operators $\Pi(K_1, K_2)$.

Theorem 6. *Let K_1 and K_2 be polyhedral cones represented by Gale pairs (X_1, Z_1) and (X_2, Z_2), respectively, and let $\mathcal{BP}$ be the corresponding cone of bipositive operators.*

(i) *If F_x is a face of $\pi_x(\mathcal{BP})$, then $F = \pi_x^{-1}(F_x) \cap \mathcal{BP}$ is a face of $\mathcal{BP}$. Moreover, $\pi_x(F) = F_x$.*

(ii) *Let F be a face of $\mathcal{BP}$.*
Then there exists a face F_x of $\pi_x(\mathcal{BP})$ such that $F = \pi_x^{-1}(F_x) \cup \mathcal{BP}$ if and only if $F = (F + \ker \pi_x) \cap \mathcal{BP}$.

(iii) $(\ker \pi_x) \cap \mathcal{BP} = \{0\}$.

Corollary 7. *Let K_1 and K_2 be polyhedral cones and let $\mathcal{BP}$ be a corresponding cone of bipositive operators (defined by suitable Gale-pair representations). Then:*

$$\operatorname{ext} \Pi(K_1, K_2) \subseteq \pi_x(\operatorname{ext} \mathcal{BP}).$$

Corollary 7 enables us to find the extremals of $\Pi(K_1, K_2)$: first find the extremals of (the relatively simply defined) $\mathcal{BP}$, or even merely a small finite set containing representatives of all the extreme rays. Then project them by means of π_x, and pick out those projections that are not non-negative combinations of others.

This result will be used in the next section to find ext $\Pi(K_1, K_2)$ in case K_1 has relatively few extremals.

3. Extreme bipositive and positive operators

In this main section use is made of the concepts and methods developed in the previous section, in order to describe the extreme rays of $\Pi(K_1, K_2)$ — briefly in the case of general polyhedral K_1 and K_2 and in a concrete constructive way for the relatively simple special cases. These results generalize and extend the one by Fiedler and Pták [4].

This structure of $\Pi(K_1, K_2)$ is determined, as explained at the end of the previous section, through the structure of a corresponding cone $\mathcal{BP}$ of bipositive operators. This, on the other hand, is determined through the Gale-pair duality, exploiting the assumed low rank of Z (that is, the small difference between the number of extreme rays of a cone and its dimension). Examples, illustrating the above-mentioned results, are given in the next section.

Let (X_1, Z_1) and (X_2, Z_2) be Gale pairs representing the polyhedral cones K_1 and K_2, respectively; $X_i \in M_{k_i \times n_i}$ and $Z_i \in M_{l_i \times n_i}$ $(i = 1, 2)$. Again, let

$$\mathcal{BP} \equiv \mathcal{BP}[(X_1, Z_1); (X_2, Z_2)] = \{B \in M_{n_2 \times n_1} \mid B \geqslant 0,\ X_2 B Z_1^{\mathrm{T}} = 0\}.$$

From the definition and properties of π_z it follows that

$$\mathcal{BP} = \{B \in M_{n_2 \times n_1} \mid B \geqslant 0,\ (\exists C \in M_{l_1 \times l_2})(Z_1 B^{\mathrm{T}} = C Z_2)\}.$$

The following two theorems summarize the information available, in general, about the extreme rays of $\mathcal{BP}$. The first one deals with the case $C = \pi_z(B) = 0$, characterizing the extremals completely (for general polyhedral cones). The second theorem deals with $C \neq 0$.

Theorem 8.

(i) *An $n_2 \times n_1$ matrix B is a nonzero extremal of $\mathcal{BP}$ with $Z_1 B^{\mathrm{T}} = 0$, if and only if:*

(a) *All the rows of B — with one exception — are zero.*

(b) *There exists a vanishing convex combination of the columns of Z_1, the coefficients of which are — up to a constant positive factor — the entries in the unique nonzero row of B.*

(c) *In the family of sets of indices of the nonzero coefficients in a vanishing convex combination of the columns of Z_1, ordered by inclusion, the set corresponding to our B (as above) is* minimal.

(ii) *Every matrix B satisfying* (a), (b) *and* (c) *of* (i) *induces (through π_x) an extremal of $\Pi(K_1, K_2)$. The extremals thus obtained are exactly the rank* 1 *extremals of $\Pi(K_1, K_2)$.*

Remark. Let $N_1 = \{1, \ldots, n_1\}$ and $N_2 = \{1, \ldots, n_2\}$. Let z_j^1 $(j = 1, \ldots, n_1)$ denote the columns of Z_1, and similarly for X_1 and X_2. Let

$$\mathcal{F}_1 = \{J \subseteq N_1 \mid 0 \in \operatorname{conv}(\{z_j^1 \mid j \in J\})\}$$

be ordered by inclusion, and let $\mathcal{M}_1$ be the set of minimal elements in $\mathcal{F}_1$. Note that by Carathéodory's Theorem (cf., for example, [9]) every member of $\mathcal{M}_1$ has at most l_1+1 elements. For every $J\in\mathcal{M}_1$ assume that $\sum_{j\in N_1}\alpha_j(J)z_j^1=0$, where $\alpha_j(J)\geqslant 0$ for all $j\in N_1$, and $\alpha_j(J)=0\Leftrightarrow j\notin J$. Let $i_0\in N_2$. Then Theorem 8 gives us the following extremals of $\mathcal{BP}$:

$$B(i_0\,|\,J)\colon b_{ij}=\begin{cases}\alpha_j(J), & i=i_0\\ 0, & \text{otherwise}\end{cases}\qquad (i_0\in N_2,\ J\in\mathcal{M}_1),$$

and projected to $\Pi(K_1,K_2)$:

$$A(i_0\,|\,J)\colon Ax_j^1=\alpha_j(J)x_{i_0}^2\qquad (i_0\in N_2,\ J\in\mathcal{M}_1).$$

Notice that the elements of $\mathcal{M}_1$ are exactly the sets of indices of minimal cofaces of $(z_j^1)_{j=1}^{N_1}$. This means that $\mathcal{M}_1$ is composed of the N_1-complements of the index-sets of extremals of the *maximal faces* of K_1. (For the definitions and theorems underlying this statement the reader is referred to [5], [10] and [2].) Thus, our result coincides with the well-known description of rank 1 extremals of $\Pi(K_1,K_2)$. ($x_2y_1^{\mathrm{T}}$, where $x_2\in\operatorname{ext}K_2$ and $y_1\in\operatorname{ext}K_1^*$, K_1^* being the cone dual to K_1; cf. [6].)

Theorem 9. *A matrix* $B=(b_{ij})\in\mathcal{BP}$, *with* $\pi_z(B)\neq 0$ *is an extremal of* $\mathcal{BP}$ *if and only if*:

(a) *The columns* z_j^1 *corresponding to the* positive *entries* b_{ij} *in any specific row* i *of* B *are linearly independent. In particular, every row of* B *contains at most* l_1 *nonzero entries.*

(b) $\dim(\bigcap_{i=1}^{n_2}\{D\in M_{l_1\times l_2}\,|\,Dz_i^2\in\operatorname{span}\{z_j^1\,|\,b_{ij}>0\}\})=1$.

The proof of this important theorem reads as follows.

(i) Suppose that B has a row i such that the elements of $\{z_j^1\,|\,b_{ij}>0\}$ are linearly dependent: $\sum_{j=1}^{n_1}\alpha_j z_j^1=0$, where $\alpha_j=0$ if $b_{ij}=0$. Let $\tilde{B}$ be an $n_2\times n_1$ matrix with zero entries, except for row i which is equal to $(\alpha_1,\alpha_2,\ldots,\alpha_{n_1})$. Then $Z_1\tilde{B}^{\mathrm{T}}=0$, and $B\pm\varepsilon\tilde{B}\in\mathcal{BP}$ for a small enough $\varepsilon>0$. Since $\tilde{B}$ is not a multiple of $B(\pi_z(B)\neq 0!)$, B is not extremal. Hence (a) is proven.

Now let W denote the intersection in (b). Since $0\neq C=\pi_z(B)\in W$, certainly $\dim W\geqslant 1$. If equality does not hold, there exists $\tilde{C}\in W$ which is linearly independent of C. There exists $\tilde{B}=(\tilde{b}_{ij})\in V$ (cf. Definition 4) such that $\tilde{C}=\pi_z(\tilde{B})$ and $\tilde{b}_{ij}=0$ whenever $b_{ij}=0$. Hence, $B\pm\varepsilon\tilde{B}\in\mathcal{BP}$ for $\varepsilon>0$ small enough, and since $\tilde{B}$ is not a multiple of B this contradicts the assumption that B is extremal. This proves (b) and completes the proof of (i).

(ii) Let $B\in\mathcal{BP}$ satisfy (a), (b), and $C=\pi_z(B)\neq 0$. Assume that $B=B_1+B_2$ for $B_1,B_2\in\mathcal{BP}$. Then

$$\pi_z(B),\pi_z(B_1),\pi_z(B_2)\in W\equiv\bigcap_{i=1}^{n_2}\{D\in M_{l_1\times l_2}\,|\,Dz_i^2\in\operatorname{span}(\{z_j^1\,|\,b_{ij}>0\})\},$$

and since $\dim W = 1$ we conclude that $\pi_z(B_1)$ and $\pi_z(B_2)$ are multiples of $\pi_z(B)$: $\pi_z(B_i) = \varepsilon_i \pi_z(B)$ $(i = 1,2)$. Thus, $\pi_z(B_i - \varepsilon_i B) = 0$, and since the entries of B_i are zero whenever the corresponding entry of B is zero, condition (a) implies that $B_i = \varepsilon_i B$. Therefore B is extremal. □

Let us now examine the consequences of the above theorems in some special cases: $l_1 = 0, 1, 2$.

We may choose the Gale pair (X_1, Z_1) representing K_1 so that columns of Z_1 will be unit-vectors or zero.

(i) $l_1 = 0$. The cone K_1 is *simplicial* (and decomposable). Z_1 and C are empty matrices, and therefore

$$\mathcal{BP} = \{B \in M_{n_2 \times n_1} \mid B \geqslant 0\}$$

and the extreme rays are represented by

$$B(i_0 \mid j_0):\ b_{ij} = \begin{cases} 1, & i = i_0, \quad j = j_0, \\ 0, & \text{otherwise.} \end{cases}$$

A projection to $\Pi(K_1, K_2)$ gives:

$$A(i_0 \mid j_0):\ Ax_j^1 = \begin{cases} x_{i_0}^2, & j = j_0, \\ 0, & \text{otherwise,} \end{cases}$$

and all these matrices (corresponding to $C = 0$) have, obviously, rank equal to 1.

(ii) $l_1 = 1$. The cone K_1 is *minimal.* The columns of Z_1 are scalars, which (according to a remark above) may be chosen to belong to $\{0, 1, -1\}$. C, likewise, has one row.

(a) $C = 0$: The unique nonzero row of B has, according to Theorem 8, either one positive entry (corresponding to $z_j^1 = 0$) or two positive entries (corresponding to $z_{j_1}^1 = 1$ and $z_{j_2}^1 = -1$). This yields:

$$B^{(1)}(i_0 \mid j_0):\ b_{ij} = \begin{cases} 1, & i = i_0, \quad j = j_0 \\ 0, & \text{otherwise} \end{cases} \qquad (\text{for } i_0 \in N_2,\ j_0 \in N_1,\ z_{j_0}^1 = 0),$$

$$B^{(2)}(i_0 \mid j_1, j_2):\ b_{ij} = \begin{cases} 1, & i = i_0, \quad j \in \{j_1, j_2\} \\ 0, & \text{otherwise} \end{cases} \qquad (\text{for } i_0 \in N_2,\ j_1, j_2 \in N_1,\ z_{j_1}^1 = -z_{j_2}^1 = 1).$$

Use of π_x leads to:

$$A^{(1)}(i_0 \mid j_0)\colon\ Ax_j^1 = \begin{cases} x_{i_0}^2, & j = j_0 \\ 0, & \text{otherwise} \end{cases} \qquad (\text{for } i_0 \in N_2,\ j_0 \in N_1,\ z_{j_0}^1 = 0),$$

$$A^{(2)}(i_0 \mid j_1, j_2)\colon\ Ax_j^1 = \begin{cases} x_{i_0}^2, & j \in \{j_1, j_2\} \\ 0, & \text{otherwise} \end{cases} \qquad (\text{for } i_0 \in N_2,\ j_1, j_2 \in N_1,$$

$$z_{j_1}^1 = -z_{j_2}^1 = 1).$$

Note that $z_{j_0}^1 = 0$ if and only if the ray generated by $x_{j_0}^1$ is a direct summand in a decomposition of K_1.

(b) $C \neq 0$: This case is possible, of course, only for $l_2 > 0$. According to Theorem 9, each row of B contains at most one positive entry, corresponding to a nonzero z_j^1. Let

$$\mathscr{F} = \{I \subseteq N_2 \mid \dim \operatorname{span}\{z_i^2 \mid i \in I\} = l_2 - 1\}$$

be ordered by inclusion, and let $\mathscr{M}$ be the set of maximal elements in $\mathscr{F}$. For every $I \in \mathscr{M}$ there is a solution $C = C(I) \in M_{1 \times l_2}$, unique up to a constant factor, to the homogeneous system of equations:

$$Cz_i^2 = 0 \qquad (i \in I),$$

and this solution also satisfies:

$$Cz_i^2 \neq 0 \qquad (i \in I).$$

For such $C(I)$, and for every $i \notin I$, choose $j_i \in N_1$ such that

$$z_{j_i}^1 = \operatorname{sign}(C(I)z_i^2)$$

and let

$$B_+^{(3)}(I \mid \{j_i \mid i \notin I\})\colon\ b_{ij} = \begin{cases} |C(I)z_i^2|, & i \notin I,\ j = j_i \\ 0, & \text{otherwise} \end{cases}$$

$(z_{j_i}^1 = \operatorname{sign}(C(I)z_i^2)$ for $i \notin I)$.

Similarly, choosing for every $i \notin I$ a $j_i \in N_1$ such that

$$z_{j_i}^1 = -\operatorname{sign}(C(I)z_i^2)$$

leads to

$$B_-^{(3)}(I \mid \{j_i \mid i \notin I\})\colon\ b_{ij} = \begin{cases} |C(I)z_i^2|, & i \notin I,\ j = j_i \\ 0, & \text{otherwise} \end{cases}$$

$(z_{j_i}^1 = -\operatorname{sign}(C(I)z_i^2)$ for $i \notin I)$.

(This corresponds to taking the solution $-C(I)$ instead of $C(I)$.)

Projecting through π_x we get:

$$A_+^{(3)}(I \mid \{j_i \mid i \notin I\}): \ Ax_j^1 = \sum_{\substack{i \notin I \\ j_i = j}} |C(I) z_i^2| x_i^2$$

$$(I \in \mathcal{M} \text{ and } z_{j_i}^1 = \operatorname{sign}(C(I) z_i^2) \text{ for } i \notin I),$$

$$A_-^{(3)}(I \mid \{j_i \mid i \notin I\}): \ Ax_j^1 = \sum_{\substack{i \notin I \\ j_i = j}} |C(I) z_i^2| x_i^2$$

$$(I \in \mathcal{M} \text{ and } z_{j_i}^1 = -\operatorname{sign}(C(I) z_i^2) \text{ for } i \notin I).$$

Although $B_\pm^{(3)}$ are extremals of $\mathcal{BP}$ (according to Theorem 9), the $A_\pm^{(3)}$ obtained above are not always extremals of $\Pi(K_1, K_2)$.

Simple criteria for the extremality of $A_\pm^{(3)}$ may easily be constructed, which can be used in each specific case. For example, let $l_2 = 1$. Obviously $\mathcal{M} = \{I_0\}$, where $I_0 = \{i \mid z_i^2 = 0\}$. $C(I_0)$ is a scalar, which may be taken to be 1. The extremality problems of $A_\pm^{(3)}(I_0 \mid \{j_i \mid i \notin I_0\})$ are solved upon requiring each of the sets

$$\{j \mid \exists i \in N_2, \text{ such that } z_i^2 = 1 \text{ and } j_i = j\}$$

and

$$\{j \mid \exists i \in N_2, \text{ such that } z_i^2 = -1 \text{ and } j_i = j\}$$

to contain at least two elements.

Paying attention to the nonexistence of type $A^{(1)}$ for an indecomposable K_1, the types $A^{(2)}$ and $A_\pm^{(3)}$ (with the restriction) constitute the well-known result of Fiedler and Pták [4].

(iii) $l_1 = 2$, $l_2 = 1$. As mentioned in the abstract, in such a case we call K_1 *almost-minimal*. The case $C = 0$ is most clearly described in Theorem 8. The construction for $C \neq 0$ is the following.

Choose $j_0 \in N_1$ such that $z_{j_0}^1 \neq 0$; let $\mathcal{M}^+(j_0)$ be the set of minimal elements in the set:

$$\mathcal{F}^+(j_0) = \{J \subseteq N_1 \mid z_{j_0}^1 \in \operatorname{cone}(\{z_j^1 \mid j \in J\})\}$$

(ordered by inclusion), and let $\mathcal{M}^-(j_0)$ be the set of minimal elements in the (similarly ordered) set:

$$\mathcal{F}^-(j_0) = \{J \subseteq N_1 \mid -z_{j_0}^1 \in \operatorname{cone}(\{z_j^1 \mid j \in J\})\}.$$

Note that every member of $\mathcal{M}^\pm(j_0)$ has one or two elements, and also that $\{j_0\} \in \mathcal{F}^+(j_0)$.

Now choose, for every $i \in N_2$ with $z_i^2 \neq 0$:

$$J_i^+ \in \mathcal{M}^+(j_0), \quad \text{if } z_i^2 = 1,$$

and

$$J_i^- \in \mathcal{M}^-(j_0), \quad \text{if } z_i^2 = -1,$$

provided that $J_i^+ = \{j_0\}$ for at least one index i. (Here we assume that all elements of the row-vector z_i^2 belong to $\{1, -1, 0\}$.)

Every $J \in \mathcal{M}^+(j_0) \cup \mathcal{M}^-(j_0)$ defines a unique set of non-negative coefficients $\{\alpha_j(J) \mid j \in N_1\}$, such that

$$\sum_{j \in N_1} \alpha_j(J) z_j^1 = \begin{cases} z_{j_0}^1, & J \in \mathcal{M}^+(j_0), \\ -z_{j_0}^1, & J \in \mathcal{M}^-(j_0), \end{cases}$$

and $\alpha_j(J) = 0$ for $j \notin J$. Then:

$$B_+(j_0, \{J_i^+ \mid z_i^2 = 1\}, \{J_i^- \mid z_i^2 = -1\})\colon b_{ij} = \begin{cases} \alpha_j(J_i), & z_i^2 = \pm 1, \\ 0, & \text{otherwise.} \end{cases}$$

The definition of $B(j_0, \{J_i^- \mid z_i^2 = 1\}, \{J_i^+ \mid z_i^2 = -1\})$ is similar, choosing

$$J_i^- \in \mathcal{M}^-(j_0), \quad \text{if } z_i^2 = 1,$$

and

$$J_i^+ \in \mathcal{M}^+(j_0), \quad \text{if } z_i^2 = -1.$$

These are all the extremal operators in $\mathcal{BP}$ with $C \neq 0$.

Projecting to $\Pi(K_1, K_2)$:

$$A_+(j_0, \{J_i^+ \mid z_i^2 = 1\}, \{J_i^- \mid z_i^2 = -1\})\colon Ax_j^1 = \sum_{\substack{i \in N_2 \\ j \in J_i}} \alpha_j(J_i) x_i^2,$$

and similarly for A_-.

The extremality of $A_\pm$ should be checked in each particular case. Note that $B_\pm$ (and $A_\pm$) depend, in fact, not upon j_0 but merely upon $z_{j_0}^1$ (so that we may avoid duplicates if Z_1 has identical columns).

4. Examples

The two examples to be given below illustrate the cases $l_1 = 1$, $l_2 = 2$, and $l_1 = 2$, $l_2 = 1$ (in the notation of the preceding section). The cones dealt with are three-dimensional, with a regular polygon as a cross section. The general cone of this form, with a regular n-gonal cross section, is conveniently represented by the extremals:

$$x_t = \begin{vmatrix} \cos(2\pi t/n) \\ \sin(2\pi t/n) \\ 1 \end{vmatrix} \qquad (t = 1, 2, \ldots, n).$$

Let these vectors form the columns of a matrix X (in order of increasing t). A short calculation shows that a matrix Z, complementing X to a Gale pair, may be formed by taking as rows vectors $(a_1, \ldots, a_t)$ of one of the forms:

$$a_t = \cos\frac{2\pi jt}{n} \qquad (t = 1, \ldots, n), \tag{1}$$

or

$$a_t = \sin\frac{2\pi jt}{n} \qquad (t = 1, \ldots, n), \tag{2}$$

and a suitable choice is $2 \leqslant j \leqslant \lfloor n/2 \rfloor$ in (1) and $2 \leqslant j \leqslant \lfloor (n-1)/2 \rfloor$ in (2). (Here, for a real number x, $\lfloor x \rfloor$ denotes the largest integer not exceeding x.) Note that

$$\left(\left\lfloor \frac{n}{2} \right\rfloor - 1\right) + \left(\left\lfloor \frac{n-1}{2} \right\rfloor - 1\right) = n - 3.$$

The specific cones we examine are those corresponding to $n = 4$ (square cross section, minimal cone) and $n = 5$ (pentagonal cross section, almost minimal cone).

(i) Let K_1 have a square cross section, and K_2 a pentagonal cross section. Use of the results in the previous section shows that $\Pi(K_1, K_2)$ has 60 extreme rays: 20 of them are of rank 1, mapping K_1 onto an extreme ray of K_2; while the other 40 are nonsingular (rank 3), mapping K_1 onto the cone generated by four out of the five extreme rays of K_2.

(ii) Let K_1 have a pentagonal, and K_1 a square cross section. It turns out that $\Pi(K_1, K_2)$ again has 60 extreme rays, of which 20 are rank 1 (mapping K_1 onto an extreme ray of K_2). The other 40 are nonsingular (rank 3), mapping three adjacent extremals of K_1 onto three (adjacent) extremals of K_2, and mapping the other two extremals of K_1 into the two-dimensional faces containing the fourth extremal of K_2.

References

[1] R. Adin, Extreme positive operators on minimal and almost-minimal cones, Linear Algebra Appl. 44 (1982) 61–86.

[2] J. Bair and R. Fourneau, Etude Géometrique des Espaces Vectoriels, II (Polyèdres et Polytopes Convexes), Lecture Notes in Mathematics No. 802 (Springer-Verlag, Berlin, 1980).

[3] A. Berman and R.J. Plemmons, Nonnegative Matrices in the Mathematical Sciences (Academic Press, New York, 1979).

[4] M. Fiedler and V. Pták, The rank of extreme positive operators on polyhedral cones, Czech. Math. J. 28 (103) (1978) 45–55.

[5] B. Grünbaum, Convex Polytopes (John Wiley and Sons, London, 1967).

[6] E. Haynsworth, M. Fiedler and V. Pták, Extreme operators on polyhedral cones, Linear Algebra Appl. 13 (1976) 163–172.

[7] R. Loewy and H. Schneider, Indecomposable cones, Linear Algebra Appl. 11 (1975) 235–245.
[8] R. Loewy and H. Schneider, Positive operators on the n-dimensional ice-cream cone, J. Math. Anal. Appl. 49 (1975) 375–392.
[9] R.T. Rockafellar, Convex Analysis (Princeton University Press, Princeton, 1970).
[10] J. Stoer and C.J. Witzgall, Convexity and Optimization in Finite Dimensions, I (Springer-Verlag, Berlin, 1970).
[11] B.-S. Tam, Some results of polyhedral cones and simplicial cones, Linear and Multilinear Algebra 4 (1977) 281–284.

Received 15 April 1981; revised August 1981

Annals of Discrete Mathematics 20 (1984) 13–23
North-Holland

MISCELLANEOUS PROPERTIES OF EQUI-ECCENTRIC GRAPHS

Jin AKIYAMA and Kiyoshi ANDO
Department of Fundamental Sciences, Nippon Ika University, Kawasaki 211, Japan

David AVIS*
School of Computer Sciences, McGill University, Montreal, Canada H3A 2K6

The eccentricity $e(v)$ of a vertex v of a connected graph G is the number $\max_{u\in V(G)} d(u, v)$, where $d(u, v)$ is the distance in G between u and v. A graph is r-equi-eccentric if $e(v) = r$ for every vertex v of G. An r-equi-eccentric graph G of order p is r-minimum if it has the least number of edges of any r-equi-eccentric graph of order p. The following results are given in this paper.

(1) A few fundamental properties of equi-eccentric graphs.

(2) Several operations for producing equi-eccentric graphs.

(3) A characterization of 2-minimum graphs and a proof that the number of edges in a 2-minimum graph of order p is $2p-5$.

1. Introduction

We deal only with connected graphs throughout this paper. The *eccentricity* $e(v)$ of a vertex v of a connected graph G is the number $\max_{u\in V(G)} d(u, v)$, where $d(u, v)$ stands for the distance between u and v. A *central vertex* of a connected graph G is a vertex v with the property that:

$$e(v) = \min_{u\in V(G)} e(u).$$

The *radius* $r(G)$ of a connected graph G is the eccentricity of its central vertices. Thus, we have:

$$r(G) = \min_{u\in V(G)} e(u) = \min_{u} \max_{v} d(u, v).$$

The subgraph induced by the set of central vertices of G is called the *center* of G. Then, a graph G is *r-equi-eccentric* (or briefly, *r-equi*) if $e(v) = r(G) = r$ for every vertex of G, that is, a graph whose center is the graph itself. An r-equi-eccentric graph G is said to be *r-minimal* if $G - e$ is no longer r-equi for any edge e of G.

* Research supported by N.S.E.R.C. Grant no. A3013.

An r-equi-eccentric graph G of order p is *r-minimum* if G has the least number of edges among all r-equi-eccentric graphs of order p. We denote by $N(v)$ the *neighborhood* of a vertex v of G consisting of the vertices of G adjacent to v. The *closed neighborhood* $N[v]$ of v is defined by $N[v] = N(v) \cup \{v\}$.

All other definitions and notation used in this paper can be found in [1] or [3].

We first present a few fundamental properties of equi-eccentric graphs.

Proposition 1. *Every equi-eccentric graph G is a block.*

Proof. Every vertex of G is a central vertex by definition. It can be verified that the set of centers of a connected graph lie in a single block [1, p. 38]. □

Proposition 2. *Let G be r-equi of order p with maximum degree Δ. Then the following inequality holds:*

$$\Delta \leqslant p - 2(r-1).$$

Proof. Let v be an arbitrary vertex of G and u be a vertex with $d(u, v) = r$. By Proposition 1, G is a block. If G is K_2, then the theorem is immediate. On the other hand, if G is a block with at least three points, there is at least one cycle containing both u and v. By C we denote the smallest of these cycles. Then note that $|V(C)| \geqslant 2r$ since $d(u, v) = r$, and $|V(C) \cap N[v]| = 3$, since C is the smallest such cycle.

Thus, the following inequalities hold:

$$|V(G)| - |N[v]| \geqslant |V(C)| - |V(C) \cap N[v]| \geqslant 2r - 3.$$

Since

$$|V(G)| - |N[v]| = p - (\deg v + 1),$$

we have

$$\deg v \leqslant p - 2(r-1) \quad \text{for every vertex } v \text{ of } G,$$

completing the proof. □

2. Operations producing equi-eccentric graphs

In this section we exhibit several interesting operations that produce equi-eccentric graphs. We omit the proofs when they are immediate from the constructions.

2.1. Mycielski's operation [4]

Generalizing Mycielski's operation to an arbitrary graph $G = (V, E)$ with p vertices and q edges, we define its (Mycielski) successor $\hat{G} = (\hat{V}, \hat{E})$ as follows.

(i) For each $x \in V$, generate its *twin* x'. Call the set of twins V'.

(ii) Join x' to $N(x)$ in G, for every $x' \in V'$.

(iii) Create a new vertex z and join it to all twin vertices $x' \in V'$.

Example 1. Let G be the graph $K_4 - e$. Then G and its Mycielski successor $\hat{G}$ are shown in Fig. 1. Note that a graph with $p+1$ vertices is 2-equi if and only if it contains no $K(1, p)$ and $\max_{u,v \in V(G)} d(u, v) = 2$.

Using this observation it is straightforward to verify the following result.

Theorem 1. *If G is 2-equi then $\hat{G}$ is 2-equi.*

2.2. The join operation

The join of two graphs G_1 and G_2 is denoted $G_1 + G_2$ and consists of $G_1 \cup G_2$ and all lines joining vertices of G_1 and vertices of G_2 [3, p. 21].

Theorem 2. *If G is 2-equi, then $G + \bar{K}_n$ $(n \geq 2)$ is 3-equi (Fig. 2).*

2.3. Operations to produce minimal 2-equi-eccentric graphs

The *corona* $G_1 \circ G_2$ of two graphs G_1 and G_2 with orders p_1 and p_2, respectively, is defined as the graph obtained by taking one copy of G_1 and p_1 copies of G_2 and joining the ith vertex of G_1 to each vertex in the ith copy of G_2. In Fig. 3 we illustrate $C_4 \circ K_2$.

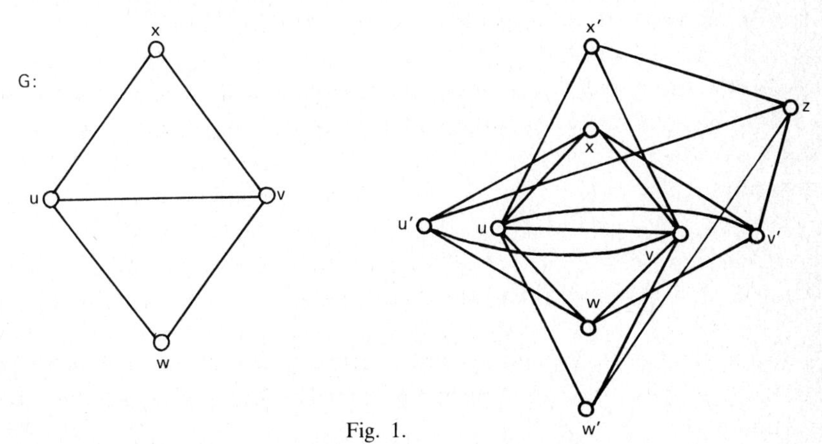

Fig. 1.

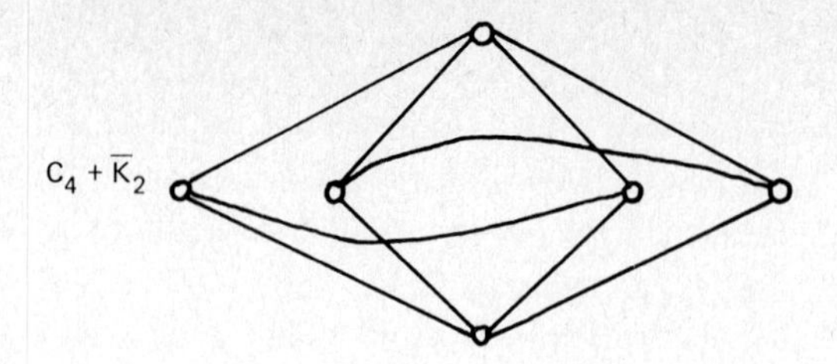

Fig. 2.

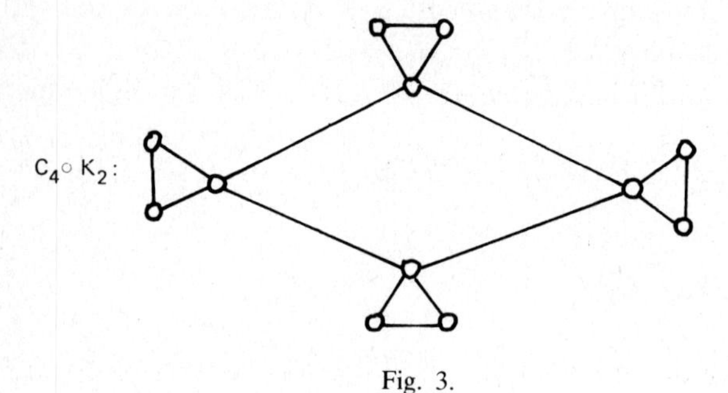

Fig. 3.

We define the graph $G_n = K_n \circ K_1 \oplus K_1$ $(n \geq 2)$ as the graph obtained from $K_n \circ K_1$ by adding a new vertex z and joining z to the vertices of degree 1 of $K_n \circ K_1$. In Fig. 4 we illustrate the graph $K_3 \circ K_1 \oplus K_1$.

Theorem 3. *The graph* $G_n = K_n \circ K_1 \oplus K_1$ $(n \geq 2)$ *obtained by the operation above is minimal 2-equi.*

2.4. *The cartesian product operation*

All of the three operations mentioned above produce 2-equi-eccentric graphs. We now present other operations to produce r-equi-eccentric graphs for an arbitrary integer $r \geq 2$.

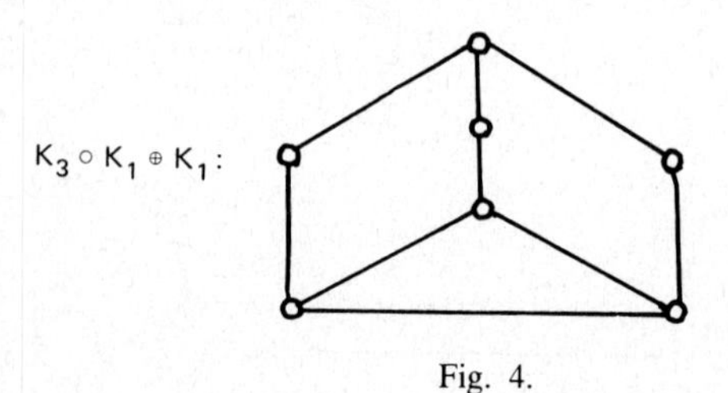

Fig. 4.

The *cartesian product* $G = G_1 \times G_2$ has $V(G) = V(G_1) \times V(G_2)$, and two vertices (u_1, u_2) and (v_1, v_2) of G are adjacent if and only if either

$$u_1 = v_1 \quad \text{and} \quad u_2v_2 \in E(G_2) \tag{1}$$

or

$$u_2 = v_2 \quad \text{and} \quad u_1v_1 \in E(G_1). \tag{2}$$

Theorem 4. *Let G_1 and G_2 be r_1, r_2-equi. Then their cartesian product $G = G_1 \times G_2$ is $(r_1 + r_2)$-equi.* □

As an immediate consequence of Theorem 4, we obtain the next result.

Corollary 1. *The r-cube $Q_r = (K_2)^r$ is r-equi.* □

2.5. The shift operation by P_n

Let F be any given graph. Define a graph $G_r(F)$ $(r \geq 2)$ consisting of F, a copy of P_{2r}, and all edges joining the end vertices of P_{2r} to the vertices of F. Fig. 5 illustrates the graph $G_2(\bar{K}_3)$.

Theorem 5. *Let F be an arbitrary graph. Then the graph $G_r(F)$ $(r \geq 2)$ is r-equi.* □

Corollary 2. *For any given nonempty graph F and an integer $r \geq 2$, there exists an r-equi eccentric graph containing F as an induced subgraph.* □

Note that the above corollary suggests that it is impossible to characterize *r*-equi-eccentric graphs in terms of forbidden subgraphs.

3. 2-equi-eccentric graphs

We denote the degree of a vertex v_i by d_i for the sake of convenience.

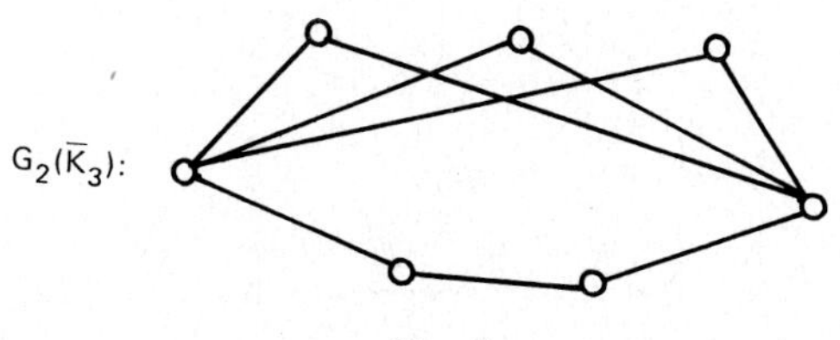

Fig. 5.

Proposition 3. *There are no 2-equi-eccentric graphs G with minimum degree $\delta = 3$ and $q \leqslant 2p - 5$, other than the Petersen graph.*

Proof. We show that G is isomorphic to the Petersen graph, if G is 2-equi with $\delta = 3$ and $q \leqslant 2p - 5$. Let v_1 be a vertex of degree 3. By v_2, v_3, and v_4 we denote vertices adjacent to v_1, and the $(p-4)$ remaining vertices in G by $v_5, v_6, \ldots, v_p$ (see Fig. 6). Each vertex v_i $(5 \leqslant i \leqslant p)$ is adjacent to at least one vertex of v_2, v_3 and v_4, otherwise $d(v_i, v_1) \geqslant 3$.

From this fact the inequality (3) follows:

$$d_2 + d_3 + d_4 \geqslant p - 1. \tag{3}$$

On the other hand, the reverse inequality of (3) follows from the fact that $\sum_{i=1}^{p} d_i = 2q \leqslant 4p - 10$ and $\sum_{i=5}^{p} d_i \geqslant 3(p-4)$ since $d_i \geqslant \delta = 3$. Therefore we have:

$$d_2 + d_3 + d_4 \leqslant p - 1. \tag{4}$$

Thus, we obtain the following equalities:

$$d_2 + d_3 + d_4 = p - 1, \tag{5}$$

$$d_i = 3 \qquad (5 \leqslant i \leqslant p). \tag{6}$$

From (5) it follows at once that

$$N(v_i) \cap N(v_j) = \{v_1\} \qquad (i \neq j, 2 \leqslant i, j \leqslant 4).$$

Applying the same argument for each vertex v_i $(5 \leqslant i \leqslant p)$ instead of v_1, we see that $d_i = 3$ for $i = 2, 3, 4$ and so G is cubic. Thus, $p = 10$, and denoting by V_i' the vertex set $N(v_i) - \{v_1\}$ for $i = 2, 3$ and 4, we have that $|V_i'| = 2$. Without loss of generality we may assume that $V_2' = \{v_5, v_6\}$, $V_3' = \{v_7, v_8\}$ and $V_4' = \{v_9, v_{10}\}$. On the basis of the fact that G is 2-equi, we see that the graph $G' = G - \{v_1, v_2, v_3, v_4\}$ is connected, which implies G' is a 6-cycle. Thus, it is easy to verify that the graph with the properties mentioned above is isomorphic to the Petersen graph illustrated in Fig. 7. □

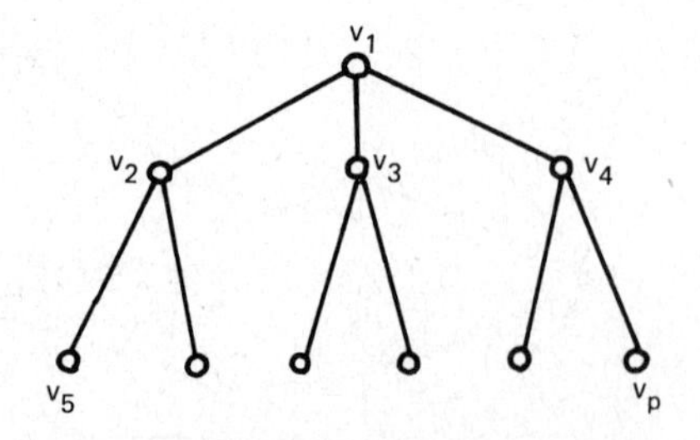

Fig. 6. A stage of the proof of Proposition 3.

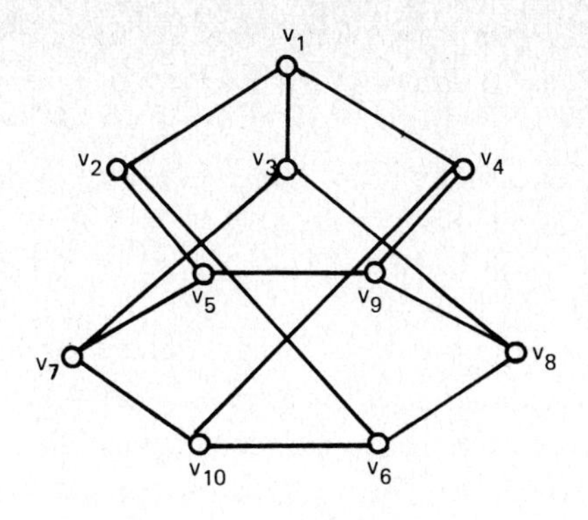

Fig. 7. The Petersen graph.

Theorem 6. *If a (p, q)-graph G is 2-equi, then $q \geqslant 2p-5$.*

Proof. Let G be 2-equi. Then G is a block by Proposition 1. Thus, $\delta(G) \geqslant 2$. If $\delta(G) \geqslant 4$, the theorem is true since $q \geqslant 2p$. If $\delta(G)=3$, then it follows from Proposition 3 that $q \geqslant 2p-5$. We may thus assume that $\delta=2$. Let v be a vertex of degree 2 and u and w be vertices adjacent to v in G. We define three vertex sets I, U, and W (see Fig. 8) and denote their cardinalities by i, j, and k, respectively:

$$I = N(u) \cap N(w) - \{v\},$$

$$U = N(u) - I - \{v\},$$

$$W = N(w) - I - \{v\}.$$

Since $d(x, y) \leqslant 2$ for any pair of vertices $x \in U$ and $y \in W$, x is connected to y in the induced subgraph $G' = \langle G - \{v, u, w\} \rangle$. Thus, the induced graph $G'' = \langle U \cup W \rangle$ is in a connected component of G', which implies that G' has at least

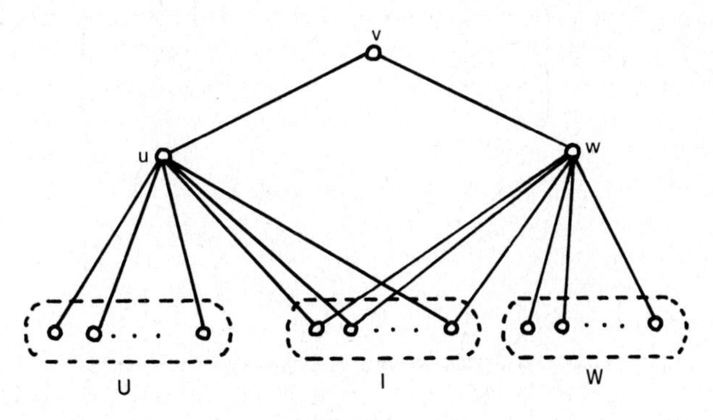

Fig. 8.

$j+k-1$ edges. Therefore, we obtain the inequality as required, since $i+j+k=p-3$:

$$q \geq 2+j+2i+k+(j+k-1)=2(i+j+k)+1$$
$$=2p-5. \qquad \square$$

Before presenting a characterization theorem for minimum 2-equi-eccentric graphs, we require two definitions.

For any tree T, we denote by T' the subtree obtained by deleting the endvertices of T. Then a *double star* is a tree T such that $T'=K_2$; it is denoted by $T(m,n)$ when m endvertices are adjacent to one vertex of this K_2 and n to the other. A set of vertices $X(G)$ is a *dominating set* of $V(G)$ if every vertex in $V(G)$ is either in $X(G)$ or adjacent to a vertex in $X(G)$.

Lemma 1. *Let T be a tree. There is a partition $\{U, W\}$ of $V(T)$ such that:*

(1) *$d(u,w) \leq 2$ for any $u \in U$ and $w \in W$, and*

(2) *both U and W are dominating sets of T,*

if and only if T is either a star or a double star.

Proof. It is easy to see that if T is either a star or a double star then there is such a partition (see Fig. 9). On the other hand, if T is neither a star nor a double star, then there exist two endvertices x and y of T connected by a path P of length at least 4. In order to satisfy (3), both x and y must belong to the same set of the partition, say U. Let x' and y' be the vertices in P adjacent to x and y, respectively (see Fig. 10). Then (2) implies that x' and y' are in W. But this is a contradiction, since $d(x,y') \geq 3$. $\square$

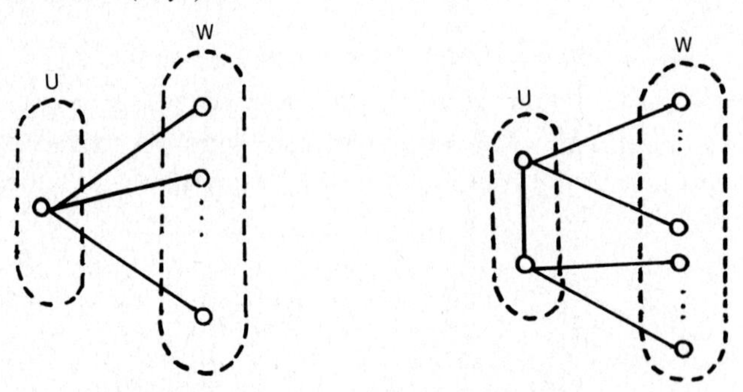

Fig. 9. A star and double star.

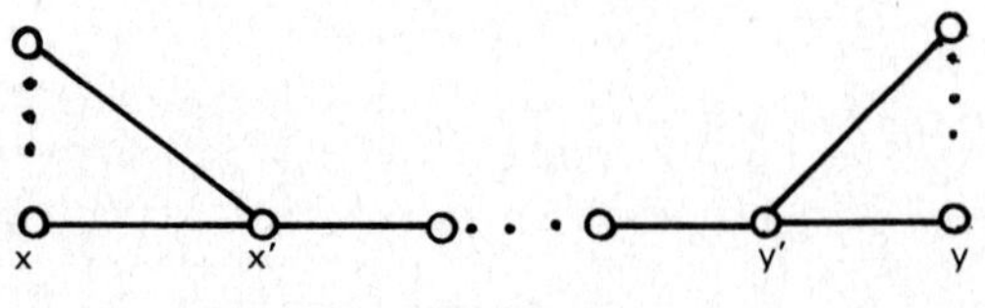

Fig. 10.

In the next theorem we use the following terminology. The graph $K_3(l, m, n)$ is the graph obtained from K_3 by adding l, m, and n pendent edges from each vertex of K_3, respectively. Fig. 11 illustrates the graph $K_3(1,2,3)$.

Theorem 7. *Let G be a minimum 2-equi-eccentric graph other than the Petersen graph. Then G is one of the following:*

(a) *The graph obtained from the double star $T(m, n)$ by adding a new vertex z and joining z to every vertex of degree 1 of $T(m, n)$, where m, n are arbitrary positive integers (see Fig. 12(a)).*

(b) *The graph obtained from $K_3(l, m, n)$, l, m, $n \geq 1$, by adding a new vertex z and joining z to every vertex of degree 1 of $K_3(l, m, n)$ (see Fig. 12(b)).*

Proof. The constructions (a) and (b) and Theorem 6 show that for every $p \geq 5$, the minimum 2-equi graphs of order p have exactly $2p-5$ edges. Furthermore, if G is a minimum 2-equi-eccentric graph other than the Petersen graph, then the minimum degree δ of G is 2 by Theorem 6. Let v be a vertex of degree 2 in G and u, w be the vertices adjacent to v. Then every vertex of $V(G)-\{u, w\}$ is adjacent to either u or w, since G is 2-equi. Define three vertex-subsets I, U and W as follows (see Fig. 13):

$$I = N(u) \cap N(w),$$

$$U = N(u) - I - \{w\},$$

$$W = N(w) - I - \{u\}.$$

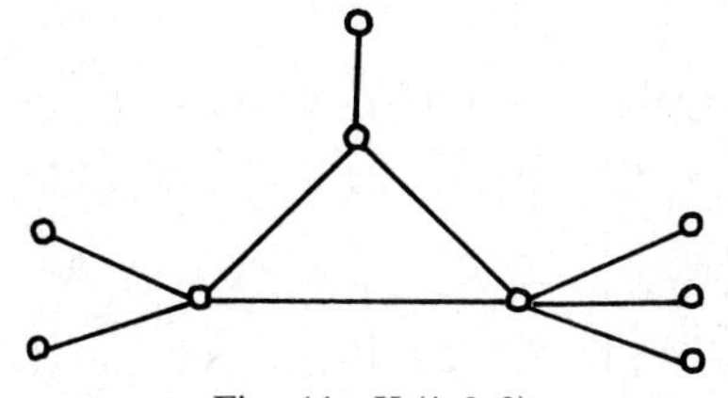

Fig. 11. $K_3(1,2,3)$.

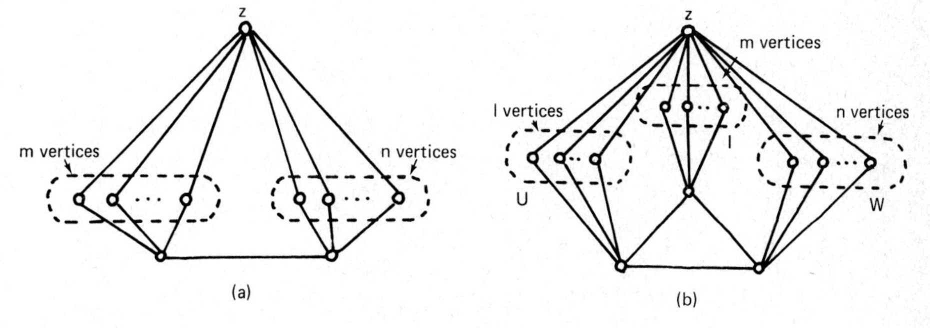

Fig. 12.

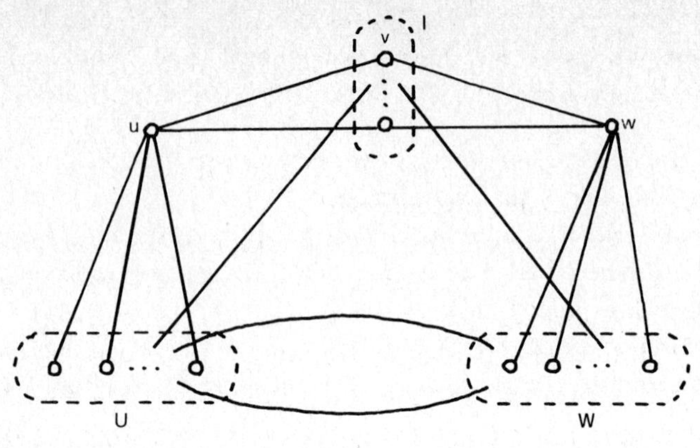

Fig. 13.

Let $|I|=p_1$, $|U|=p_2$ and $|W|=p_3$. Observe that if both p_2 and p_3 are 0, then

$$\begin{aligned} q &\geqslant 2p_1 > 2p_1 - 1 \\ &= 2p - 5, \end{aligned}$$

contradicting the hypothesis that G is minimum 2-equi. Now suppose $p_2=0$ and $p_3>0$. (The case $p_2>0$ and $p_3=0$ is similar.) Since G is 2-equi, u and w cannot be adjacent. (Otherwise $e(w)=1$.) Furthermore, for all $y \in W$, $d(u,y) \leqslant 2$. This implies that every vertex in W is adjacent to a vertex in I. Therefore

$$q \geqslant 2p_1 + 2p_3 = 2p - 4,$$

again contradicting the hypothesis that G is minimum 2-equi. Thus, both p_2 and p_3 are greater than zero. Set

$$\begin{aligned} G' &= G - \{u, w\} \\ &= \langle I \cup U \cup W \rangle_G \end{aligned}$$

and

$$\begin{aligned} T &= \langle U \cup W \rangle_G \\ &= \langle U \cup W \rangle_{G'}. \end{aligned}$$

Then since G is 2-equi, $d'(x,y) \leqslant 2$ for any $x \in U$ and $y \in W$, where $d'(x,y)$ stands for the distance between x and y in G' (see Fig. 13). So T lies in a connected component H of G'. On the other hand, we have:

$$\begin{aligned} q(H) \leqslant q' &\leqslant q - (\deg u + \deg w) \\ &\leqslant 2p - 5 - (2p_1 + p_2 + p_3) \\ &= p_2 + p_3 - 1. \end{aligned}$$

So $p(H) \leqslant p_2 + p_3$ since H is connected. The fact that $H \cong T$ follows immediately from the inequality $p(H) \leqslant p_2 + p_3 = p(T)$ and that $H \supseteq T$. Thus, we obtain the following facts:

(i) $T = \langle U \cup W \rangle_G$ is a tree;

(ii) the vertices u and w are not adjacent in G; and

(iii) $\langle I \rangle = G' - T$ is totally disconnected.

For any $y \in W$, it follows from the fact that $d(u, y) = 2$ and condition (ii) that $N_G(y) \cap U = N_T(y) \cap U \neq \emptyset$. Similarly, we see that $N_T(x) \cap W \neq \emptyset$ for any $x \in U$. We thus obtain:

(iv) both U and W are dominating sets of $T = \langle U \cup W \rangle_G$.

Applying Lemma 1 and (iv), we have that T is either a star or a double star. Then we obtain the graphs illustrated in Fig. 12(a) and (b) according to whether T is a star or a double star. □

Acknowledgements

The authors gratefully acknowledge a communication from F.B. Buckley and M.F. Capabianco pointing out ref. [2]. The term "equi-eccentric" is identical to "self-center" in this reference. The result of Theorem 6 is also stated in [2].

References

[1] M. Behzad, G. Chartrand and L. Lesniak-Foster, Graphs and Digraphs (Prindle, Weber & Schmidt, Reading, 1979).

[2] F. Buckley, Self-centered graphs with a given radius, Proc. 10th S–E Conf. Combinatorics, Graph Theory, and Computing, Congressus Numerantum XXIII (Utilitas Mathematica, Winnipeg, 1979) pp. 211–215.

[3] F. Harary, Graph Theory (Addison-Wesley, Reading, 1969).

[4] J. Mycielski, Sur le coloriage des graphes, Coll. Math. 3 (1955) 161–162.

Received March 1981

Annals of Discrete Mathematics 20 (1984) 25–36
North-Holland

ON THE NUMBER OF SUBGRAPHS OF PRESCRIBED TYPE OF PLANAR GRAPHS WITH A GIVEN NUMBER OF VERTICES

N. ALON and Y. CARO
Department of Mathematics, Tel Aviv University, Tel Aviv, Israel

For a planar graph H and a positive integer n we study the maximal number $f = f(n, H)$, such that there exists a planar graph on n vertices containing f subgraphs isomorphic to H. We determine $f(n, H)$ precisely if H is either a complete bipartite (planar) graph or a maximal planar graph without triangles that are not faces, and estimate $f(n, H)$ for every maximal planar graph H.

1. Introduction

All graphs considered are finite, undirected, with no loops and no multiple edges, and unless otherwise specified, they are all planar. For every two graphs, G and H, $N(G, H)$ is the number of subgraphs of G isomorphic to H. G^n is a graph on n vertices. A *triangulation* G^n is a maximal planar graph on n vertices, i.e. a planar graph all of whose faces (including the unbounded face) are triangles. For every graph H and for every $n \geqslant 1$ put $f(n, H) = \max N(G^n, H)$, where the maximum is taken over all planar graphs G^n. (Clearly, this maximum is attained for some triangulation G^n.) Hakimi and Schmeichel [2] investigated $f(n, H)$, where $H = C_k$ is a cycle of length k. They found that:

$$f(n, C_3) = 3n - 8, \qquad \text{for } n \geqslant 3, \tag{1}$$

$$f(n, C_4) = (n^2 + 3n - 22)/2, \quad \text{for } n \geqslant 4, \tag{2}$$

and that for $k \geqslant 5$ there exist positive constants $c_1(k)$ and $c_2(k)$ such that:

$$c_1(k) \cdot n^{[k/2]} \leqslant f(n, C_k) \leqslant c_2(k) \cdot n^{[k/2]}, \quad \text{for all } n \geqslant k.$$

Here we determine $f(n, H)$ precisely if H is either a complete bipartite (planar) graph or a triangulation without triangles that are not faces, and estimate $f(n, H)$ for every triangulation H.

2. Notation and definitions

$V(G)$ is the set of vertices of the graph G. $G \simeq H$ denotes that the graphs G and H are isomorphic.

If G is a triangulation, a *cut* of G is a triangle in G that is not a face. G is *cut-free* if it includes no cuts. A subgraph of G that is a cut-free triangulation on more than three vertices is called a *block* of G. $b(G)$ is the set of all blocks of G and $c(G)$ is the set of all cuts of G.

A triangulation G is *stacked* if it is C_3 or if every block of G is isomorphic to K_4. (A stacked triangulation is in fact the graph of a stacked 3-polytope, as defined in [3].)

K_k is the complete graph on k vertices, $K_{1,k}$ is the star with k edges, and $K_{2,k}$ is the complete bipartite graph consisting of k independent vertices with two common nonadjacent neighbours. $I(k)$ is the graph consisting of k independent edges. W_k ($k \geq 3$) is the wheel obtained by joining a new vertex to the k vertices of the cycle C_k.

3. The complete bipartite (planar) graphs

When one considers the problem of determining $f(n, H)$ for various graphs H, it seems natural to begin with the complete planar graphs $K_3 = C_3$ and K_4. However, as was remarked, Hakimi and Schmeichel determined $f(n, K_3)$ for all $n \geq 3$. We shall determine $f(n, K_4)$ in Remark 3 of Section 4 as a special case of Theorem 6. Therefore we begin here with the complete bipartite graphs. The main results of this section are the following two theorems:

Theorem 1. *For every $k \geq 2$ and $n \geq 4$:*

$$f(n, K_{1,k}) = g(n, k), \tag{3}$$

where

$$g(n, k) = 2 \cdot \binom{n-1}{k} + 2 \cdot \binom{3}{k} + (n-4) \cdot \binom{4}{k}.$$

Theorem 2. *For every $k \geq 2$ and $n \geq 4$:*

$$f(n, K_{2,k}) = h(n, k), \tag{4}$$

where

$$h(n,k) = \begin{cases} \binom{n-2}{k}, & \text{if } k \geq 5 \text{ or if } k = 4 \text{ and } n \neq 6, \\ 3 & \text{if } k = 4 \text{ and } n = 6, \\ \binom{n-2}{3} + 3(n-4), & \text{if } k = 3 \text{ and } n \neq 6, \end{cases}$$

$$h(n,k)=\begin{cases}12 & \text{if } k=3 \text{ and } n=6,\\ \binom{n-2}{2}+4n-14, & \text{if } k=2.\end{cases}$$

We begin with a simple lemma.

Lemma 3. *If u, v and w are the degrees of three different vertices of a planar graph G^n, then*

$$u+v+w\leqslant 2n+2.$$

Proof. Let $x_1, x_2, \ldots, x_n$ be the vertices of G^n and suppose that u, v and w are the degrees of x_1, x_2 and x_3, respectively. Since G^n contains no $K_{3,3}$, there are at most two vertices x_i with $i\geqslant 4$ that are adjacent to x_1, x_2 and x_3. Thus, the number of edges that join x_1, x_2 and x_3 to some x_i, $i\geqslant 4$, is at most $2\cdot 3+(n-5)\cdot 2=2n-4$, and we obtain:

$$u+v+w\leqslant 6+(2n-4)=2n+2. \qquad \square$$

Proof of Theorem 1. Note that if $d_1, d_2, \ldots, d_n$ are the degrees of the vertices of a graph G^n, then

$$N(G^n, K_{1,k})=\sum_{i=1}^{n}\binom{d_i}{k}, \quad \text{for all } k\geqslant 2.$$

Therefore $f(n, K_{1,k})$ is just

$$\max\sum_{i=1}^{n}\binom{d_i}{k},$$

where the maximum is taken over all degree sequences of planar graphs on n vertices.

For every $n\geqslant 3$ let S^n be the graph obtained by joining each of two adjacent vertices to each of the $n-2$ vertices of a path of length $n-3$. (Note that $S^3=K_3$ and $S^4=K_4$.)

As is easily checked, for every $k\geqslant 2$ and $n\geqslant 4$

$$f(n, K_{1,k})\geqslant N(S^n, K_{1,k})=g(n,k). \tag{5}$$

In order to complete the proof we have to show that for every $k\geqslant 2$ and every graph G^n, where $n\geqslant 4$,

$$N(G^n, K_{1,k})\leqslant g(n,k). \tag{6}$$

We prove (6) for every fixed k by induction on n. If $n=4$, (6) is trivial. Assuming it holds for $n-1$, let us prove it for n $(n\geqslant 5)$. Let G^n be a graph. Clearly, we may assume that G^n is a triangulation. Let $d_1 \leqslant d_2 \leqslant \cdots \leqslant d_n$ be the degrees of the vertices of G^n. Euler's formula implies that $\sum_{i=1}^n d_i = 6n - 12$, and clearly $3 \leqslant d_1 \leqslant d_n \leqslant n-1$. As remarked above

$$N(G^n, K_{1,k}) = \sum_{i=1}^{n} \binom{d_i}{k}. \tag{7}$$

If $\bar{y} = (y_1, y_2, \ldots, y_n)$ and $\bar{z} = (z_1, \ldots, z_n)$ are nondecreasing sequences of positive integers, and if there exist i and j, $1 \leqslant i < j \leqslant n$, such that $z_i = y_i - 1$, $z_j = y_j + 1$ and $z_l = y_l$, for all $l \neq i, j$, then we say that $\bar{z}$ is obtained from $\bar{y}$ by a simple improvement. It is easily checked that in this case

$$\sum_{i=1}^{n} \binom{y_i}{k} \leqslant \sum_{i=1}^{n} \binom{z_i}{k}, \quad \text{for all } k \geqslant 2, \tag{8}$$

and the inequality is strict iff $y_j \geqslant k-1$.

Returning to our G^n we consider two possible cases.

Case 1. $d_1 \geqslant 4$.

In this case:

$$4 \cdot n \leqslant \sum_{i=1}^{n} d_i = 6n - 12,$$

and thus $n \geqslant 6$. It is easily checked that the vector of length n $(3, 3, 4, \ldots, 4, n-1, n-1)$ can be obtained from $(d_1, \ldots, d_n)$ by a finite sequence of simple improvements. By (7) and (8) we obtain:

$$N(G^n, K_{1,k}) = \sum_{i=1}^{n} \binom{d_i}{k} \leqslant 2\binom{3}{k} + (n-4) \cdot \binom{4}{k} + 2 \cdot \binom{n-1}{k} = g(n,k),$$

as needed.

Case 2. $d_1 = 3$.

Let x be a vertex of degree 3 in G^n, and let u, v and w be the degrees of its three neighbours, where $3 \leqslant u \leqslant v \leqslant w \leqslant n-1$. The number of copies of $K_{1,k}$ in G^n that contain x is precisely

$$\binom{3}{k} + \binom{u-1}{k-1} + \binom{v-1}{k-1} + \binom{w-1}{k-1}.$$

By Lemma 3 $u + v + w \leqslant 2n + 2$. It is easily checked that there exist u', v', and w', $4 \leqslant u' \leqslant v' \leqslant w' \leqslant n-1$, such that $u \leqslant u'$, $v \leqslant v'$, $w \leqslant w'$ and $u' + v' + w' = 2n + 2$. The vector $(3, n-2, n-2)$ can be obtained from $(u'-1, v'-1, w'-1)$ by a finite number of simple improvements. Thus, the number of copies of $K_{1,k}$ in G^n that contain x is:

$$\binom{3}{k}+\binom{u-1}{k-1}+\binom{v-1}{k-1}+\binom{w-1}{k-1}$$
$$\leq \binom{3}{k}+\binom{u'-1}{k-1}+\binom{v'-1}{k-1}+\binom{w'-1}{k-1}$$
$$\leq \binom{3}{k}+\binom{3}{k-1}+2\cdot\binom{n-2}{k-1}. \tag{9}$$

By the induction hypothesis:

$$N(G^n - x, K_{1,k}) \leq g(n-1,k). \tag{10}$$

Combining (9) and (10) with the definition of $g(n,k)$ we obtain:

$$N(G^n, K_{1,k}) \leq g(n-1,k)+\binom{3}{k}+\binom{3}{k-1}+2\cdot\binom{n-2}{k-1}=g(n,k).$$

This completes the proof for Case 2 and establishes the theorem. □

Remark 1. Theorem 1 states that for every $k \geq 2$ and $n \geq 4$ and for every graph G^n:

$$N(G^n, K_{1,k}) \leq g(n,k), \tag{11}$$

and equality holds if G^n is the graph S^n appearing in the proof of the theorem. One can easily check that the proof implies that for $k = 2,3,4$ and $n \geq k+1$ equality holds in (11) iff $G^n = S^n$.

Remark 2. The proof of Theorem 1 implies that if $n \geq 12$ and $d_1 \leq \cdots \leq d_n$ are the degrees of the vertices of a triangulation G^n, then:

$$N(G^n, K_{1,2}) = \sum_{i=1}^{n}\binom{d_i}{2} \geq 12\cdot\binom{5}{2}+(n-12)\cdot\binom{6}{2},$$

since $(d_1,\ldots,d_n)$ can be obtained by a finite sequence of simple improvements from the vector $\bar{c}=(c_1,\ldots,c_n)$, where

$$c_i = \begin{cases} 5, & \text{if } i \leq 12, \\ 6, & \text{if } i > 12. \end{cases}$$

Since every triangulation G^n contains $\binom{3n-6}{2}$ pairs of edges, and each such pair is either $K_{1,2}$ or $I(2)$, we conclude that:

$$N(G^n, I(2)) \leq \binom{3n-6}{2} - 12\binom{5}{2} - (n-12)\cdot\binom{6}{2} = (9n^2-69n+162)/2,$$

with equality iff the degree sequence of G^n is $\bar{c}$. In [1] it is proved that such a triangulation G^n exists whenever $n \geq 12$, except for $n = 13$, and thus we obtain:

$$f(n, I(2)) = (9n^2 - 69n + 162)/2,$$

for all $n \geq 12$, except $n = 13$.

The proof of Theorem 2 is similar to that of Theorem 1, although somewhat more complicated. The result of Hakimi and Schmeichel that appears as equation (2) in this paper proves Theorem 2 for $k = 2$. We prove the theorem here for $k \geq 5$ and give only an outline for $k = 3, 4$, since the proof in these cases is rather lengthy and quite similar.

We need two simple lemmas.

Lemma 4. *Let G^n be a (planar) graph that has a vertex x of degree $n-1$, (i.e. x is adjacent to every other vertex of G^n). If $n \geq 5$, then G^n contains two nonadjacent vertices, each of degree ≤ 3.*

Proof.[1] Note that we may assume that G^n is a triangulation. We prove the lemma by induction on n. For $n = 5$ it is trivial. Assuming it holds for all n', $5 \leq n' < n$, let us prove it for n. Let G^n be a triangulation, and let x be a vertex of G^n of degree $n-1$. Since G^n is a triangulation, there is a Hamiltonian cycle C in $G^n - x$. If no edge of G is a chord of C, then all vertices of C have degree 3 and the assertion of the lemma follows. Thus, we may assume that there is a diagonal joining the vertices y and z of C. This diagonal splits C into two cycles, C_1 and C_2, with a common edge yz. For $i = 1, 2$ let H_i be the induced subgraph of G^n with vertex set $\{x\} \cup C_i$. We claim that H_1 contains a vertex t of degree ≤ 3 in H_1, $t \neq x, y, z$. Indeed, if $|V(H_1)| = 4$ this is trivial, and if $|V(H_1)| \geq 5$ this follows from the induction hypothesis. Similarly H_2 contains a vertex r of degree ≤ 3 in H_2, $r \neq x, y, z$. However, the degree of t in H_1 equals its degree in G^n and the degree of r in H_2 equals its degree in G^n. Thus, t and r are two nonadjacent vertices of G^n, each of degree ≤ 3 in G^n, which completes the proof. □

Lemma 5. *Let G^n be a triangulation and let $d_1 \leq d_2 \leq \cdots \leq d_n$ be its degree sequence. If $4 \leq d_1 \leq d_n \leq n-2$, then for every $k \geq 5$:*

$$N(G^n, K_{2,k}) \leq \binom{n-2}{k}.$$

Proof. Since G^n includes no $K_{3,3}$, every $K_{1,k}$ in G^n is included in at most one

[1] **Editorial remark.** The following shorter proof was suggested by a referee. $G^n \setminus \{x\}$ is outerplanar, hence it has two nonadjacent vertices ($n \geq 5$) of valences ≤ 2 in $G^n \setminus \{x\}$; they are nonadjacent vertices in G^n, of valences ≤ 3.

$K_{2,k}$ in G^n. Clearly, every $K_{2,k}$ in G^n includes exactly two copies of $K_{1,k}$, and thus

$$N(G^n, K_{2,k}) \leqslant \tfrac{1}{2} N(G^n, K_{1,k}) = \tfrac{1}{2} \sum_{i=1}^{n} \binom{d_i}{k}. \tag{12}$$

It is easily checked that the vector of length n $(4, 4, \dots, 4, n-2, n-2)$ can be obtained from the vector $(d_1, d_2, \dots, d_n)$ by a finite sequence of simple improvements. Therefore (12) implies

$$N(G^n, K_{2,k}) \leqslant \tfrac{1}{2} \sum_{i=1}^{n} \binom{d_i}{k} \leqslant \tfrac{1}{2} \left((n-2) \binom{4}{k} + 2 \binom{n-2}{k} \right) = \binom{n-2}{k},$$

as needed. □

Proof of Theorem 2 for $k \geqslant 5$. As is easily checked, for every $k \geqslant 5$ and $n \geqslant 4$:

$$f(n, K_{2,k}) \geqslant N(S^n, K_{2,k}) = \binom{n-2}{k} = h(n, k).$$

In order to complete the proof we have to show that for every $k \geqslant 5$ and every triangulation G^n, where $n \geqslant 4$,

$$N(G^n, K_{2,k}) \leqslant \binom{n-2}{k}. \tag{13}$$

We prove (13) for every fixed k by induction on n. If $n \leqslant k+1$, (13) is trivial. Assuming it holds for $n-1$, let us prove it for n $(n \geqslant k+2)$. Let G^n be a triangulation, and let $d_1 \leqslant d_2 \leqslant \cdots \leqslant d_n$ be the degrees of its vertices. If $d_1 \geqslant 4$, then by Lemma 4 $d_n \leqslant n-2$ and by Lemma 5 (13) holds, as needed. Thus, we may assume that $d_1 = 3$. Let t be a vertex of degree 3 in G^n, and let x, y and z be its three neighbours. Let k_1, k_2, and k_3 denote the numbers of common neighbours of x and y, y and z, and z and x, respectively, in $V(G^n) \setminus \{x, y, z, t\}$. Since G^n includes no $K_{3,3}$, it can be easily verified that $k_1 + k_2 + k_3 \leqslant n-2$, and if $k_1 = 0$, then $k_1 + k_2 + k_3 \leqslant n-4$. Clearly, $0 \leqslant k_1, k_2, k_3 \leqslant n-4$ and we may assume that $k_1 \leqslant k_2 \leqslant k_3$. It is easily checked that there exist k_1', k_2', and k_3', $1 \leqslant k_1' \leqslant k_2' \leqslant k_3' \leqslant n-4$, such that $k_i \leqslant k_i'$ for $i = 1, 2, 3$ and $k_1' + k_2' + k_3' = n-2$.

The number of $K_{2,k}$'s in G^n that contain t is clearly at most

$$\sum_{i=1}^{3} \binom{k_i+1}{k-1} \leqslant \sum_{i=1}^{3} \binom{k_i'+1}{k-1},$$

and since $(1, 1, n-4)$ can be obtained from (k_1', k_2', k_3') by a finite number of simple improvements, this number is at most

$$2 \cdot \binom{2}{k-1} + \binom{n-3}{k-1} = \binom{n-3}{k-1}. \tag{14}$$

By the induction hypothesis:

$$N(G^n - t, K_{2,k}) \leq \binom{n-3}{k}. \tag{15}$$

Combining (14) and (15) we obtain:

$$N(G^n, K_{2,k}) \leq \binom{n-3}{k-1} + \binom{n-3}{k} = \binom{n-2}{k}.$$

This completes the proof and establishes Theorem 2 for $k \geq 5$. □

An outline of the proof of Theorem 2 for $k = 3, 4$**.** For $n \leq 7$ one can easily prove the theorem by checking all the possible triangulations G^n. Clearly, for $k = 3, 4$ and $n \geq 8$:

$$f(n, K_{2,k}) \geq N(S^n, K_{2,k}) = h(n, k).$$

Thus, we have to show that for $k = 3, 4$ and for every triangulation G^n, where $n \geq 7$,

$$N(G^n, K_{2,k}) \leq h(n, k). \tag{16}$$

We prove (16) for each of the two possible values of k by induction on n. For $n = 7$, (16) holds. Assuming it holds for $n - 1$, let G^n be a triangulation and let $d_1 \leq d_2 \leq \cdots \leq d_n$ be its degree-sequence. If $d_1 = 3$, we proceed exactly as in the proof for $k \geq 5$. Thus, we may assume that $d_1 \geq 4$. By Lemma 4 $d_n \leq n - 2$. If $d_{n-1} = d_n = n - 2$, we can show that G^n must be the graph obtained by joining each of two nonadjacent vertices to each of the $n - 2$ vertices of the cycle C_{n-2} and, as is easily checked in this case, (16) holds. Therefore we may assume that $d_1 \geq 4$, $d_{n-1} \leq n - 3$ and $d_n \leq n - 2$. It is easily seen that in this case the vector of length n $(4, 4, \ldots, 4, 5, n-3, n-2)$ can be obtained from the vector $(d_1, \ldots, d_n)$ by a finite sequence of simple improvements, and using the same argument as in the proof of Lemma 5 we conclude that for $k = 3, 4$ and for every $n \geq 8$:

$$\begin{aligned} N(G_n, K_{2,k}) &\leq \tfrac{1}{2} N(G^n, K_{1,k}) \\ &= \tfrac{1}{2}\left[(n-3)\binom{4}{k} + \binom{5}{k} + \binom{n-3}{k} + \binom{n-2}{k}\right] \\ &\leq h(n, k), \end{aligned}$$

as needed. □

4. The triangulations

The main results of this section are the following two theorems.

Theorem 6. *If H is a cut-free triangulation on k vertices, $k \geq 4$, then*

$$f(n, H) = [(n-3)/(k-3)], \quad \text{for all } n \geq 3.$$

Theorem 7. *For every triangulation H that contains a cut and for every $n \geq 4$:*

$$f(n, H) \leq 12(n-4)/|\mathrm{Aut}\, H|,$$

where $|\mathrm{Aut}\, H|$ is the number of automorphisms of H.

(Considering Theorem 6, one can easily verify that Theorem 7 holds for all triangulations with four exceptions: the graphs of the triangule, the tetrahedron, the octahedron, and the icosahedron.)

In order to prove Theorems 6 and 7 we need a few simple lemmas concerning the blocks and the cuts of a triangulation. Since the contents of these lemmas seem to be well known, we shall leave most of the proofs to the reader.

Lemma 8. *Let $G = G^n$ be a triangulation, $n \geq 4$.*

(i) *If T is a cut of G that splits G into two triangulations, A and B, having T as a common face, then $c(G)$ is the (disjoint) union of $c(A)$, $c(B)$ and $\{T\}$, and $b(G)$ is the (disjoint) union of $b(A)$ and $b(B)$.*

(ii) *Every face of G is contained in a unique block of G and every cut of G is contained in precisely two blocks of G.*

(iii) *Let $cb(G)$ denote the graph whose vertex set is $b(G)$, and $B_1, B_2 \in b(G)$ are joined iff their intersection is a cut of G. Then $cb(G)$ is a tree.*

(iv) $|c(G)| = |b(G)| - 1$.

Proof. Most assertions of part (i) can be easily verified. In showing that $b(G) \subset b(A) \cup b(B)$, use the fact that every block of G is a 3-connected graph. Part (ii) and part (iii) are proved by induction on n, using the assertions of the preceding part(s). Part (iv) follows immediately from (iii). □

Lemma 9. *Let G^n be a triangulation, $n \geq 4$.*

(i) *If G^n has q blocks $H_1^{n_1}, H_2^{n_2}, \ldots, H_q^{n_q}$, then*

$$n - 3 = \sum_{i=1}^{q} (n_i - 3).$$

(ii) *The number of cuts in G^n is at most $n-4$, and equality holds iff G^n is a stacked triangulation.*

Proof. Part (i) is proved by induction on q, using part (i) of Lemma 8. Part (ii) follows easily from part (i), using part (iv) of Lemma 8. (Note that a block has at least four vertices, and a block with four vertices is K_4.) □

Lemma 10. *Let $T_1, T_2, \ldots, T_q$ be q cut-free triangulations, each containing more than three vertices. Then there exists a triangulation G with precisely q blocks $H_1, \ldots, H_q$ such that $H_i \simeq T_i$ for $1 \leq i \leq q$.*

Proof. By induction on q. The case $q=1$ is trivial. If $q>1$, and F is a triangulation with $q-1$ blocks $H_1, \ldots, H_{q-1}$, isomorphic to $T_1, \ldots, T_{q-1}$, respectively, then the required triangulation G is obtained by gluing together F and an isomorphic copy of T_q along a common face. □

Lemma 11. *Let $H = H^n$ and $F = F^n$ be two cut-free triangulations, $n \geq 4$. Let x_1, x_2, and x_3 be the vertices of a face of H and let y_1, y_2, and y_3 be the vertices of a face of F. Then there exists at most one isomorphism $g: H \to F$ that satisfies $g(x_i) = y_i$, for $1 \leq i \leq 3$.*

Proof. Let g be such an isomorphism. The edge x_1x_2 is included in precisely two triangles of H, one of them is $x_1x_2x_3$. Let the other triangle be x_1x_2a. Similarly, y_1y_2 is included in precisely two triangles of F, $y_1y_2y_3$ and y_1y_2b. Clearly, g must satisfy $g(a) = b$. By repeated application of this argument one can easily show that $g(c)$ is uniquely determined for all $c \in V(H)$. □

Proof of Theorem 6. Let G^n be a triangulation. By definition, every copy of H in G^n is a block of G^n. By part (i) of Lemma 9:

$$n-3 \geq N(G^n, H) \cdot (k-3).$$

Therefore

$$f(n, H) \leq [(n-3)/(k-3)].$$

Conversely, put $r = [(n-3)/(k-3)]$. By Lemma 10 there is a triangulation $G = G^{r(k-3)+3}$ with r blocks, each isomorphic to H. Thus:

$$f(n, H) \geq f(r(k-3)+3, H) \geq N(G, H) = r = [(n-3)/(k-3)]. \qquad \square$$

Remark 3. The proof of Theorem 6 implies that if H is a cut-free triangulation on k vertices, $k \geq 4$, and if $k-3$ divides $n-3$, then for every triangulation G^n:

$$N(G^n, H) \leq (n-3)/(k-3),$$

and equality holds iff every block of G^n is isomorphic to H. In particular, for every $n \geq 3$ and for every triangulation G^n:

$$N(G^n, K_4) \leq n-3,$$

and equality holds iff G^n is a stacked triangulation.

Note also that Euler's formula and part (ii) of Lemma 9 imply that for every triangulation G^n, $n \geq 3$:

$$N(G^n, K_3) \leq (2n-4)+(n-4) = 3n-8, \tag{17}$$

and equality holds iff G^n is a stacked triangulation. This is just the result of Hakimi and Schmeichel, quoted in equation (1) of this paper.

Proof of Theorem 7. Let H be a triangulation that contains a cut and let G^n be a triangulation, $n \geq 5$. We must show that

$$N(G^n, H) \leq 12(n-4)/|\mathrm{Aut}\, H|. \tag{18}$$

Let C be a cut of H and let B be a block of H that contains C. Every isomorphism of H into G clearly maps C onto some cut T of G and maps B onto a block of G that includes T. But T is included in precisely two blocks of G, say A_1 and A_2. The number of possible maps of C onto T is six. By repeated application of Lemma 11 it is easily shown that there are at most six isomorphisms of H into G that map C onto T and B onto A_1, and there are at most six isomorphisms of H into G that map C onto T and B onto A_2. Therefore there are at most 12 isomorphisms of H into G that map C onto a given cut T of G. By part (ii) of Lemma 9 the number of cuts in G is at most $n-4$, and thus there are at most $12(n-4)$ isomorphisms of H into G. However, the number of such isomorphisms is exactly

$$N(G^n, H) \cdot |\mathrm{Aut}\, H|,$$

which implies (18). □

Remark 4. Recall that S^5 is the graph obtained from K_5 by deleting an edge. By Theorem 7:

$$N(G^n, S^5) \leq 12(n-4)/|\mathrm{Aut}(S^5)| = n-4, \tag{19}$$

for every triangulation G^n, $n \geq 5$. The proof of Theorem 7 implies that equality holds in (19) iff G^n is a stacked triangulation and thus $f(n, S^5) = n-4$, for all $n \geq 5$. This shows that Theorem 7 is, in a sense, the best possible. However, by a slight modification of the proof of Theorem 7 it is not difficult to obtain a better upper bound for $f(n, H)$, if H is a triangulation that has a cut but is not stacked.

Remark 5. For every fixed graph H, the function $\varphi_H(n) = f(n, H)$ is clearly super-additive, and therefore $f(n, H)/n$ tends to a (finite or infinite) limit as $n \to \infty$. By Theorem 7 this limit is finite for every triangulation H.

We conclude the paper with the following conjecture of M.A. Perles.

Conjecture. For every 3-connected (planar) graph H there is a constant $c(H)$ such that

$$f(n, H) \leqslant c(H) \cdot n, \quad \text{for all } n.$$

(One should note that if $H \neq K_3$ is planar and not 3-connected, then $f(n, H) \geqslant c(H) \cdot n^2$ for a suitable positive constant $c(H)$ and for all $n \geqslant |V(H)|$.)

By Theorem 7 the conjecture holds for every triangulation H. We can prove the conjecture if H is any wheel W_k ($k \geqslant 3$). It is worth noting that unlike the case of the triangulations, the constant $c(H)$ in the conjecture cannot be chosen independently of H, since it can be easily shown that for every $k \geqslant 2$ and $m \geqslant 1$:

$$f(m \cdot (4k+1), W_{3k}) \geqslant m \cdot \binom{2k}{k}.$$

Acknowledgement

Thanks are due to Prof. M.A. Perles from the Hebrew University of Jerusalem for many useful suggestions.

References

[1] B. Grünbaum, Convex Polytopes (Interscience Publishers, London, New York and Sydney, 1967) pp. 271–272.

[2] S.L. Hakimi and E.F. Schmeichel, On the number of cycles of length k in a maximal planar graph, J. Graph Theory 3 (1979) 69–86.

[3] G.C. Shephard, Subpolytopes of stack polytopes, Israel J. Math. 19 (1974) 292–296.

Received 25 April 1981

Annals of Discrete Mathematics 20 (1984) 37–41
North-Holland

SOME CONDITIONS FOR DIGRAPHS TO BE HAMILTONIAN

D. AMAR, I. FOURNIER and A. GERMA
Laboratoire de Recherche en Informatique, Université de Paris-Sud, 91405 Orsay Cedex, France

We answer a question of B. Bollobas and we determine the minimum number of arcs in a diagraph with given minimum indegree and outdegree guaranteeing this digraph to be Hamiltonian or strongly Hamilton-connected.

1. Introduction

The notation we use can be found in [1]. In what follows, $D=(X,E)$ denotes a 1-graph with order $|X|=n\geqslant 3$.

Lewin [4] proved the following two theorems:

If D is strongly connected and has no less than n^2-3n+5 arcs, then D is Hamiltonian.

If D has no less than n^2-2n+3 arcs, then D is strongly Hamilton-connected (i.e. $\forall a\in X$, $\forall b\in X$, *there exists a Hamiltonian path from a to b*).

Lemma 1. *Let D be a 1-graph satisfying (P_k):*

$$(P_k)\text{: } \forall x\in X,\ d^+(x)\geqslant k \text{ and } d^-(x)\geqslant k, \qquad k \text{ an arbitrary integer,}$$

where $d^+(x)$ and $d^-(x)$ denote the outdegree and the indegree of x. If D has no less than $n(n-1)-(k+1)(n-k-1)+1$ arcs, then D is strongly connected.

Proof. If D is not strongly connected, we can find $A\subset X$, $A\neq\emptyset$, such that there exist no arcs from A to $X\setminus A$. (Fig. 1). Let

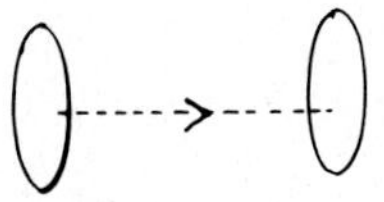

A X\A

Fig. 1.

$$p = |A|;\ \forall x \in A,\ d^+(x) \geq k \Rightarrow p \geq k+1,$$

$$\forall y \in X \setminus A,\ d^-(y) \geq k \Rightarrow n-p \geq k+1.$$

The number of arcs of D is no more than

$$p(p-1)+(n-p)(n-p-1)+p(n-p) = n(n-1)-p(n-p).$$

The maximum of this function for $k+1 \leq p \leq n-k-1$ is $n(n-1)-(k+1)\cdot(n-k-1)$.

Remark 1. If a digraph D has no less than n^2-2n+3 arcs, then D satisfies:

$$(P_1)[\text{i.e. } \forall x \in X,\ d^+(x) \geq 1 \text{ and } d^-(x) \geq 1].$$

So the theorem we prove here is in a way a generalization of the Lewin's theorem:

Theorem. *Let $D = (X, E)$ be a 1-graph with order n satisfying (P_k).*
If $|E| > f(n,k) = n(n-1)-(k+1)\ (n-k-1)$, D is Hamiltonian.
If $|E| > g(n,k) = n(n-1)-k(n-k-1)$, D is strongly Hamilton-connected.

2. Remarks

A theorem of Ghouila Houri [3] has the following corollaries [1, pp. 188–195]:

If for each vertex x of D, $d^+(x) \geq n/2$, $d^-(x) \geq n/2$, then D is Hamiltonian.

If for each vertex x of D, $d^+(x) \geq (n+1)/2$, $d^-(x) \geq (n+1)/2$, then D is strongly Hamilton-connected.

For small values of k ($k < n/2$ or $k < (n+1)/2$) the bounds given by the theorem are the best possible, as proved below. Indeed, if $k < n/2$, we can find 1–graphs D with $f(n,k)$ arcs, satisfying (P_k), which are not Hamiltonian. In Fig. 2, $n-k-1 \geq k$, i.e. $k \leq (n-1)/2$; S_{k+1} is an independent set with $k+1$ vertices. For $p \in N$, K_p^* is the complete symmetric diagraph with p vertices.

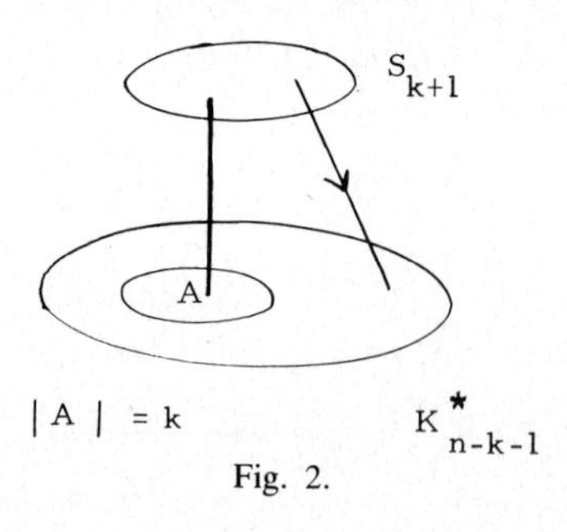

Fig. 2.

3. Notations

Fig. 3(a) shows every arc with origin in C and end in B, and Fig. 3(b) shows every arc between B and C. If k is strictly less than $(n+1)/2$, the bound $g(n, k)$ is the best possible: we can find 1-graphs D, with $g(n, k)$ arcs, satisfying (P_k), and which are not strongly Hamilton-connected.

In Fig. 4, $n-k \geqslant k$, i.e. $k \leqslant n/2$; S_k is an independent set with k vertices. A 1-graph D is k-strongly connected if for each $(x, y) \in X^2$, there exist k vertex-disjoint paths with beginning point x and end point y.

In the examples of Fig. 2 and Fig. 4, the 1-graphs considered are k-strongly connected. So the function $f(n, k)$ (resp. $g(n, k)$) is the best possible for $k < n/2$ (resp. $k < (n+1)/2$) even if we replace the condition (P_k) by the condition to be k-strongly connected.

Proof of the theorem. We prove the theorem by induction on n. If $n=3$, we can easily verify it. To pass from n to $n+1$, we use Lemmas 1 and 2.

Notations. For $n \in N^*$ and $k \leqslant n-1$, let $\mathcal{F}(n, k)$ (resp. $\mathcal{G}(n, k)$) be the set of 1-graphs D, with order n, satisfying (P_k) and having no less than $f(n, k)+1$ (resp. $g(n, k)+1$) arcs.

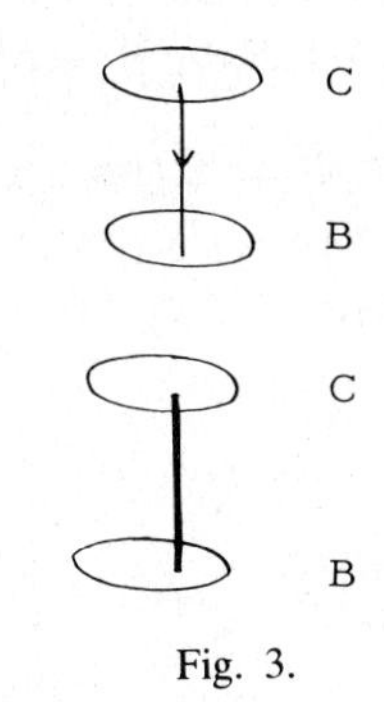

Fig. 3.

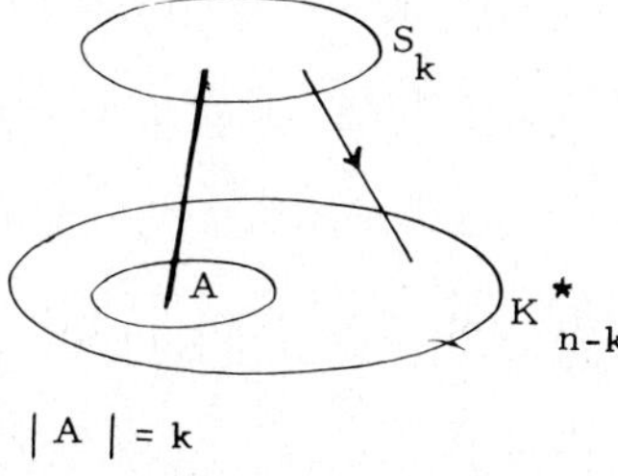

Fig. 4.

Lemma 2. *If for each $k \leqslant n-1$, every D in $\mathscr{F}(n,k)$ is Hamiltonian, then for each $k \leqslant n$, every D in $\mathscr{G}(n+1,k)$ is strongly Hamilton-connected.*

Proof. Let D be in $\mathscr{G}(n+1,k)$ and let a and b be two vertices of X. We consider the following 1-graph $\tilde{D}=(\tilde{X},\tilde{E})$:

$$\tilde{X}=X\setminus\{a,b\}\cup\{c\}, \text{ where } c \text{ is not in } X,$$

$$\tilde{E}=\{(x,y), x\in\tilde{X},\ y\in\tilde{X},\ (x,y)\in E\}\cup\{(x,c),\ x\neq c,\ (x,b)\in E\}$$

$$\cup\{(c,y),\ y\neq c,\ (a,y)\in E\}$$

(the 1-graph $\tilde{D}$ has been considered by Ghouila Houri [3]).

$\tilde{D}$ satisfies (P_{k-1}) and

$$|E(\tilde{D})|\geqslant|E(D)|-2(n-1),$$

$$|E(\tilde{D})|>n(n+1)-k(n-k)-2(n-1),$$

$$|E(\tilde{D})|>f(n,k-1).$$

So $\tilde{D}$ is in $\mathscr{F}(n,k-1)$: it is Hamiltonian. In a Hamiltonian circuit in $\tilde{D}$, there is an arc (x,c) and an arc (c,y); we deduce from it a Hamiltonian path in D, with a beginning point a and an end point b.

Lemma 3. *If for each $k\leqslant n$, every D in $\mathscr{G}(n+1,k)$ is strongly Hamilton-connected, then for each $k\leqslant n$, every D in $\mathscr{F}(n+1,k)$ is Hamiltonian.*

We prove the lemma by induction on decreasing k. By the corollaries of the theorem of Ghouila Houri, if $k\geqslant(n+1)/2$ and if D is in $\mathscr{F}(n+1,k)$, D is Hamiltonian. Let us assume every D in $\mathscr{F}(n+1,k')$ is Hamiltonian for $k+1\leqslant k'\leqslant n$. Let D be in $\mathscr{F}(n+1,k)$.

First case. For each vertex x in X, $d^+(x)\geqslant k+1$, and $d^-(x)\geqslant k+1$:
if $n+1\leqslant 2k+2$, D is Hamiltonian;
if $n+1\geqslant 2k+3$, we have:

$$f(n+1,k)-f(n+1,k+1)=n+1-(2k+3)\geqslant 0,$$

so that D is in $\mathscr{F}(n+1,k+1)$ and is Hamiltonian by assumption.

Second case. There is a vertex $x_0\in X$ such that $d^+(x_0)=k$ or $d^-(x_0)=k$; we assume $d^+(x_0)=k$.

We consider the following 1-graph $D_1=(X_1,E_1)$:

$$X_1=X,$$

$$E_1=E\cup\{(x_0,y),\ y\neq x_0,\ (x_0,y)\notin E\}.$$

So $|E_1| = |E| + n - k$. Then:

$$|E_1| > n(n+1) - (k+1)(n-k) + (n-k),$$

and hence

$$|E_1| > g(n+1, k).$$

By our hypothesis, D_1 is strongly Hamilton-connected. Let x_1 be a vertex such that $(x_0, x_1) \in E$. We consider a Hamiltonian path in D_1 from x_1 to x_0. It is a Hamiltonian path in D and we can deduce, with the arc (x_0, x_1), a Hamiltonian circuit in D.

References

[1] C. Berge, Graphes et Hypergraphes (Dunod, Paris, 1973).
[2] J.C. Bermond, A. German, M.C. Heydemann and D. Sotteau, Chemins et circuits dans les graphes orientés, Ann. Discr. Math. 8 (1980) 293–309.
[3] A. Ghouila Houri, Flots et tensions dans un graphe, in: Annales Scientifiques de l'Ecole Normale Supérieure (1964) pp. 317–327.
[4] M. Lewin, On maximal circuits in directed graphs, J. Combinat. Theory (B) 18 (1975) 175–179.

Received March 1981

Annals of Discrete Mathematics 20 (1984) 43–45
North-Holland

COVERING OF VERTICES OF A SIMPLE GRAPH WITH GIVEN CONNECTIVITY AND STABILITY NUMBER

D. AMAR, I. FOURNIER and A. GERMA
Laboratoire de Recherche en Informatique, Université de Paris-Sud, 91405 Orsay Cedex, France

R. HÄGGKVIST
Institut Mittag Leffler, S 182 62 Djursholm, Sweden

We prove the following lemma. Let G be a k-connected simple graph, $k \geqslant 2$, with stability number $\alpha = k + 1$. Let C be the longest cycle in G, then $G \setminus C$ is a complete subgraph.

Chvátal and Erdös [1] have proved the following theorem:

Theorem. *If G is a k-connected simple graph with stability number $\alpha \leqslant k$, then G is Hamiltonian.*

We extend this theorem by the following result:

Theorem. *Let G be a k-connected simple graph, $k \geqslant 2$, with stability number $\alpha = k + 1$. Let C be the longest cycle in G; then the subgraph $G \setminus C$ is complete.*

We shall prove the theorem by contradiction. For that we use three lemmas which are the basis of the proof of the Chvátal–Erdös theorem.

Lemma 1. *In a simple graph G, let C be the longest cycle. If x is not in $V(C)$ and if $[x, a_1]$ and $[x, a_2]$ are two paths from x to C which are disjoint except in x, then a_1 and a_2 are not consecutive on C.*

Notation. For an arbitrary orientation of C, if y is a vertex of C, we denote by y^+ the successor of y.

Lemma 2. *With the hypothesis of Lemma 1 there does not exist a path from a_1^+ to a_2^+ which does not intersect C and the paths $[x, a_1]$ and $[x, a_2]$.*

C would not be the longest cycle, as we can see in Fig. 1.

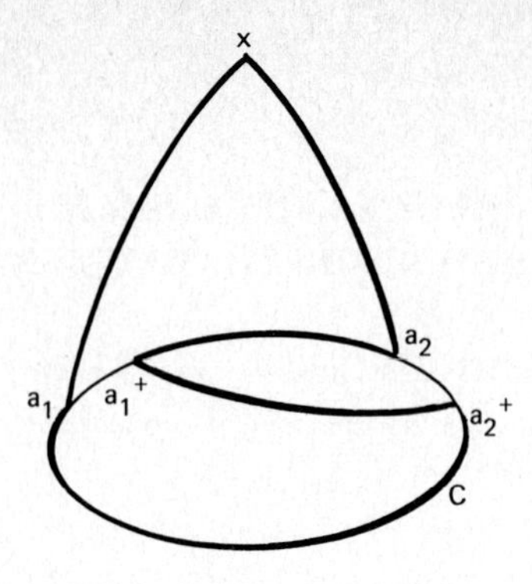

Fig. 1.

Lemma 3. *Let G be a graph with stability number α. If C is the longest cycle in G and if H is a component of $G \setminus C$, then there exist at most $(\alpha - 1)$ paths from H to C which are vertex disjoint except in H.*

Proof. If $a_1, a_2, \ldots, a_p$ are the end points on C of these paths, the set $\{x, a_1^+, a_2^+, \ldots, a_p^+\}$ is a stable set.

Proof of the theorem. Let G be a k-connected graph with stability number $\alpha = k + 1$. Let C be the longest cycle in G. Suppose $G \setminus C$ is not a complete graph.

(i) Let u and v be two vertices of $G \setminus C$ which are not adjacent.

G being k-connected, $|V(C)|$ is not less than $(k+1)$ and there exist k paths $[u, a_1] \cdots [u, a_k]$ from u to C, disjoint except in u.

The set $\{u, v, a_1^+, \ldots, a_k^+\}$ has $(k+2)$ vertices and so is not a stable set. By Lemmas 1 and 2 there exists an edge (v, a_i^+): we suppose it is (v, a_1^+).

If u and v were in the same component of $G \setminus C$, C would not be the longest cycle, as we can see in Fig. 2. So, u and v are in different components of $G \setminus C$, and the components of $G \setminus C$ are complete subgraphs.

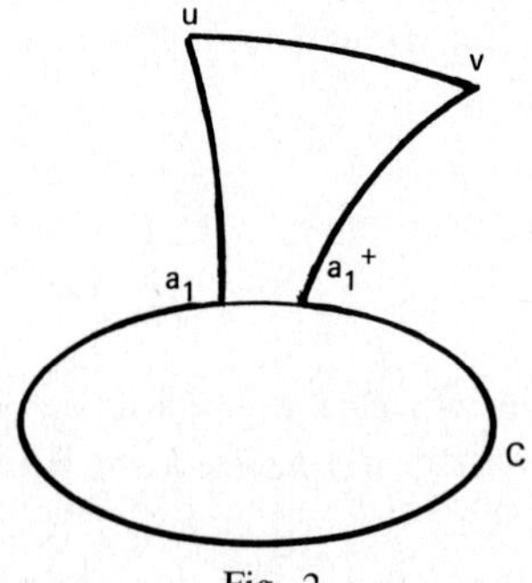

Fig. 2.

Lemma 1 implies that $\{a_1^{++}, a_2^+, \ldots, a_k^+\}$ are pairwise distinct, therefore:

(ii) The set $\{u, v, a_1^{++}, a_2^+, \ldots, a_k^+\}$ is not a stable set, so there exists an edge (a_1^{++}, a_i^+) (see Fig. 3) or an edge (u, a_1^{++}) (see Fig. 4). In both cases we can find a cycle C' such that $|V(C')| \geq |V(C)|$ and which is drawn on Figs. 3 and 4.

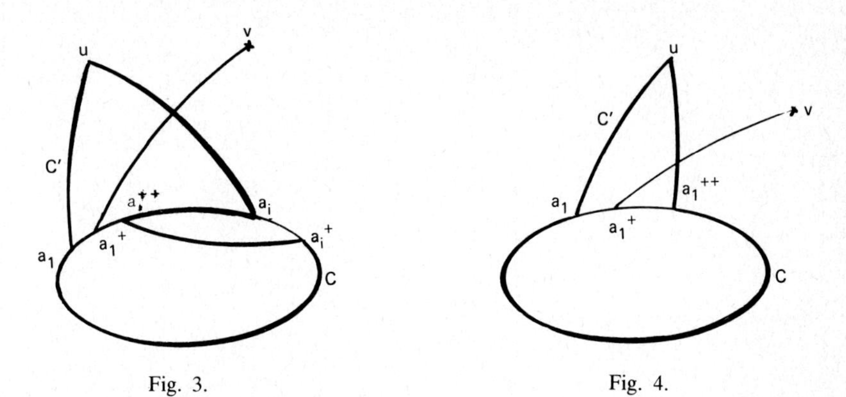

Fig. 3.

Fig. 4.

There exist k vertex-disjoint paths from v to C, one of them is the edge (v, a_1^+); we denote the others by $[v, b_i]$, $2 \leq i \leq k$. Let H be the component of v in $G \setminus C'$. Between H and C' there exist $(k+1)$ paths which are vertex-disjoint except in H, (a_1^+, a_1), (a_1^+, a_1^{++}) and the $(k-1)$ paths $[v, b_i]$, $2 \leq i \leq k$. From Lemma 3 we have a contradiction.

Reference

[1] V. Chvátal and P. Erdös, A note on hamiltonian circuits, Discr. Math. 2 (1972) 111–113.

Received June 1981

Annals of Discrete Mathematics 20 (1984) 47–54
North-Holland

EMBEDDING LATIN SQUARES IN STEINER QUASIGROUPS, AND HOWELL DESIGNS IN TRIPLE SYSTEMS

L.D. ANDERSEN
Mathematics Institute, Aarhus, Denmark

E. MENDELSOHN*
University of Toronto, Toronto, Ontario, Canada

This paper shows that any Latin square of side u can be a subsquare of the multiplication table of a Steiner triple system of size $3u + k$, where k depends only on the residue class of u modulo 6, and of Steiner triple systems all of sizes at least $6u + 2k + 1$. A generalization of "conjugate" to subsquares is given and it is then shown that if a Latin square is a subsquare of Steiner quasigroup then so are all its subsquare conjugates. Finally, as an application it is shown that if a square is part of a family of $s + 1$ mutually orthogonal Latin squares of side u then a Generalized Howell Design $(s, 3, u, 3u)$ exists which is embeddable in a Steiner quasigroup of size $3u + k$.

1. Introduction

Steiner triple systems and Latin squares appear in many guises and we shall assume the reader is familiar with these facts and refer him to [1].

1.1. *Latin squares*

A Latin square is an $n \times n$ matrix with entries from a set N of size n so that every row and column is a permutation of N; a quasigroup is an algebra with a binary operation whose multiplication table is a Latin square; an orthogonal array $\mathrm{OA}(n, 3)$ is a set of ordered triples on an n-set so that every projection to $n \times n$ is a bijection. An $\mathrm{OA}(n, 3)$ is equivalent to a quasigroup in which $x * y$ is defined by the ordered triple $(x, y, x * y)$. An $\mathrm{OA}(n, k)$ is equivalent to a set of $k - 2$ mutually orthogonal Latin squares of side n.

Here we must note carefully the distinction between a *subsquare* of a quasigroup and a *subquasigroup* of a quasigroup. A subquasigroup of a Latin

*The authors wish to thank the referee for valuable suggestions.

square on a set N is a subset $U \subset N$ such that $U * U = U$. A subsquare of a Latin square is a triple of subsets $(R, C, E) \subset N^3$ such that $R * C = E$, and $|R| = |C| = |E|$, where $A * B = \{a * b \mid a \in A, b \in B\}$. (We think of R as a subset of the rows, C the columns, E the entries.) For example, a subquasigroup of a group is a subgroup but a subsquare could be, for example, (xH, yH, xyH), where H is a normal subgroup.

We now extend the definition of conjugate from Latin square to Latin subsquare. We recall from [1] that a p-conjugate of a Latin square is a Latin square obtained by applying a permutation p of S_3 (the symmetric group on three symbols) to the coordinates of the $\mathrm{OA}(n, 3)$ to obtain a new $\mathrm{OA}(n, 3)$. The result is always a new Latin square. If we extend the definition to subsquare and obtain, for example, (E, R, C), we obtain a conjugate-fragment which may or may not be a subsquare of our original square. If it is, we say that the subsquare has a p-conjugate subsquare.

Lemma 1. *If a Latin square L has a subsquare H with a p-conjugate subsquare then an isotope of the p-conjugate of H is a subsquare of L.*

1.2. Steiner triple systems

A Steiner triple system (S.T.S.) is a pair $(S, \mathcal{B})$, where $\mathcal{B}$ is a set of 3-subsets of S such that every pair of elements of S is in exactly one member of $\mathcal{B}$. In other words, an S.T.S. is a decomposition of the complete graph on $|S|$ vertices, $K_{|S|}$ into K_3's. A necessary and sufficient condition for this is $n \equiv 1, 3 \pmod 6$ [1]. A Steiner quasigroup is a quasigroup satisfying $x*(x*y) = y$, $x*y = y*x$ and $x*x = x$. A Steiner $\mathrm{OA}(n, 3)$ [6] is an $\mathrm{OA}(n, 3)$, $\mathcal{O}$, such that all of its conjugates are identical, i.e. if $(a_1, a_2, a_3) \in \mathcal{O}$, then $(a_{p(1)}, a_{p(2)}, a_{p(3)}) \in \mathcal{O}$, for all $p \in S_3$. Steiner $\mathrm{OA}(n, 3)$'s are coextensive with Steiner quasigroups. A subsystem of an S.T.S. is a subset U of S such that $(U, \mathcal{B} \cap U^3)$ is an S.T.S.

Lemma 2. *U is a subsystem of a Steiner triple system iff U is a subquasigroup of the corresponding Steiner quasigroup.*

Lemma 3. *If $H = (R, C, E)$ is a subsquare of a Steiner quasigroup and R, C, E are* not *mutually disjoint, then $R = C = E$ and thus H is a subquasigroup.*

Proof. Let $x \in R \cap C$, then as $*$ is idempotent $x \in E$. If $|R| = 1$ we are done; if not, let $y \neq x$, $y \in R$. Now $y*x = a \in E$ and as $x*y = y*x$, $y \in R \cap C$ and so $y \in R \cap C \cap E$, yielding $R = C = E$. Thus, (R, C, E) is a subquasigroup. The cases of the other intersections are similar.

Corollary 1. *If a non-Steiner quasigroup of side u is a subsquare of a Steiner quasigroup of side n, $n \geq 3u$.*

2. Embedding Latin squares

In this section we will regard the subquasigroup as arbitrary and will always think of an embedding of a Latin square in a Steiner quasigroup as an *off-diagonal* embedding. In Table 1 we show the Steiner quasigroup on 7 points, one of its 2×2 subsquares "circled" and one of its conjugates "squared".

Table 1

	1	2	[3]	4	(5)	[6]	(7)
(1)	1	4	7	2	(6)	5	(3)
(2)	4	2	5	1	(3)	7	(6)
3	7	5	3	6	2	4	1
4	2	1	6	4	7	3	5
[5]	6	3	[2]	7	5	[1]	4
6	5	7	4	3	1	6	2
[7]	3	6	[1]	5	4	[2]	7

Theorem 1. *If a Latin square of side u, and denoted (R, C, E), is (off-diagonal) embedded in a Steiner quasigroup S of size n, then:*

(1) *every p-conjugate of (R, C, E) is embedded in S; and*

(2) *any Latin square of side u can be embedded in a Steiner quasigroup of side n.*

Proof. Since S is equivalent to a Steiner OA$(n, 3)$, then if (R, C, E) is a subset of this array so is (C, R, E), etc. Thus, every p-conjugate of (R, C, E) is a subsquare of S for all $p \in S_3$.

To prove (2) we let $\langle R, C, E\rangle$ be the set of unordered triples $\langle r, c, e\rangle$ such that $(r, c, e) \in R \times C \times E \cap \mathrm{OA}(n, 3)$. This set of triples covers exactly the pairs $\{w, v\}$, where w is in one of R, C or E and v is in a different one. This set of pairs is covered by $\langle R, C, E\rangle$ independently of which Latin square of side u was chosen (provided it is on the same set of symbols). Removing $\langle R, C, E\rangle$ from the set of triples of S and replacing it by $\langle R', C', E'\rangle$ from another square of side u thus gives a Steiner quasigroup of side $|S|$ containing the new square.

Note. Any Steiner-quasigroup of side u can be a subquasigroup of a Steiner quasigroup of every side $\geq 2u+1$ [3].

Theorem 2. *A Latin square* (R, C, E) *of side* u *can be embedded in a Steiner quasigroup of side* $3u+k$, *where* k *depends only on the residue class of* u modulo 6. *In particular,* $k=0$ *if* $u \equiv 1$ *or* 3 (mod 6); $k=1$ *if* $u \equiv 0$ *or* 2 (mod 6); $k=3$ *if* $u \equiv 4$ (mod 6); *and* $k=6$ *if* $u \equiv 5$ (mod 6). *Moreover, these values of* k *are as small as possible when the embedding is off-diagonal.*

Proof. In the following the S.T.S. will be constructed and hence the result on Steiner quasigroups will follow.

Case 1: $u \equiv 1$ or 3 (mod 6). There exists an S.T.S. on each of the sets R, C and E. Adding these triples to the set $\langle R, C, E\rangle$ gives a Steiner triple system.

Case 2: $u \equiv 0$ or 2 (mod 6). In this case $3u \equiv 0$ (mod 6) so the minimal value would be of size $3u+1$. Let us denote the extra point by ∞. Then $|R \cup \{\infty\}| \equiv 1$ or 3 (mod 6) and thus there is a Steiner triple system on each of $R \cup \{\infty\}$, $C \cup \{\infty\}$, and $E \cup \{\infty\}$ which when added to the set $\langle R, C, E\rangle$ give an S.T.S. on $R \cup C \cup E \cup \{\infty\}$.

Case 3: $u \equiv 4$ (mod 6). This implies $3u \equiv 0$ (mod 6). The smallest possible system would be of size $3u+1$. Let the extra point be ∞. Since no new triple involves a pair of elements, one from R and one from C or E, $\{\infty\} \cup R$ must be a subsystem. But we have $|\{\infty\} \cup R| \equiv 5$ (mod 6). Thus, the minimal size must be at least $3u+3$. Let the extra points be given by $X = \{\infty_1, \infty_2, \infty_3\}$ and add to the $\langle R, C, E\rangle$ on S.T.S. on each of $R \cup X$, $C \cup X$ and $E \cup X$, in which each of these contains the triple X. We note we could also do this in the case $u \equiv 0$ (mod 6) and achieve an embedding into the next largest possible size, $3u+3$.

(Case 1 is a special case of the generalized direct product, [1]; Case 2 is a special case of the singular direct product [5].)

Case 4: $u \equiv 5$ (mod 6). In this case the first possible values for k are 0, 4, and 6. We shall show that 0 and 4 are impossible and give a recursive construction for $k=6$.

$k=0$. This is impossible since $u \not\equiv 1$ or 3 (mod 6).

$k=4$. Let $S = R \cup C \cup E \cup X$, where $X = \{\infty_1, \infty_2, \infty_3, \infty_4\}$. Consider the graph K_4 with vertices ∞_1, ∞_2, ∞_3, ∞_4. Clearly, the edges represent the unordered pairs from X. We want to partition the edges of K_4 into three subgraphs, S_R, S_C and S_E, so that the edges of each of the graphs $G_Y = K_{|Y|,|X|} \cup K_{|Y|} \cup S_Y$,

$Y \in \{R, C, E\}$, can be partitioned into triples. (Note that $K_{a,b}$ is the complete bipartite graph with bipartition of sizes a and b.) Thus, in G_Y every vertex must have even degree and the number of edges must be divisible by 3. Simple counting shows that for each Y, S_Y must have each vertex of odd degree and 0, 3 or 6 edges. This is impossible.

$k = 6$. A similar argument on edge counting forces the edges of K_6 to be similarly partitioned. It *is* possible to divide K_6 into three copies of H (each vertex is of odd degree). The graph H is the graph on six vertices which looks like the letter H [4]. The partition is given by:

30	01	02	34	35
14	12	13	45	40
25	23	24	50	51

where the edges in each copy of H are given by the rows of the array (Fig. 1).

What is necessary to complete the proof is the following proposition.

Proposition 3. *If* $t \equiv 5 \pmod 6$, $t > 5$, *then there exists a pair* $S_t = (X_t, \mathcal{B}_t)$ *such that*:

(1) $|X_t| = t$;

(2) $\mathcal{B}_t$ *is a collection of three subsets of* X_t *such that every pair is in at most one such triple*; *and*

(3) *the graph of the uncovered pairs is* $\bar{K}_{t-6} \cup \bar{H}$, *where for any graph* G, $\bar{G}$ *denotes the complement of* G.

Proof. We give an explicit listing for S_{11} and S_{17}. The $\bar{H}$ graph left uncovered is the complement of AF, BE, CF, DE, and EF:

$$S_{11}\colon X_{11} = \{A, B, C, D, E, F, \infty, 0, 1, 2, 3\},$$

$$B_{11} = \{EF\infty,\ ED1,\ EB0,\ E32,\ FA3,\ FC2,\ F01,\ D03,\ D\infty 2,\ C0\infty,\ C13,\ B12,\ B\infty 3,\ A02,\ A\infty 1\},$$

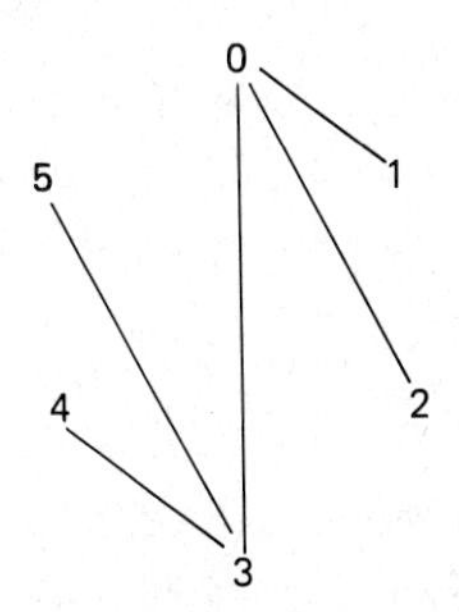

Fig. 1.

S_{17}: $X_{17} = \{A, B, C, D, E, F, \infty, 0, 1, 2, 3, 4, 5, 6, 7, 8, 9\}$,

$B_{17} = \{\infty EF, \infty A0, \infty B2, \infty C1, \infty D3, \infty 45, \infty 67, \infty 89, EB3, ED2, CF0, AF1, C26, C38, C47, C59, A27, A39, A48, A56, F29, F35, F46, F87, B07, B16, B49, B58, D04, D18, D57, D69, E14, E05, E68, E79, 019, 234, 215, 036, 137, 028\}$.

We now note (as has been noted by Schönheim [13], and especially Severn [14]) that the fundamental recursive embedding of an S.T.S. on v vertices into one on $2v+1$ and $2v+7$ vertices (the first using complete one factorizations and the second Skolem sequences) of Rosa (cf., for example, [10, 7, 8]) also works for embedding partial triple systems of size $v \equiv 5 \pmod 6$ in which the original missing pairs are exactly those missing after the embedding. Thus, with the partial systems S_{11} and S_{17} we can obtain S_{6k+5}, $k \geqslant 1$.

With the aid of this we now complete the proof of Theorem 2 by taking a copy of S_{u+6} on each of $R \cup X$, $C \cup X$ and $E \cup X$, making sure that we use a different H of K_6 in each copy. We give an example for a Latin square of order 5 being embedded in an S.T.S. of order 21.

The square:

	a	b	c	d	e
α	0	1	2	3	4
β	1	2	3	4	0
γ	2	3	4	0	1
δ	3	4	0	1	2
ε	4	0	1	2	3

gives rise to the following triples:

$a\alpha 0$	$e\delta 2$	$d\beta 4$	$c\varepsilon 1$	$b\gamma 3$
$b\beta 2$	$a\varepsilon 4$	$e\gamma 1$	$d\alpha 3$	$c\delta 0$
$c\gamma 4$	$b\alpha 1$	$a\delta 3$	$e\beta 0$	$d\varepsilon 2$
$d\delta 1$	$c\beta 3$	$b\varepsilon 0$	$a\gamma 2$	$e\alpha 4$
$e\varepsilon 3$	$d\gamma 0$	$c\alpha 2$	$b\delta 4$	$a\beta 1$

Let the extra points be $ABCDEF$. Then the three sets of $R \cup X$, $C \cup Y$, and $E \cup X$ triples are:

$R \cup X$:	$CD\varepsilon$	$AC\beta$	$CE\alpha$	$C\gamma\delta$	$FD\delta$
	$BD\gamma$	$D\alpha\beta$	$A\alpha\delta$	$A\gamma\varepsilon$	$B\alpha\varepsilon$
	$B\beta\delta$	$E\beta\gamma$	$E\delta\varepsilon$	$F\alpha\gamma$	$F\beta\varepsilon$
$C \cup X$:	ABe	EAb	ADa	Acd	CBd
	FBc	Bab	Ead	Ece	Fae
	Fbd	Dbc	Dde	Cac	Cbe

$E \cup X$: $EF4$	$ED1$	$EB0$	$E23$	$AF3$
$FC2$	$F01$	$D03$	$D24$	$C04$
$C13$	$B12$	$B34$	$A02$	$A14$

If we look at the triples coming from the square itself, we notice that every row and every column contains each element of $R \cup C \cup E$ exactly once. This is no coincidence but has to do with the fact that the Latin square is one of a set of at least 3 mutually orthogonal Latin squares. This is a maximal partial doubly resolvable set (compare with [9]). This leads us to a discussion of arrays of this type similar to Rosa's generalized Howell designs.

This completes the proof of Theorem 2.

Corollary 2.1. *A partial Latin square of side u can be embedded in a Steiner quasigroup of side* $6u + l$, where $l = 1,1,3,1,1,3$ *as* $u \equiv 0,1,2,3,4,5 \pmod 6$.

3. Embedding generalized Howell designs and Rosa's MRDs

We define a d-dimensional generalized Howell design (GHD) following Rosa [11, 12] as follows.

A GHD(d, k, s, n) is a d-dimensional array of size s such that:

(a) every cell is either empty or contains a k-subset of an n-set, X;

(b) each element of X appears in every file exactly once (a file is a set of all cells whose uth coordinate is fixed); and

(c) each k-subset appears at most once in the array.

Theorem 4. *If a set of l mutually orthogonal Latin squares of side n exist then so does a* GHD$(d, l+2-d, n, n(l+2-d))$, *for all* $1 < d \leqslant l$.

Proof. If a set of l mutually orthogonal Latin squares exists then an OA$(n, l+2)$ exists [1]. Let us reinterpret the columns as follows: if $(i_1, i_2, \ldots, i_d, i_{d+1}, i_{d+2}, \ldots, i_{l+2})$ is a row then in cell $(i_1, i_2, \ldots, i_d)$ of a d-dimensional array put the subset $\{(i_{d+1}, d+1), (i_{d+2}, d+2), \ldots, (i_{l+2}, l+2)\}$. This is easily seen to be a GHD$(d, l+2-d, n, n(l+2-d))$. (What we have in the example of the previous section is a GHD$(2,3,5,15)$ based on mutually orthogonal Latin squares of order 5.)

Corollary 5. *If* $d+1$ *mutually orthogonal Latin squares exist, then there is a* $GHD(d,3,n,3n)$ *which can be embedded in an S.T.S. of size* $3n+k$.

Finally, we note that the converse of Theorem 4 is not true. For example: there do not exist 3 M.O.L.S. of order 2, but a GHD$(2,3,3,9)$ exists:

123	456	789
467	189	235
589	237	146

4. Some open questions

Corollary 6. *If two Latin squares have the same Latin square graphs, then the triple systems generated by Theorem 2 are isomorphic.*

Work point 1: Can we modify the construction to obtain the converse of this corollary?

Work point 2: Can any 3-quasigroup be embedded in an S.Q.S.

Work point 3: In the case $u \equiv 5 \pmod 6$, is the S.T.S. ever resolvable?

References

[1] J. Dénes and A.D. Keedwell, Latin Squares and their Applications (Academic Press, New York, 1976).

[2] J. Doyen and A. Rosa, An updated bibliography and survey of Steiner systems, Ann. Discr. Math. 7 (1980) 317–349.

[3] J. Doyen and R.M. Wilson, Embeddings of Steiner triple systems, Discr. Math. 5 (1973) 229–239.

[4] S. Fiorini and R.J. Wilson, Edge-colorings of Graphs, Research Notes in Mathematics (Pittman, 1977).

[5] C.C. Lindner, Construction of quasigroups using the singular direct product, Proc. Amer. Math. Soc. 29 (1971) 263–266.

[6] C.C. Lindner and E. Mendelsohn, On the conjugates of $n^2 \times 4$ orthogonal array, Discr. Math. 20 (1977) 123–132.

[7] C.C. Lindner and A. Rosa, On the existence of automorphism free Steiner triple systems, J. Algebra 34 (1975) 430–443.

[8] C.C. Lindner, E. Mendelsohn and A. Rosa, On the number of 1-factorizations of the complete graph, J. Combinat. Theory (B) 20 (1976) 265–282.

[9] R. Mathon and S.A. Vanstone, On the existence of doubly resolvable Kirkman systems and equidistant permutation arrays, Discr. Math. (to appear).

[10] A. Rosa, On the chromatic number of Steiner triple systems, in: Combinat. Structures and their Applications (Proc. Conf. Calgary, 1969) (Gordon and Breach, New York, 1970) pp. 369–371.

[11] A. Rosa, Generalized room squares and multidimensional room designs, in: Proc. Carribean Conference on Combinatorics & Computing (1977) pp. 209–214.

[12] A. Rosa, Generalized Howell Designs, Ann. N.Y.A.S. 3193 (1979) 484–489.

[13] J. Schönheim, On the number of mutually disjoint triples in Steiner systems and related maximal packing and minimal covering systems, in: Recent Progress in Combinatorics (Proc. 3rd Waterloo Conf., 1968) (Academic Press, New York, 1969) pp. 311–318.

[14] E. Severn, Maximal Partial Triple Systems (to appear).

Received March 1981

Annals of Discrete Mathematics 20 (1984) 55–59
North-Holland

CONVEXITY, GRAPH THEORY AND NON-NEGATIVE MATRICES

Abraham BERMAN*
Department of Mathematics, Technion — Israel Institute of Technology, Haifa 32000, Israel

Problems involving non-negative matrices in which both convexity and graph theory play an important role are surveyed. First, we describe work of Barker and Tam who study the irreducibility of a matrix which leaves a cone invariant via a directed graph of the faces of the cone. The second part of the paper is devoted to graph theoretical characterizations, due mainly to Brualdi, of the extreme points of various polytopes of non-negative matrices.

1. Introduction

Since the subject of this volume is "Convexity and Graph Theory", it may be appropriate to point out how these two disciplines meet in problems concerning non-negative operators.

Two such examples are given. The first involves a graph of faces of a convex cone which is left invariant by a matrix A. This graph is used in studying the irreducibility of A.

The second example describes graph theoretical characterizations of extreme points of polytopes of non-negative matrices.

2. Cone irreducibility

One of the active research areas in matrix theory is the study of operators which leave a cone invariant.

A set K in $\mathbb{R}^n$ is:

a cone, if $\alpha \geqslant 0 \Rightarrow \alpha K \subseteq K$;
convex, if $K + K \subseteq K$;
solid, if $\operatorname{int} K \neq \emptyset$; and
pointed, if $K \cap (-K) = \{0\}$.

All cones in this paper are closed, convex, solid and pointed.

A subcone F of K is a *face* of K if $x \in F$, $y \in K$, $x - y \in K$, $\Rightarrow y \in F$. An

* Research supported in part by Technion's V.P.R., Grant no. 100–473.

extreme ray of K is a one-dimensional face of K. The intersection $\phi(S)$ of all faces of K containing S is a face called the *face generated by S*.

An $n \times n$ real matrix A is *K-non-negative* if $AK \subseteq K$. Such a matrix is *K-irreducible* if the only faces of K left invariant by A are $\{0\}$ and K; *K-primitive* if the only subset of bd K left invariant by A is $\{0\}$; and *K-positive* if $A(K - \{0\}) \subseteq \text{int } K$.

There are several relations between these definitions, for example:

(a) A is K-irreducible if and only if $(I + A)^{n-1}$ is K-positive [11].

(b) A is K-primitive if and only if for some natural number l, A^l is K-positive [1].

Let $\rho(A)$ denote the spectral radius of A and let $\nu(\lambda)$ denote the multiplicity of λ in the minimal polynomial of A. ($\nu(\lambda) = 0$ if λ is not an eigenvalue of A.)

K-non-negative matrices have nice spectral properties [3, 5, 9, 11]:

(a) If A is K-non-negative, then it has an eigenvector in K which corresponds to $\rho(A)$ and if $|\lambda| = \rho(A)$, then $\nu(\lambda) \leq \nu(\rho(A))$.

(b) If A is K-irreducible, then $\rho(A)$ is simple and A has only one (up to scalar multiples) eigenvector in K and this eigenvector (which corresponds to $\rho(A)$) lies in int K.

(c) If A is (K-positive or) K-primitive and $\lambda \neq \rho(A)$ is an eigenvalue of A, then $|\lambda| < \rho(A)$.

When K is $\mathbb{R}^n_+$ — the non-negative orthant in $\mathbb{R}^n$ — A is K-non-negative if and only if all its entries are non-negative, and the above spectral results are part of the classical Perron–Frobenius Theorem. In this case the irreducibility and primitivity of an $n \times n$ non-negative matrix A have elegant graph theoretical characterizations [12]. Let $G(A)$ be a directed graph with vertices $v_1, \ldots, v_n$ and an arc from v_i to v_j if and only if $a_{ij} > 0$. Then A is irreducible if and only if $G(A)$ is strongly connected. In this case let S_i be the set of all lengths m_i of circuits in $G(A)$ through v_i and

$$h_i = \underset{m_i \in S_i}{\text{g.c.d.}} \{m_i\}.$$

Then $h_1 = h_2 = \cdots = h_n$ $(= h)$ and A is primitive if and only if $h = 1$.

We now describe work of Barker and Tam [2], on a similar characterization of K-irreducible matrices.

With a cone K and a matrix A we associate a directed graph $G(A, K)$. The vertices of $G(A, K)$ are the non-zero faces of k and there is an arc from a face F to a face G if and only if G is a face of $\phi((I + A)F)$.

Example. Let

$$K = \mathbb{R}^2_+, \qquad A_1 = \begin{pmatrix} 0 & 1 \\ 1 & 0 \end{pmatrix}, \qquad A_2 = \begin{pmatrix} 1 & 0 \\ 0 & 1 \end{pmatrix}.$$

The corresponding graphs are:

Note that in $G(A, K)$ there are arcs from k to all other faces. Let F be a face of K and consider the chain of faces $\{F_l\}$, where $F_l = \phi((I + A)^l F)$. In $G(A, K)$ there is an arc from F_{l-1} to F_l, so if A is K-irreducible there is a path from F to $K = F_{n-1}$ (and to its faces) and the graph is strongly connected.

Conversely, if $F \neq K$ is a non-zero invariant face of A, there is a face (an extremal ray) G such that $G \cap F = \{0\}$ and there is no path from F to G. This proves

Theorem [2]. *A K-non-negative matrix is K-irreducible if and only if $G(A, K)$ is strongly connected.*

Note that when $K = \mathbb{R}^n_+$, $G(A, K)$ is different from $G(A)$. To have a better analogy, one should consider a graph with extremal rays as vertices. If this graph is strongly connected, then A is K-irreducible, but an example in [2] shows that the converse is not true.

Subgraphs of $G(A, K)$ are used in [2] to study K-primitivity of A.

3. Extreme points of matrix polytopes

The famous Van der Waerden Conjecture, proved only recently [8], states that the permanent of an $n \times n$ doubly stochastic matrix is greater than or equal to $n!/n^n$, with equality only when $(a_{ij}) = (1)$. Here A is doubly stochastic if it is non-negative and all its row sums and column sums are equal to 1 and the permanent of A is

$$\sum_{\sigma \in S_n} \prod_{i=1}^{n} a_{i\sigma(i)}.$$

Let $r = (r_i)$ and $c = (c_j)$ be non-negative vectors in $\mathbb{R}^n$ and $\mathbb{R}^m$, respectively, such that

$$\sum_{i=1}^{m} r_i = \sum_{j=1}^{n} c_j > 0.$$

The polytope of doubly stochastic matrices is a special case of $U(r, c)$ = the polytope of non-negative matrices with row sums r_i, $i = 1, \ldots, m$, and column sums c_j, $j = 1, \ldots, n$.

An important problem in transportation research [10] is scaling a non-negative matrix to one in $U(r, c)$.

To determine the extreme points of $U(r, c)$ we associate with an $m \times n$ non-negative matrix A a bipartite graph $\mathrm{BG}(A)$ with vertex sets $x_1, \ldots, x_m$; $y_1, \ldots, y_n$ and with an edge between x_i and y_j if and only if $a_{ij} > 0$.

Theorem [6]. *A matrix $A \in U(r, c)$ is an extreme point of $U(r, c)$ if and only if $\mathrm{BG}(A)$ is a forest.*

In the special case of doubly stochastic matrices one obtains as a corollary the well-known result of Birkhoff [4] that the permutation matrices of order n are the vertices of the polytope of the $n \times n$ doubly stochastic matrices.

Consider now the following polytopes.

(a) $U(\leqslant r, \leqslant c)$ — the set of non-negative matrices A for which there exists a non-negative matrix B such that $A + B \in U(r, c)$.

(b) $U(r)$ — the set of symmetric $m \times m$ non-negative matrices whose row sums are r_i, $i = 1, \ldots, m$.

(c) $U(\leqslant r)$ — the set of non-negative symmetric A for which there exists a non-negative matrix B such that $A + B \in U(r)$.

Recall that a *simple cactus* in a graph G consists of a simple cycle without chords along with a tree (possibly empty) rooted at each vertex. The cactus is odd if the length of the cycle is odd.

For a symmetric non-negative matrix A, consider the graph $G(A)$ as a non-directed graph (there is an edge between v_i and v_j if and only if $a_{ij} > 0$).

Theorem [7]. *A matrix $A \in U(r)$ is an extreme point of $U(r)$ if and only if the components of $G(A)$ are trees or simple odd cacti.*

Example.

$$A = \begin{bmatrix} 0 & 0 & 0 & 1 \\ 0 & 0 & 1 & 1 \\ 0 & 1 & 0 & 2 \\ 1 & 1 & 2 & 0 \end{bmatrix}$$

is an extreme point of $U(1, 2, 3, 4)$. Here $G(A)$ is the 3-cactus:

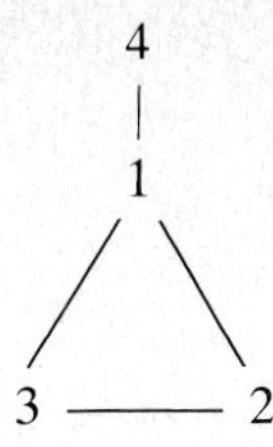

Similar conditions are given in [7] for $U(\leqslant r, \leqslant c)$ and for $U(\leqslant r)$. Additional references are given in [3].

References

[1] G.P. Barker, On matrices having an invariant cone, Czech. Math. J. 22 (97) (1972) 49–68.
[2] G.P. Barker and B.S. Tam, Graphs for cone preserving maps, Linear Algebra Appl. 37 (1981) 199–204.
[3] A. Berman and R.J. Plemmons, Nonnegative Matrices in the Mathematical Sciences (Academic Press, New York, 1979).
[4] G. Birkhoff, Tres observaciones sobre el algebra lineal, Univ. Nac. Tucuman Rev. Ser. A5 (1946) 147–150.
[5] G. Birkhoff, Linear transformations with invariant cones, Amer. Math. Monthly 72 (1967) 274–276.
[6] R.A. Brualdi, Convex sets of non-negative matrices, Canad. J. Math. 20 (1968) 144–157.
[7] R.A. Brualdi, Combinatorial properties of symmetric non-negative matrices, in: Proc. Intern. Conf. Combinatorial Theory (Rome, 1973) (Academia Nazionale dei Lincei, Rome, 1976) pp. 99–120.
[8] G.P. Egorichev, Solution of the van der Waerden permanent conjecture, Preprint, Kirenski Institute of Physics, Kransojarsk (1980).
[9] M.G. Krein and M.A. Rutman, Linear operators leaving invariant a cone in a Banach space, Usp. Mat. Nauk (N.S.) 3 (1948) 3–95.
[10] P. Robillard and N.F. Stewart, Iterative numerical methods for trip distribution problems, Transp. Res. 8 (1974) 575–582.
[11] J.S. Vandergraft, Spectral properties of matrices which have invariant cones, SIAM J. Appl. Math. 16 (1968) 1208–1222.
[12] R.S. Varga, Matrix Iterative Analysis (Prentice-Hall, Englewood Cliffs, New Jersey, 1962).

Received 28 April 1981

Annals of Discrete Mathematics 20 (1984) 61–64
North-Holland

LATTICE POINTS AND LATTICE POLYTOPES

Ulrich BETKE*
Universität Siegen, D 5900 Siegen, Fed. Rep. Germany

1. Introduction

In recent years the number of lattice points in convex bodies has been studied by several authors. In particular, the number of lattice points in lattice polytopes or, more generally, in lattice manifolds has been found to be of some interest. For a bibliography see [1]. Our aim is to describe the connection between the number of lattice points in dilatates of lattice balls and lattice spheres with the combinatorial structure of a given triangulation and some properties of the simplexes in the triangulation. Furthermore, we give some applications of this connection.

2. Notation

In what follows we denote by $\mathbb{Z}^d$ the ordinary lattice of points with integer coordinates in E^d, and by $\mathcal{P}^d$ the set of lattice polytopes in E^d: $\mathcal{P}^d = \{P \mid P = \operatorname{conv}\{x_0, \dots, x_n\},\ x_i \in \mathbb{Z}^d\}$. Furthermore, we understand a lattice manifold to be a geometrical cell-complex C, where all cells are lattice polytopes, and set C is a manifold.

By the face-figure M/P of a d-dimensional manifold M at one of its j-dimensional faces P, we understand the $(d-j-1)$-dimensional cell-complex whose k-faces are the $(k+j+1)$-dimensional faces of M containing P, $k = -1, \dots, d-j-1$.

The number of j-faces of a cell-complex C is denoted by $f_j(C)$, and the g-numbers of a d-dimensional cell-complex C as introduced in [4] are defined by:

$$g_k^{(d+1)}(C) = \sum_{j=-1}^{k} (-1)^{k-j} \binom{d-j}{d-k} f_j(C),$$

for $-1 \leq k \leq d$, where we adopt the convention $f_{-1}(C) = 1$.

* This paper contains material from joint work with Dr. Peter McMullen (see [1a]).

$G(M) = \text{card}\{M \cap \mathbb{Z}^d\}$ is the number of lattice points in M, and nM is the dilatate of M by $n \in \mathbb{N}$.

3. Properties of the lattice-point enumerator

The lattice point enumerator G of lattice manifolds has several interesting properties (a more detailed list of results can be found in [1]):

(1) The inclusion-exclusion principle holds, i.e. for $M = \bigcup_{i=1}^n P_i$:

$$G(M) = \sum_{i=1}^{n} G(P_i) - \sum_{i<j} G(P_i \cap P_j) + \sum_{i<j<k} G(P_i \cap P_j \cap P_k) - \cdots + (-1)^{n-1} G\left(\bigcap_{i=1}^{n} P_i\right).$$

(2) G is invariant under unimodular transformations, i.e. for any transformation $\beta(x) = (a_{ij})x + (b_i)$, a_{ij}, $b_i \in \mathbb{Z}$, $\det(a_{ij}) = 1$:

$$G(\beta(M)) = G(M).$$

(3) For any dilatation by integers, $G(nM)$ is a polynomial of degree d in n [2]:

$$G(nM) = \sum_{i=0}^{d} n^i G_i(M), \quad \dim M = d.$$

The proof of (3) is a consequence of (1) and the following fundamental lemma of Ehrhart.

Lemma 1. *Let $S = \text{conv}\{x_0, \ldots, x_d\}$ be a lattice simplex, then:*

$$G(nS) = \sum_{s=0}^{d} C_s(S) \binom{d-s+n}{d}$$

with

$$C_s(S) = \text{card}\left\{x \in \mathbb{Z}^d \mid x = \sum_{i=0}^{d} \lambda_i x_i,\ 0 \leq \lambda_i < 1,\ \sum_{i=0}^{d} \lambda_i = s\right\}.$$

The following theorem describes the situation in (3) more explicitly:

Theorem. *Let M be a triangulated lattice ball or a lattice sphere, $\dim M = d$, and let $S_1^{(j)}$ be the simplexes of the triangulation (j denoting the dimension of the simplex $S_1^{(j)}$). Then:*

$$G(nM) = \sum_{S_1^{(j)}} \sum_{k=0}^{d-j} \sum_{s=0}^{j} C_s^*(S_1^{(j)}) g_{k-1}^{(d-j)}(M/S_1^{(j)}) \binom{d-k-s+n}{d},$$

with

$$C_s^*(S_1^{(j)}) = \text{card}\left\{x \in \mathbb{Z}^d \mid x = \sum_{i=0}^{j} \lambda_i x_i,\ 0 < \lambda_i < 1,\ \sum_{i=0}^{j} \lambda_i = s\right\},$$

for $S_1^{(j)} = \text{conv}\{x_0, \ldots, x_j\}$.

The theorem has some consequences:

Corollary 1.

$$G(nM) = \sum_{s=0}^{d+1} C_s(M)\binom{d-s+n}{d},$$

with $C_s(M) \in \mathbb{N} \cup \{0\}$, $s = 0, \ldots, d+1$, *for any* d*-dimensional lattice ball or lattice sphere* M, *and*

$$\sum_{s=0}^{d} sC_s(M) \leq (d+1)/2 \sum_{s=0}^{d} C_s(M) - 1; \qquad C_{d+1}(M) = 0,$$

for lattice balls, and

$$C_s(M) = C_{d+1-s}(M)$$

for lattice spheres.

Corollary 2. *Let* $P \in \mathcal{P}^d$, *P having at least one lattice point in its interior. Then*:

$$C_s(P) = A_s(P) + B_s(P),$$

with $A_s(P)$, $B_s(P) \in \mathbb{N} \cup \{0\}$ *and* $A_s(P) = A_{d-s}(P)$, $B_s(P) = B_{d-s+1}(P)$.

Corollary 3. *Let* M *be a* d*-dimensional lattice ball and* $V(M)$ *be its volume. Then*:

(a) $$G(nM) \leq \binom{d+n-1}{d} d!\,V(M) + \binom{n+d-1}{d-1},$$

(b) $$G(nM) \geq \binom{d+n}{d} + (d!\,V(M) - 1)\binom{(d-1)/2+n}{d}, \qquad d = 2k+1,$$

$$G(nM) \geq \binom{d+n}{d} + \tfrac{1}{2}(d!\,V(M) - 1)\left\{\binom{d/2+n}{d} + \binom{d/2+n-1}{d}\right\},$$

$$d = 2k.$$

(a) and (b) are the best possible, since equality holds for certain simplexes.

For $n = 1$, (a) is a result of Blichfeldt, compare Lekkerkerker [3]. The inequalities (b) are rather interesting since there are polytopes with arbitrarily large volume containing just $d+1$ lattice points.

The proof of the theorem depends on Lemma 1, the results of McMullen–Walkup [4] on g-numbers, and the following:

Lemma 2. *Let M be a lattice manifold,* $\dim M = d$, *and $S_1^{(j)}$ the j-dimensional cells of M. Then*:

$$G(M) = \sum_{S_1^{(j)} \not\subset \operatorname{bd} P} (-1)^{d-j} G(S_1^{(j)}).$$

References

[1] U. Betke and J.M. Wills, Stetige und diskrete Funktionale konvexer Körper, in: Contributions to Geometry (Birkhäuser, 1979) pp. 226–237.

[1a] U. Betke and P. McMullen, Lattice points in lattice polytopes, to appear.

[2] E. Ehrhart, Sur un problème de géométrie diophantienne linéaire, J. Reine Angew. Math. 226 (1967) 1–29; 227 (1967) 25–49.

[3] C.G. Lekkerkerker, Geometry of Numbers (Wolters-Noordhoff, 1969).

[4] P. McMullen and D.W. Walkup, A generalized lower-bound conjecture for simplicial polytopes, Mathematika 18 (1971) 264–273.

Received 27 April 1981

Annals of Discrete Mathematics 20 (1984) 65–66
North-Holland

A NEW UPPER BOUND FOR THE CARDINALITY OF 2-DISTANCE SETS IN EUCLIDEAN SPACE

A. BLOKHUIS

Department of Mathematics, Eindhoven University of Technology, Eindhoven 5600 MB, The Netherlands

It is proved that the cardinality of a 2-distance set S in Euclidean d-dimensional space satisfies

$$\mathrm{card}(S) \leq \tfrac{1}{2}(d+1)(d+2).$$

A set S in Euclidean d-space, E^d, is called a 2-distance set if the distance between distinct points of S assumes only two values. The maximum size of such a set is 5 in E^2 (Kelly), and 6 in E^3 (Croft). Delsarte, Goethals and Seidel [1] treated the case where the points of S lie on a sphere. Their argument can be modified to obtain the bound $\mathrm{card}(S) \leq \frac{1}{2}(d+1)(d+4)$ for general 2-distance sets as was established by Larman, Rogers and Seidel [2]. Bannai and Bannai [3] showed that equality does not occur in this case. The proof of Larman, Rogers and Seidel can be modified again to obtain $\mathrm{card}(S) \leq \frac{1}{2}(d+1)(d+2)$.

Theorem. *Let S be a 2-distance set in E^d, then*

$$\mathrm{card}(S) \leq \tfrac{1}{2}(d+1)(d+2).$$

Proof. Let a and b be the distances in S. For each point s in S and $x \in E^d$ we define

$$F_s(x) = \frac{1}{a^2 b^2}(\|x-s\|^2 - a^2)(\|x-s\|^2 - b^2).$$

These functions form an independent set of functions since $F_s(t) = \delta_{s,t}$ for all $s, t \in S$. They are linear combinations of the following functions:

$$\|x\|^4; \qquad \|x\|^2 x_i; \qquad x_i x_j; \qquad x_i; \qquad 1, \quad \text{where } 1 \leq i \leq j \leq d.$$

Hence, the total number of functions F_s cannot exceed

$$1 + d + \tfrac{1}{2}d(d+1) + d + 1 = \tfrac{1}{2}(d+1)(d+4).$$

We proceed to show that in fact the set

$$\{F_s(x), x_i, 1 \mid s \in S, 1 \leq i \leq d\}$$

is linearly independent, which implies

$$\text{card}(S) + d + 1 \leq \tfrac{1}{2}(d+1)(d+4),$$

and hence

$$\text{card}(S) \leq \tfrac{1}{2}(d+1)(d+2).$$

Now suppose we have:

$$\sum_{s \in S} c_s F_s(x) + \sum_{i=1}^{d} c_i x_i + c = 0. \tag{1}$$

Inserting s into relation (1) we get:

$$c_s + \sum_i c_i s_i + c = 0. \tag{2}$$

Inserting ke_i into (1), where e_i is the ith column of the unit matrix, we get:

$$\frac{1}{a^2 b^2} \sum_s c_s (k^2 - 2ks_i + \|s\|^2 - a^2)(k^2 - 2ks_i + \|s\|^2 - b^2) + kc_i + c = 0. \tag{3}$$

Comparing the coefficients of k^4 and of k^3 we obtain:

$$\sum_s c_s = 0 \quad \text{and} \quad \sum_s c_s s_i = 0, \tag{4}$$

for $i = 1, \ldots, d$.

Multiply relation (2) by c_s and sum over all $s \in S$:

$$\sum_s c_s^2 + \sum_i c_i \sum_s c_s s_i + c \sum_s c_s = 0. \tag{5}$$

Now (4) and (5) yield $c_s = 0$ for all $s \in S$, whence also $c = c_i = 0$ for $i = 1, \ldots, d$. This completes the proof of the theorem.

References

[1] Ph. Delsarte, J.M. Goethals and J.J. Seidel, Spherical codes and designs, Geometrica Dedicata 6 (1977) 363–388.

[2] D.G. Larman, C.A. Rogers and J.J. Seidel, On two-distance sets in Euclidean space, Bull. London Math. Soc. 2 (1977) 261–267.

[3] E. Bannai and E. Bannai, An upper bound for the cardinality of an s-distance subset in Euclidean space (to appear in Combinatorica).

Received 28 April 1981

Annals of Discrete Mathematics 20 (1984) 67–73
North-Holland

GEODESICS IN ORIENTED GRAPHS

Béla BOLLOBÁS
University of Cambridge, Cambridge, U.K.

A *geodesic* from a to b in a directed graph is the shortest directed path from a to b. R.C. Entringer made the beautiful conjecture that every graph has an orientation in which every geodesic is unique. Unfortunately this is not true; the aim of this note is to present two different ways of disproving it. First, we show that for a large enough q the Paley graph Q_q has no orientation in which every geodesic is unique, and then we apply a probabilistic argument to show that almost no graph has an orientation with unique geodesics. *I* am grateful to Adrian Bondy for bringing Entringer's conjecture to my attention.

Let q be a prime power and write $\mathbb{F}_q$ for the field of order q. If q is a power of a prime congruent to 1 modulo 4, then the Paley graph Q_q is defined on $\mathbb{F}_q$ by joining a to b if $a-b$ is a quadratic residue. It is well known that Q_q is a strongly regular graph of degree $(q-1)/2$, in which two adjacent vertices have $(q-1)/4$ common neighbours. For the sake of completeness we give detailed proofs of the following folklore lemmas concerning Paley graphs. I am grateful to J.W.S. Cassels for the slick proof of the first.

Lemma 1. *Let χ be a non-principal (multiplicative) character of $\mathbb{F}_q$ and let $S \subset \mathbb{F}_q$ be an arbitrary set of s elements. Then*:

$$\left| \sum_{a,b \in S} \chi(a-b) \right| \leq q^{1/2} s.$$

Proof. Let ψ be a non-principal additive character. It is well known (see [4, p. 47]) that the Gaussian sum $G(\bar{\chi}, \psi)$ has modulus $q^{1/2}$:

$$|G(\bar{\chi}, \psi)| = \left| \sum_x \bar{\chi}(x)\psi(x) \right| = q^{1/2}, \tag{1}$$

where the summation is over all values in $\mathbb{F}_q$. Note that for $c \in \mathbb{F}_q$,

$$\chi(c)G(\bar{\chi}, \psi) = \sum_y \bar{\chi}(y)\psi(cy).$$

Indeed, this is trivial for $c = 0$, and if $c \neq 0$, then

$$\sum_x \chi(c)\bar{\chi}(x)\psi(x) = \sum_x \bar{\chi}(x/c)\psi(x) = \sum_y \bar{\chi}(y)\psi(cy).$$

Consequently, by (1):

$$\begin{aligned}
q^{1/2}\Big|\sum_{a,b\in S}\chi(a-b)\Big| &= \Big|G(\bar{\chi},\psi)\sum_{a,b\in S}\chi(a-b)\Big| \\
&= \Big|\sum_{a,b\in S}\sum_y \bar{\chi}(y)\psi((a-b)y)\Big| \\
&= \Big|\sum_y \bar{\chi}(y)\sum_{a,b\in S}\psi((a-b)y)\Big| \\
&= \Big|\sum_y \bar{\chi}(y)\Big|\sum_{c\in S}\psi(cy)\Big|^2\Big| \\
&\leqslant \sum_y \Big|\sum_{c\in S}\psi(cy)\Big|^2 = \sum_y\sum_{a,b\in S}\psi((a-b)y) \\
&= \sum_{a,b\in S}\sum_y \psi((a-b)y) = \sum_{c\in S}\sum_y \psi(0.y) = sq,
\end{aligned}$$

implying the lemma. □

Lemma 2. *Let $S\subset \mathbb{F}_q$ be an arbitrary set of s vertices of the Paley graph Q_q and denote by $e(S)$ the number of edges in the subgraph spanned by S. Then*:

$$\left|e(S)-\frac{1}{2}\binom{s}{2}\right| \leqslant \tfrac{1}{4}sq^{1/2}.$$

Proof. Let χ be the quadratic residue character. Then:

$$\Big|\sum_{a,b\in S}\chi(a-b)\Big| = \left|2e(S)-2\left\{\binom{s}{2}-e(S)\right\}\right| = \left|4e(S)-2\binom{s}{2}\right|,$$

so the assertion follows from Lemma 1. □

Lemma 3. *Let χ be a non-principal character of $\mathbb{F}_q$ and let U and W be disjoint subsets of $\mathbb{F}_q$. Then*:

$$\begin{aligned}
&\Big|\sum_{x\notin U\cup W}\prod_{u\in U}(1+\chi(x-u))\prod_{w\in W}(1-\chi(x-w))-q\Big| \\
&\qquad\leqslant \{(m-2)2^{m-1}+1\}q^{1/2}+m2^{m-1},
\end{aligned}$$

where $m=|U|+|W|$.

Proof. As above, we write Σ_x to denote summation over the whole of $\mathbb{F}_q$. Clearly,

$$\left| q - \sum_x \sum_{Z \subset U \cup W} (-1)^{|Z \cap W|} \chi\left(\prod_{z \in Z} (x - z) \right) \right|$$

$$= \left| \sum_x \sum_{Z \subset U \cup W}' (-1)^{|Z \cap W|} \chi(P_Z(x)) \right| = \left| \sum_{Z \subset U \cup W}' (-1)^{|Z \cap W|} \sum_x \chi(P_Z(x)) \right|,$$

where $\sum'$ denotes summation over the non-empty subsets of $U \cup W$ and $P_Z(x)$ denotes the polynomial $\Pi_{z \in Z} (x - z)$.

By Weil's inequality concerning the Riemann hypothesis for algebraic curves over finite fields (see [4, p. 43]) we have:

$$\left| \sum_x \chi(P_Z(x)) \right| \leq (|Z| - 1)q^{1/2}.$$

Hence, the sum above is bounded by

$$(m2^{m-1} - 2^m + 1)q^{1/2}.$$

Since

$$\left| \sum_{x \in U \cup W} \prod_{u \in U} (1 + \chi(x - u)) \prod_{w \in W} (1 - \chi(x - w)) \right| \leq m2^{m-1},$$

the lemma follows. □

Lemma 4. *Let U and W be disjoint subsets of vertices of the Paley graph Q_q and denote by $v(U, W)$ the number of vertices joined to each vertex of U and no vertex of W. Then*:

$$|v(U, W) - 2^{-m}q| \leq \tfrac{1}{2}(m - 2 + 2^{-m+1})q^{1/2} + m/2,$$

where $m = |U| + |W|$.

Proof. Note that if χ is the quadratic residue character, then:

$$\prod_{u \in U} (1 + \chi(x - u)) \prod_{w \in W} (1 - \chi(x - w)) = 2^m v(U, W).$$

Apply Lemma 3. □

Now we are ready to prove the main result of the note. Given a directed graph $\vec{G}$, write $E(\vec{G})$ for the edge set of $\vec{G}$, and $d^+(x)$ for the outdegree of a vertex x. If there is an edge from x to y, we say that x *dominates* y. Four vertices, a, b, c and d, are said to form a *diamond* of $\vec{G}$ if $ab, ac, bd, cd \in E(\vec{G})$ and $ad \notin E(\vec{G})$. Clearly, the geodesics of length 2 are unique iff there is no diamond. Thus, the following theorem disproves the conjecture in a considerably stronger sense than mentioned above.

Theorem 5. *Let $q \geqslant 6^4$ be a power of a prime congruent to* 1 *modulo* 4. *Then every orientation of the Paley graph Q_q contains a diamond.*

Proof. Suppose $G = Q_q$ has an orientation $\vec{G}$ without a diamond. Denote by S the set of vertices of outdegree at least $(q-1)/4 + q^{1/2}$ and put $s = |S|$. By Lemma 2 the subgraph of G spanned by S has at least

$$\tfrac{1}{4}s(s - 1 - q^{1/2})$$

edges so there are vertices $x_0, x_1, \ldots, x_t \in S$,

$$t \geqslant \tfrac{1}{4}s(s-1-q^{1/2})/s = \tfrac{1}{4}(s - 1 - q^{1/2}), \tag{2}$$

such that x_0 dominates each x_i, $1 \leqslant i \leqslant t$. Now each x_i is joined to $(q-5)/4$ neighbours of x_0 so each x_i dominates at most $(q-5)/4$ of the vertices dominated by x_0. Consequently, each x_i dominates at least

$$\frac{q-1}{4} + q^{1/2} - \frac{q-5}{4} = q^{1/2} + 1$$

vertices not dominated by x_0. Since $\vec{G}$ contains no diamond, a vertex not dominated by x_0 is dominated by at most one x_i, $1 \leqslant i \leqslant t$. Consequently,

$$t(q^{1/2} + 1) \leqslant q - 1 - \frac{q-1}{2},$$

so

$$t \leqslant \tfrac{1}{2}(q^{1/2} - 1).$$

Comparing this with (2) we see that

$$s \leqslant 3q^{1/2} - 1.$$

Since by reversing the orientations of the edges we obtain another orientation without a diamond, there are at most $3q^{1/2} - 1$ vertices of indegree at least $(q-1)/4 + q^{1/2}$. Therefore if we denote by M the set of vertices whose outdegrees and indegrees lie strictly between $(q-1)/4 - q^{1/2}$ and $(q-1)/4 + q^{1/2}$, then

$$m = |M| \geqslant q - 6q^{1/2} + 2.$$

Analogously to (2) we find that there are vertices $y_0, y_1, \ldots, y_u \in M$ such that

$$u \geqslant \tfrac{1}{4}(m - 1 - q^{1/2}) \geqslant \tfrac{1}{4}(q - 7q^{1/2} + 1),$$

and y_0 dominates each y_i, $1 \leqslant i \leqslant u$. Let $z_1, z_2, \ldots, z_w$ be the other vertices dominated by y_0 and put $U = \{y_1, y_2, \ldots, y_n\}$ and $W = \{z_1, z_2, \ldots, z_w\}$. Each vertex not in $\{y_0\} \cup U \cup W$ is dominated by at most one y_i. Hence, if $e(U, W)$

denotes the number of $U-W$ edges directed into W and $e(U)$ denotes the number of edges spanned by U, we have:

$$\sum_{i=1}^{u} d^{+}(y_i) \leq e(U) + e(U, W) + q - 1 - u - w. \tag{3}$$

By Lemma 2:

$$e(U) \leq \tfrac{1}{4}u(u - 1 + q^{1/2}).$$

By Lemma 4 Q_q does not contain a $K(3, r)$ with $r = \frac{1}{8}q + \frac{5}{8}q^{1/2} + \frac{3}{2}$, so by [1, Lemma 2.1, Ch. VI]:

$$e(U, W) \leq ud,$$

where

$$\binom{d}{3} = \frac{r-1}{u}\binom{w}{3}.$$

However, instead of this estimate we shall be satisfied with the crude bound

$$e(U, W) \leq uw.$$

Hence, (3) gives

$$u\left(\frac{q-1}{4} - q^{1/2}\right) \leq \tfrac{1}{4}u(u - 1 + q^{1/2}) + uw + q - 1 - u - w, \tag{4}$$

where

$$\tfrac{1}{4}(q - 7q^{1/2} + 1) \leq u \leq u + w \leq \tfrac{1}{4}(q-1) + q^{1/2}.$$

Rearranging (4) we find that

$$u(q - 1 - 4q^{1/2} - u + 1 - q^{1/2} - 4w + 4) \leq 4q - 4 - 4w.$$

Since $q \geq 6^4$, this inequality has to hold when u has the minimal value permitted by (5) and w has the maximal value, so

$$(q - 1 - 4q^{1/2})(3q - 60q^{1/2} + 25) \leq 16(4q - 11q^{1/2} - 2).$$

Since this inequality does not hold if $q \geq 6^4$, the theorem is proved. □

We say that almost no graph has a certain property if the proportion of all graphs with n labelled vertices having the property tends to 0 as $n \to \infty$.

Theorem 6. *Almost no graph has an orientation in which every geodesic is unique.*

Proof. The proof is based on the simple fact that if α, β and γ are arbitrary positive constants, then almost every graph has the following two properties.

(a) If W is any set of $w \geqslant \alpha n$ vertices, then

$$\left| e(W) - \frac{1}{2}\binom{w}{2} \right| < \frac{\beta}{2}\binom{w}{2},$$

where $e(W)$ denotes the number of edges joining vertices in W.

(b) Any two vertices have at most $\frac{1}{2}n(1+\gamma)$ common neighbours.

These assertions are rather far from being best possible; see [3] for proofs of considerably stronger properties.

Now choose positive constants α, β, γ and δ so that $\alpha < \frac{1}{3}$, $\gamma < \delta$ and $2(1-2\alpha)(1-\beta)(1-\delta) > (1+\beta)(1+\delta)^2$. Let G be a graph satisfying (a) and (b). To prove the theorem it suffices to show that if n is sufficiently large, then every orientation of G contains a diamond. Since the proof is analogous to the proof of Theorem 5 we do not give all the details.

Suppose there is an orientation $\vec{G}$ without a diamond. Let S be the set of vertices of outdegree at least $(n/4)(1+\delta)$. If $s = |S| \geqslant \alpha n$, then by (a) some vertex of S dominates at least $[(s-1)/2](1-\beta)$ vertices of S and then by (b) we must have:

$$\frac{s-1}{2}(1-\beta)\frac{n}{4}(\delta-\gamma) < n,$$

which fails for large n. Hence, if n is large, $s < \alpha n$. By reversing the orientation of the edges we find that if M is the set of vertices whose outdegree differs from $n/4$ by less than $(\delta/4)n$, then $m = |M| \geqslant (1-2\alpha)n$. By (a) some vertex y_0 of M dominates at least $\frac{1}{4}(1-2\alpha)(1-\beta)n$ other vertices of M. Since a vertex not dominated by y_0 is dominated by at most one of these vertices, by (a) we find that:

$$\tfrac{1}{4}(1-2\alpha)(1-\beta)n\frac{n}{4}(1-\delta) \leqslant (1+\beta)(1+\delta)^2\frac{n^2}{32} + n.$$

Since this fails for large n, the proof is complete. □

In conclusion let us note that Theorem 6 can be extended to other classes of random graphs. As customary, denote by $\mathcal{G}(n, P(\text{edge}) = p)$ the probability space consisting of all graphs with a fixed set of n labelled vertices, in which the edges are chosen independently and with probability p. (For the basic properties of random graphs see [2, Ch. VII].) Clearly, $\mathcal{G}(n, P(\text{edge}) = \frac{1}{2})$ is exactly the space of all graphs with n labelled vertices, in which all graphs have the same probability. By taking a little more care with the calculations we can show that Theorem 6 holds for all $\mathcal{G}(n, P(\text{edge}) = p)$, provided $p = p(n)$ satisfies $Cn^{-1/2} \leqslant p(n) \leqslant 1-\varepsilon$, where C is a suitable constant and ε is any positive constant.

References

[1] B. Bollobás, Extremal Graph Theory (Academic Press, London, New York and San Francisco, 1978).
[2] B. Bollobás, Graph Theory — An Introductory Course, Graduate Texts in Mathematics, Vol. 63 (Springer-Verlag, New York, Heidelberg and Berlin, 1979).
[3] B. Bollobás, Degree sequences of random graphs, Discrete Mathematics 33 (1981) 1–19.
[4] W.M. Schmidt, Equations over Finite Fields — An Elementary Approach, Lecture Notes in Mathematics, Vol. 536 (Springer-Verlag, Berlin–Heidelberg–New York, 1976).

Received March 1981

Annals of Discrete Mathematics 20 (1984) 75–82
North-Holland

RECENT RESULTS ON THE STRONG PERFECT GRAPH CONJECTURE*

Mark A. BUCKINGHAM**
Courant Institute, New York University, New York, New York 10012, U.S.A.

Martin Charles GOLUMBIC***
Bell Laboratories, Murray Hill, New Jersey 07974, U.S.A.

A graph G is called *perfect* if for every induced subgraph H of G, the minimum number of colors needed to color the vertices of H equals the size of the largest clique in H. In the early 1960s, Claude Berge conjectured that a graph G is perfect if and only if neither G nor its complement contain an odd length chordless cycle, and this has become known as the Strong Perfect Graph Conjecture (SPGC). A circle graph is one whose vertices can be put into one-to-one correspondence with the chords of a circle such that two vertices are adjacent if and only if their corresponding chords intersect. In this work we will prove that the SPGC is true for the class of circle graphs.

1. Introduction

A graph G is called *perfect* if for every induced subgraph H of G, the minimum number of colors needed to color the vertices of H equals the size of the largest clique in H. Perfect graphs have been of particular interest in recent years since certain classes of perfect graphs arise naturally in a variety of applications, including optimization of computer storage, analysis of genetic structure, some routing problems for VLSI chips, synchronization of parallel processes, and certain scheduling problems. Golumbic [9] summarizes many such applications. A graph G is *p-critical* if it is minimally imperfect, that is, G itself is not perfect but every proper induced subgraph of G is perfect. The only known p-critical graphs are the odd holes C_{2k+1} and the odd antiholes $\bar{C}_{2k+1}$ on $2k+1 \geq 5$ vertices. The Berge Strong Perfect Graph Conjecture can be stated as follows:

SPGC. *If G is imperfect, then it contains an odd hole or an odd antihole.*

* This work was supported in part by the National Science Foundation under Grant no. MCS78–03820.
** Current affiliation: Arthur Anderson & Co., New York, NY 10105, U.S.A.
*** Current affiliation: IBM Israel Scientific Center, Technion City, Haifa, Israel.

Attempts at proving the SPGC have followed in two main directions. The first direction has been to determine properties of minimally imperfect graphs and has led to the study of partitionable graphs [2, 4–6]. The second direction has been to prove that the SPGC holds when restricted to special classes of graphs such as planar graphs, circular-arc graphs, and others [11, 16, 17, 19, 20, 21]. In this paper we will prove the SPGC for circle graphs. The proof is self-contained and uses only Properties 1–5 of Section 2. An alternative proof relying on the SPGC for $K_{1,3}$-free graphs [16] is given in [3].

2. Partitionable graphs

Bland, Huang and Trotter [2] introduced a notion that is closely related to p-critical graphs. We give a seemingly weaker, but equivalent, formulation. Let G be an undirected graph, and let

$\alpha(G)$ be the size of a maximum stable set,
$\omega(G)$ be the size of the maximum clique,
$\chi(G)$ be the size of the minimum stable set cover, and
$\theta(G)$ be the size of a minimum clique cover.

A graph G having n vertices is *partitionable* if there exist integers $\alpha, \omega \geq 2$, such that $n = \alpha\omega + 1$, and $\alpha \geq \theta(G-x)$ and $\omega \geq \chi(G-x)$ for all vertices x. Buckingham [3] has shown that this implies that, for all vertices x, $G-x$ can be partitioned into α ω-cliques and also into ω α-stable sets, which was the definition for partitionable graphs originally given in [2]. It is an immediate result of the Perfect Graph Theorem [12, 13] that *every p-critical graph is partitionable.* Examples of graphs which are partitionable but not p-critical have been given in [2, 5, 6]. It can be further shown that for partitionable graphs $\alpha = \alpha(G)$ and $\omega = \omega(G)$, and, more importantly, that *G is partitionable if and only if*:

(i) $n = \alpha\omega + 1$,
(ii) G contains exactly n ω-cliques and n α-stable sets,
(iii) every vertex of G is contained in exactly ω ω-cliques and α α-stable sets, and
(iv) each ω-clique intersects all but one α-stable set, and vice versa.

(See [2–4, 9].)

Let $\Theta(G)$ denote a minimum clique cover of G and $X(G)$ denote a minimum stable set cover. Let Adj(x) be the set of vertices adjacent to x, and let $N(x) = x \cup \mathrm{Adj}(x)$ denote the neighborhood of a vertex x. From conditions (i)–(iv) we may deduce the following three properties of partitionable graphs as given in [21]. Let x be any vertex of a partitionable graph G.

Property 1. The clique cover $\Theta(G-x)$ and the stable set cover $X(G-x)$ are unique. $\Theta(G-x)$ consists of α ω-cliques, and $X(G-x)$ consists of ω α-stable sets.

Property 2. $S \cup S'$ is connected for any two α-stable sets S, $S' \in X(G-x)$.

Property 3. $G - N(x)$ is connected.

The following new properties of partitionable graphs are proved in [3, 4] and will be used in the next section.

Property 4. $S \cup S' \cup x$ is biconnected for any two α-stable sets S, $S' \in X(G-x)$.

Property 5. $2\omega - 2 \leqslant |\mathrm{Adj}(x)| \leqslant n - 2\alpha + 1$.

Partitionable graphs can be used to prove the SPGC for certain classes of graphs as follows. Let EXTRA be a graph theoretic property which is hereditary, that is, if G is an EXTRA graph, then every induced subgraph of G is also an EXTRA graph. For example, planarity is a hereditary property while connectivity is not. If G_1 is an imperfect graph which satisfies the property EXTRA, then G_1 contains a p-critical subgraph G_2 which is therefore a partitionable EXTRA graph. Hence, the SPGC reduces to the following.

SPGC for EXTRA graphs. *If G is a partitionable EXTRA graph, then G contains an odd hold or odd antihole.*

In the remainder of this paper the property EXTRA will be that of being a circle graph.

3. Preliminary results

A *circle graph* is a graph G whose vertices can be put into one-to-one correspondence with the chords of a circle such that two vertices of G are adjacent if and only if their corresponding chords intersect. We say that the circle with its chords *derives* G. Circle graphs have been studied by Even and Itai [7], Gavril [8], Golumbic [9, Ch. 11], and Buckingham [3]. One application in which circle graphs arise involves optimizing the use of storage in micro computers [10].

In this section we will prove some properties of partitionable circle graphs which will be used in the next section to establish the main theorem.

Lemma 1. *If S is a stable set of a partitionable circle graph G, then in any circle with chords that derives G, S corresponds to a collection of chords that go 'around' the circle (see Fig.* 1).

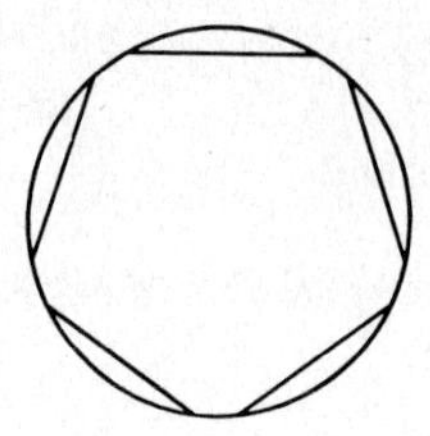

Fig. 1. Chords going 'around' the circle.

Proof. Suppose not, then there will be a triple of chords with one sandwiched between the other two (Fig. 2). Removing the middle chord of the triple and all chords intersecting it results in a circle with chords that derives $G - N(x)$, where x corresponds to the middle chord. Furthermore, the two vertices corresponding to the other two chords of our triple will be in different connected components of $G - N(x)$. However, by Property 3, $G - N(x)$ must be connected, a contradiction.

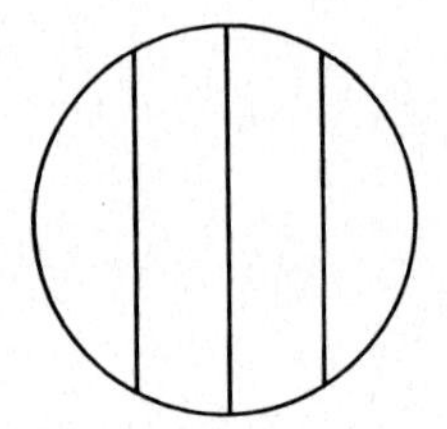

Fig. 2. A chord sandwich.

An immediate consequence of Lemma 1 is the following.

Corollary 2. *Let G be a partitionable circle graph. For any vertex x and any two stable sets $S_1, S_2 \in X(G - x)$, the union $S_1 \cup S_2$ corresponds to a collection of chords that forms either*

(i) *a chordless path going 'around' the circle, or*

(ii) *a chordless cycle (see Fig.* 3).

Lemma 3. *If G is a partitionable circle graph, then $|\mathrm{Adj}(x)| = 2\omega - 2$ for all vertices x.*

Proof. Choose a vertex x and let y be a vertex not adjacent to x. Consider the

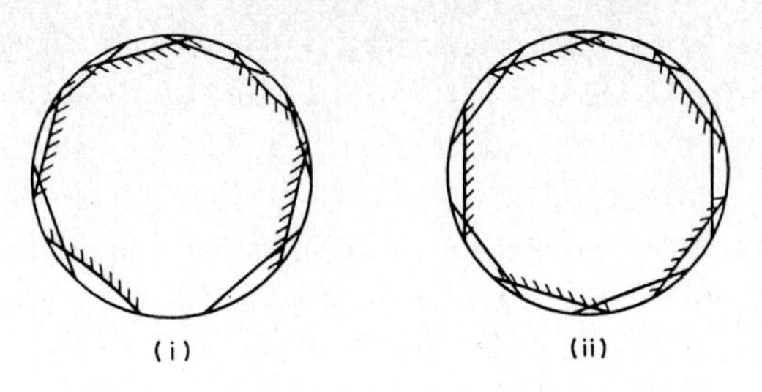

Fig. 3. (i) A chordless path going 'around' the circle, and (ii) a chordless cycle. The chords alternate between stable sets S_1 and S_2.

stable set partition $X(G-y)$. By Lemma 1, $|T \cap \mathrm{Adj}(x)| \leqslant 2$ *for each* $T \in X(G-y)$. *Since one member of* $X(G-y)$ contains x, the remaining $\omega - 1$ members cover $\mathrm{Adj}(x)$. Hence, $|\mathrm{Adj}(x)| \leqslant 2\omega - 2$. By Property 5, $|\mathrm{Adj}(x)| \geqslant 2\omega - 2$, which proves the lemma.

Lemma 4. *Let G be a partitionable circle graph. For any vertex x, the cover* $X(G-x)$ *contains two* α*-stable sets,* S_1, S_2, *such that* $|S_i \cap \mathrm{Adj}(x)| = 1$ and $\omega - 2$ α*-stable sets* $S_3, \ldots, S_\omega$ *such that* $|S_i \cap \mathrm{Adj}(x)| = 2$.

Proof. By Lemma 1, $|S \cap \mathrm{Adj}(x)| \leqslant 2$ for all $S \in X(G-x)$ and since $|\mathrm{Adj}(x)| = 2\omega - 2$, a simple counting argument yields the result.

4. The SPGC and circle graphs

In this section we prove that the Berge Strong Perfect Graph Conjecture holds for the class of circle graphs.

Theorem. *If G is a partitionable circle graph, then G contains an odd hole or an odd antihole.*

Proof. Every partitionable graph having $\omega = 2$ is an odd hole, and every partitionable graph having $\alpha = 2$ is an odd antihole. Therefore, we may assume that α, $\omega \geqslant 3$. Let x be any vertex of G and let the members of $X(G-x)$ be numbered (as in Lemma 4) $S_1, S_2, \ldots, S_\omega$ such that $|S_i \cap \mathrm{Adj}(x)| = 1$ for $i = 1,2$ and $|S_i \cap \mathrm{Adj}(x)| = 2$ for $i = 3, \ldots, \omega$. There are two cases to consider, and each case uses Corollary 2.

Case 1. $\omega = 3$. (Here we could appeal to the result of Tucker [21] on 3-chromatic graphs, but the following is more direct.) Suppose $S_1 \cup S_2$ corresponds to a collection of chords that forms a chordless path going 'around' the circle. Since $S_1 \cup S_2 \cup x$ is biconnected (Property 4), the chord corresponding to x must intersect the two end chords. There is only one way to do this (Fig. 4(i)) since $|(S_1 \cup S_2) \cap \mathrm{Adj}(x)| = 2$, and it results in $S_1 \cup S_2 \cup x$ being an odd hole.

Suppose $S_1 \cup S_2$ corresponds to a collection of chords that forms a chordless cycle. If the chord corresponding to x intersects two nonadjacent chords of the cycle, then we would have an odd hole. Therefore, we may assume the configuration in Fig. 4(ii). Consider the third α-stable set, S_3, in $X(G-x)$. Let $\{y, z\} = S_3 \cap \mathrm{Adj}(x)$. Since $S_1 \cup S_2 \cup S_3 \cup x$ covers all vertices of G, both $\mathrm{Adj}(y)$ and $\mathrm{Adj}(z)$ are contained in $S_1 \cup S_2 \cup x$. Furthermore, by Lemma 3, $|\mathrm{Adj}(y)| = |\mathrm{Adj}(z)| = 2\omega - 2 = 4$, and there is only one way up to symmetry that the chords corresponding to y and z could intersect the chords of $S_1 \cup S_2 \cup x$ without forming an odd hole keeping $\omega = 3$ (Fig. 5). But removing the four chords whose end points are circled clearly reveals an odd hole.

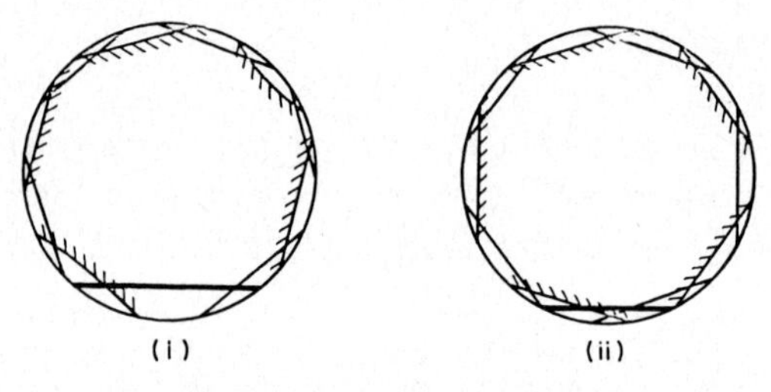

Fig. 4. $S_1 \cup S_2 \cup x$ for Case 1.

Fig. 5. $S_1 \cup S_2 \cup x \cup y \cup z$.

Case 2. $\omega \geq 4$. As before, we first suppose that $S_3 \cup S_4$ corresponds to a chordless path going 'around' the circle. Since $S_3 \cup S_4 \cup x$ is biconnected and $|(S_3 \cup S_4) \cap \mathrm{Adj}(x)| = 4$, the chord corresponding to x must intersect the two end chords and their neighbors (Fig. 6). This yields an odd hole of size $2\alpha - 1 \geq 5$.

Finally, suppose that $S_3 \cup S_4$ corresponds to a chordless cycle. Again, there is only one way, up to symmetry, that the chord corresponding to x could intersect those of $S_3 \cup S_4$ such that $|S_3 \cap \mathrm{Adj}(x)| = |S_4 \cap \mathrm{Adj}(x)| = 2$ without immediately

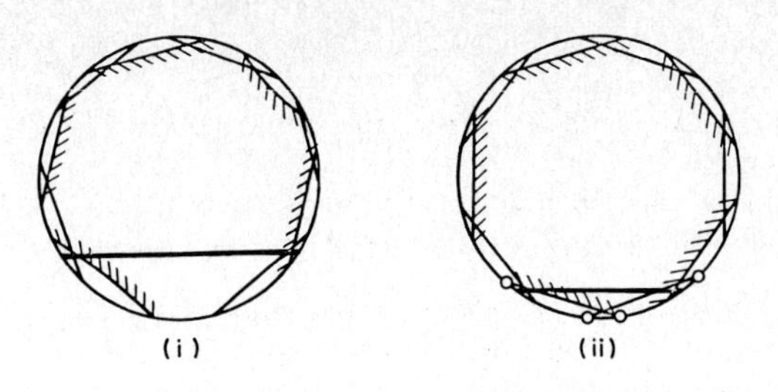

Fig. 6. $S_3 \cup S_4 \cup x$ for Case 2.

forming an odd hole. But, removing the two chords whose end points are circled reveals an odd hole.

Therefore, in all cases, G contains an odd hole or odd antihole.

References

[1] C. Berge, Färbung von Graphen, deren sämtliche bzw. deren ungerade kreise starr sind (Zusammenfassung), Wiss. Z. Martin-Luther-Univ. Halle-Wittenberg Math.-Natur. Reihe (1961) 114–115.

[2] R.B. Bland, H.C. Huang and L.E. Trotter, Jr., Graphical properties related to minimal imperfection, Discr. Math. 27 (1979) 11–22.

[3] M.A. Buckingham, Circle graphs, Ph.D. Thesis, New York University (1981). (Also available as: Courant Computer Science Report No. 21, Courant Institute of Mathematical Sciences, New York University, New York, New York, Oct. 1980).

[4] M.A. Buckingham and M.C. Golumbic, Partitionable graphs, circle graphs, and the Berge Strong Perfect Graph Conjecture, Discr. Math. 44 (1983) 45–54.

[5] V. Chvátal, On the Strong Perfect Graph Conjecture, J. Combinat. Theory (B) 20 (1976) 139–141.

[6] V. Chvátal, R.L. Graham, A.F. Perold and S.H. Whitesides, Combinatorial designs related to the Strong Perfect Graph Conjecture, Discr. Math. 26 (1979) 83–92.

[7] S. Even and A. Itai, Queues, stacks and graphs, in: Z. Kohavi and A. Paz, Eds., Theory of Machines and Computations (Academic Press, New York, New York, 1971) pp. 71–86.

[8] F. Gavril, Algorithms for a maximum clique and a maximum independent set of a circle graph, Networks 3 (1973) 261–273.

[9] M.C. Golumbic, Algorithmic Graph Theory and Perfect Graphs (Academic Press, New York, New York, 1980).

[10] M.C. Golumbic, C.F. Goss and R.B.K. Dewar, Macro substitutions in MICRO SPITBOL — a combinatorial analysis, in: Proc. 11th Southeastern Conf. on Combinatorics, Graph Theory, and Computing, Congressus Numerantium, Vol. 29 (1980) pp. 485–495.

[11] C.M. Grinstead, Toroidal graphs and the Strong Perfect Graph Conjecture, Ph.D. Thesis, University of California at Los Angeles (1978).

[12] L. Lovász, Normal hypergraphs and the Perfect Graph Conjecture, Discr. Math. 2 (1972) 253–267.

[13] L. Lovász, A characterization of perfect graphs, J. Combinat. Theory (B) 13 (1972) 95–98.

[14] M.W. Padberg, Perfect zero–one matrices, Math. Programming 6 (1974) 180–196.

[15] M.W. Padberg, Almost integral polyhedra related to certain combinatorial optimization problems, Linear Algebra Appl. 15 (1976) 69–88.

[16] K.R. Parthasarathy and G. Ravindra, The Strong Perfect Graph Conjecture is true for $K_{1,3}$-free graphs, J. Combinat. Theory (B) 21 (1976) 212–223.
[17] K.R. Parthasarathy and G. Ravindra, The validity of the Strong Perfect Graph Conjecture for $(K_4 - e)$-free graphs, J. Combinat. Theory (B) 26 (1979) 98–100.
[18] H. Sachs, On the Berge conjecture concerning perfect graphs, in: R. Guy et al., Eds., Combinatorial Structures and Their Applications (Gordon and Breach, New York, 1970) pp. 377–384.
[19] A. Tucker, The Strong Perfect Graph Conjecture for planar graphs, Can. J. Math. 25 (1973) 103–114.
[20] A. Tucker, Coloring a family of circular arcs, SIAM J. Appl. Math. 29 (1975) 493–502.
[21] A. Tucker, Critical perfect graphs and perfect 3-chromatic graphs, J. Combinat. Theory (B) 23 (1977) 143–149.

Received March 1981

Annals of Discrete Mathematics 20 (1984) 83–91
North-Holland

ITERATIVE METHODS FOR THE CONVEX FEASIBILITY PROBLEM

Yair CENSOR

Department of Mathematics, University of Haifa, Mount Carmel, Haifa 31999, Israel

The problem of finding a point in the intersection of a finite family of closed convex sets in the Euclidean space is considered here. Several iterative methods for its solution are reviewed and some connections between them are pointed out.

1. Introduction

The *convex feasibility problem* is to find a point in the nonempty intersection $Q \triangleq \bigcap_{i=1}^{m} Q_i \neq \emptyset$ of a finite family of closed and convex sets $Q_i \subseteq \mathbb{R}^n$, $i = 1, 2, \ldots, m$, in the Euclidean n-dimensional space $\mathbb{R}^n$. This fundamental problem has many applications in mathematics as well as in other fields. When the sets are given in the form

$$Q_i = \{x \subset \mathbb{R}^n \mid g_i(x) \leq 0,\ g_i \text{ is a convex function}\}, \tag{1}$$

then we deal with the problem of solving a system of inequalities with convex functions, of which the linear case is an important special case.

In this paper we review several iterative methods for the convex feasibility problem. A thorough study of the intimate relations between these methods has not yet been done. Therefore, we aim only at a description of the methods along with some of the apparent connections between them. The following methods are discussed.

(i) The method of successive orthogonal projections (SOP) of Gubin, Polyak and Raik [17] (Section 3).

(ii) The cyclic subgradient projections method (CSP) of Eremin [12], Raik [23], and Censor and Lent [6] (Section 4).

(iii) The interior points algorithm (IP) and the (δ, η)-algorithm of Aharoni, Berman and Censor [2] (Section 5).

(iv) The general schemes of Oettli [21] and Eremin [11, 14] (Section 6).

(v) D-projections and Bregman's [4] generalization of the SOP-method (Section 7).

In contrast with the constructive approach taken here, the reader will find existence results for systems of (mainly strict) inequalities involving convex functions in Fan et al. [15], and in [24, Section 21].

2. Control sequences and relaxation parameters

A *control sequence* $\{i_k\}_{k=0}^{\infty}$ is a sequence of indices according to which individual sets Q_i may be chosen for the execution of an iterative algorithm. Here are some important controls.

(1) *Cyclic control.* $i_k = k \pmod{m}+1$, where m is the number of sets in the convex feasibility problem.

(2) *Almost cyclic control* (Lent [18], see also Lent and Censor [19]). Denoting $I=\{1,2,\ldots,m\}$, $\{i_k\}_{k=0}^{\infty}$ is almost cyclic on I if $i_k \in I$ for all $k \geqslant 0$, and there exists an integer C such that for all $k \geqslant 0$, $I \subseteq \{i_{k+1}, i_{k+2}, \ldots, i_{k+C}\}$.

Almost cyclic controls are less restrictive than the cyclic control and therefore add an important option as to how the application of a method to a particular problem may be carried out (see, for example, Censor and Lent [7]).

(3) *Repetitive control* (Aharoni et al. [2]). The control $\{i_k\}_{k=0}^{\infty}$ is called repetitive on I if $i_k \in I$ for all $k \geqslant 0$ and

$$\forall l \in I,\ \forall k,\ \exists k' > k \text{ such that } i_{k'} = l.$$

(4) *Remotest set control.* This is obtained by determining i_k such that:

$$d(x^k, Q_{i_k}) = \max_{i \in I} d(x^k, Q_i),$$

where $d(x^k, Q_i)$ is the Euclidean distance between x^k and the set Q_i.

Other controls, i.e. the *approximately remotest set control*, the *most violated constraint control*, and the *threshold control* are mentioned in Censor [8].

The sequence $\{\lambda_k\}_{k=0}^{\infty}$ of *relaxation parameters*, appearing in several methods, allows us, loosely speaking, to overdo or underdo the move prescribed in an iterative step. Relaxation parameters add an extra degree of freedom to the way a method might actually be implemented and have important practical consequences.

3. The method of successive orthogonal projections

A well-known iterative algorithm for solving the convex feasibility problem is the *method of successive orthogonal projections* (SOP, for short) of Gubin, Polyak and Raik [17]. Starting from an arbitrary $x^0 \in \mathbb{R}^n$, the method generates a sequence $\{x^k\}_{k=0}^{\infty}$ that converges to a point in Q, by performing successive, possibly relaxed, orthogonal projections onto the individual convex sets Q_i. A typical iterative step has the form:

$$x^{k+1} = x^k + \lambda_k (P_{Q_{i_k}}(x^k) - x^k), \tag{2}$$

where $P_{Q_{i_k}}(x^k)$ stands for the orthogonal projection of x^k onto the set Q_{i_k} and $\{\lambda_k\}_{k=0}^{\infty}$ are relaxation parameters confined to $\eta \leqslant \lambda_k \leqslant 2-\eta$ for all $k \geqslant 0$, with some $\eta > 0$.

Gubin et al. [17] proved convergence of the SOP method under cyclic control and their proof carries over easily to the almost cyclic case. Convergence under any repetitive control follows from a general result of Aharoni et al. [2, Remark 1]. Gubin et al. [17] proved convergence also for the remotest set control.

The SOP method is particularly useful when the projections onto the individual sets are easily calculated. In general, however, application of the method would require at each step the solution of a subsidiary minimization problem associated with the projection onto the current set, namely

$$\min_{y \in Q_{i_k}} \|x^k - y\|, \tag{3}$$

where $\|\cdot\|$ stands for the Euclidean norm. An instance of easy enough projections is the linear case in which the SOP method coincides with the relaxation method of Agmon [1] and Motzkin and Schoenberg [20].

For the convex feasibility problem with sets of the form (1) the typical step of the SOP method requires that a move will be made in a direction determined by $\nabla g_{i_k}(x^{k+1})$, the gradient of g_{i_k} calculated at the *next* iterate x^{k+1}. Methods which circumvent this difficulty in various ways are described below.

4. The cyclic subgradient projections method

The cyclic subgradient projections method (CSP, for short) differs from the SOP method in that it requires at each iterative step a move in the direction of the gradient (or subgradient) calculated at the *current* iterate, i.e. $\nabla g_{i_k}(x^k)$. In this way the computational difficulties associated with the SOP method discussed in the previous section are circumvented. The CSP method presented by Censor and Lent [6] is as follows.

CSP method

Initialization. $x^0 \in \mathbb{R}^n$ arbitrary.

Typical step.

$$x^{k+1} = x^k - \lambda_k \frac{g_{i_k}^{+}(x^k)}{\|t_k\|^2} t_k,$$

where $g_i^+(x) = \max\{0, g_i(x)\}$, $t_k \in \partial g_{i_k}^+(x^k)$ is a subgradient (see, for example, [24, p. 214]) of $g_{i_k}^+$ at the point x^k and the relaxation parameters $\{\lambda_k\}_{k=0}^{\infty}$ are confined to the interval $\varepsilon_1 \leq \lambda_k \leq 2 - \varepsilon_2$ for all $k \geq 0$, with $\varepsilon_1, \varepsilon_2 > 0$.

Control. $\{i_k\}_{k=0}^{\infty}$ is almost cyclic on $\{1, 2, \ldots, m\}$.

Remotest-set-controlled, or other noncyclically controlled methods with the same typical step can be derived from the schemes of Oettli and of Eremin, see Section 6. It was recently pointed out to us that the CSP method is actually a slight generalization of the method of Eremin [12] and of Raik [23]. An extension of the CSP method to the reverse convex feasibility problem was proposed and experimented with, though not proved, by Censor et al. [5].

5. The interior points algorithm

An orthogonal projection of a point x onto a set Q amounts to an orthogonal projection of x onto the particular hyperplane H which separates x from Q and supports Q at the closest point to x. In view of the simplicity of an orthogonal projection onto a hyperplane, it is natural to ask whether one could use other separating supporting hyperplanes instead of the particular hyperplane through the closest point to x. Aside from theoretical interest, this approach leads to algorithms useful in practice, provided that the computational load of finding such other hyperplanes will favourably compete with the work involved in performing orthogonal projections onto the given sets.

Below we give a short account of some recent results of Aharoni, Berman and Censor [2]; the complete study will appear elsewhere.

A general framework for the design of algorithms for the convex feasibility problem is given by the so-called (δ, η)-algorithm. With a given point $x \in \mathbb{R}^n$ and a given closed and convex set $Q \subseteq \mathbb{R}^n$ associate a set $A_Q(x)$ in the following manner. Choose $0 \leq \delta \leq 1$ and $0 \leq \eta$ and let $B = B(x, \delta d(x, Q))$ be the ball centered at x with radius $\delta d(x, Q)$, where $d(x, Q)$ is the Euclidean distance between x and Q. For $x \notin Q$ denote by $\mathcal{H}_{x,Q}$ the set of all hyperplanes which separate (see, for example, [24, p. 95]) B from Q.

Define:

$$A_Q(x) \triangleq \begin{cases} \{x\}, & \text{if } x \in Q, \\ \{x + \lambda(P_H(x) - x) \mid H \in \mathcal{H}_{x,Q}, \eta \leq \lambda \leq 2 - \eta\}, & \text{if } x \notin Q. \end{cases} \tag{4}$$

The (δ, η)-algorithm is given by:

$$x^0 \in \mathbb{R}^n, \text{ arbitrary},$$

$$x^{k+1} \in A_{Q_{i_k}}(x^k), \tag{5}$$

$\{i_k\}_{k=0}^{\infty}$ is repetitive (see Section 2, above).

Theorem. *A sequence $\{x^k\}_{k=0}^{\infty}$ generated by a (δ, η)-algorithm with $\delta > 0$ and $\eta > 0$ converges to a point in $Q = \bigcap_{i=1}^{m} Q_i$.*

Observation. For $\delta = 1$ and any $\eta > 0$ the (δ, η)-algorithm coincides with the SOP-method (Section 3) because then all hyperplanes in $\mathcal{H}_{x,Q}$ are also supporting Q at the closest point to x.

Another realization of the (δ, η)-algorithm is the *interior points* (IP) *algorithm* of [2]. Here is a concrete version of this algorithm for the convex feasibility problem with sets of the form (1). A prerequisite of this algorithm is the availability of interior points of the individual sets, thus the sets Q_i have to be solid (i.e. have non-empty interior).

Algorithm IP.

Step 0. Find interior points y^i such that $g_i(y^i) < 0$, $i = 1, 2, \ldots, m$.

Step 1. Choose $x^0 \in \mathbb{R}^n$ arbitrary.

Step 2. Given x^k, pick Q_{i_k} by $i_k = k \pmod{m} + 1$.

Step 3. If $g_{i_k}(x^k) \leq 0$, set $x^{k+1} = x^k$ and return to Step 2.

Step 4. If $g_{i_k}(x^k) > 0$, define $z(\theta) = \theta y^{i_k} + (1 - \theta)x^k$, where $0 \leq \theta \leq 1$.

Step 5. Solve the single equation $g_{i_k}(z(\theta)) = 0$, denote by θ_k the smallest value of θ for which $z(\theta)$ solves this equation, and set $z^k = z(\theta_k)$.

Step 6. Calculate a subgradient $t_k \in \partial g_{i_k}(z^k)$.

Step 7. Let

$$x^{k+1} = x^k + \lambda_k \frac{\langle t_k, z^k \rangle - \langle t_k, x^k \rangle}{\| t_k \|^2} t_k,$$

where $0 < \eta \leq \lambda_k \leq 2 - \eta$ are relaxation parameters, and return to Step 2.

6. The schemes of Oettli and Eremin

The methods of Oettli [21] and of Eremin [11, 14] are algorithmic schemes for the solution of the convex feasibility problem with sets of the form (1).

6.1. Oettli's scheme

Initialization. $x^0 \in \mathbb{R}^n$, arbitrary.

Typical step.

$$x^{k+1} = x^k - \lambda_k \frac{\varphi(x^k)}{\|t_k\|^2} t_k, \quad \text{if } x^k \notin Q.$$

Here $Q = \bigcap_{i=1}^m Q_i$ and $\varphi: \mathbb{R}^n \to \mathbb{R}$ is defined by $\varphi(x) \stackrel{\Delta}{=} p(g_1^+(x), g_2^+(x), \ldots, g_m^+(x))$, where $g_i^+(x) = \max\{0, g_i(x)\}$. In the definition of the auxiliary function φ, we are free to choose any *monotonic norm* $p: \mathbb{R}^m \to \mathbb{R}$ (see, for example, [22, p. 52]). $t_k \in \partial\varphi(x^k)$ is any subgradient of φ at x^k and the relaxation parameters $\lambda_k \in (0, 2)$ are such that $\sum_{k=0}^{\infty} \lambda_k(2 - \lambda_k) = +\infty$.

Control. The control of the method is determined by the choice of the monotonic norm p.

Various choices of the monotonic norm p in Oettli's scheme give rise to concrete algorithms. Choosing p to be the l_∞-norm in $\mathbb{R}^m$ yields the remotest-set-controlled method of subgradient projections for solving the convex feasibility problem. However, we believe that there is no monotonic norm p that upon substitution into Oettli's scheme yields the CSP method of Section 4.

The Scheme of Eremin [11, 14] resembles that of Oettli but is not quite identical to it.

6.2. Eremin's scheme

Initialization. $x^0 \in \mathbb{R}^n$, arbitrary.

Typical step.

$$x^{k+1} = \begin{cases} x^k, & \text{if } d(x^k) \leq 0, \\ x^k - \lambda_k \dfrac{d(x^k)}{\|e_k\|^2} e_k, & \text{otherwise,} \end{cases}$$

where $d: \mathbb{R}^n \to \mathbb{R}$ is a suitably chosen continuous convex function and $e_k = e(x^k)$ with $e: \mathbb{R}^n \to \mathbb{R}^n$ a suitably chosen mapping. $\{\lambda_k\}_{k=0}^{\infty}$ are relaxation parameters confined to $\varepsilon_1 \leq \lambda_k \leq 2 - \varepsilon_2$ for all $k \geq 0$, with $\varepsilon_1, \varepsilon_2 > 0$.

Control. Determined by the choice of the functions d and e.

We do not here go into any details and direct the reader to Eremin's paper for the precise conditions on the functions d and e involved in the scheme. Various concrete methods may be derived from Eremin's algorithmic scheme through a proper choice of the functions d and e. In particular, the choice $d(x) = \max_{1 \leq i \leq m} g_i(x)$ and $e_k = \nabla g_{i_k}(x^k)$ (∇ stands for gradient), with $i_k \in \{i_k \mid g_{i_k}(x^k) = \max_{1 \leq i \leq m} g_i(x^k)\}$, again gives rise to a remotest-set-controlled gradient projections method, and is, therefore, equivalent to Oettli's method with the l_∞-norm. Other choices of d and e, which are equivalent to choices of the l_2-norm and the weighted l_1-norm in Oettli's scheme, are given in [11, 13]. To the best of our knowledge it is not possible to choose d and e such that the CSP method will be obtained; thus, the cyclic case has to be studied separately, as done in Section 4.

7. D-projections and Bregman's generalization of the SOP method

Let S be a nonempty, open, convex set such that $\bar{S} \subseteq \Lambda$ (the bar denotes closure), where $\Lambda \subseteq \mathbb{R}^n$ is the domain of a function $f\colon \Lambda \to \mathbb{R}^n$. Assume that $f(x)$ has continuous first partial derivatives at every $x \in S$ and denote by $\nabla f(x)$ its gradient at x. Following Bregman [4], Censor and Lent [7] define the function $D(x, y)$, which maps $\bar{S} \times S$ into $\mathbb{R}$, as:

$$D(x, y) = f(x) - f(y) - \langle \nabla f(y), x - y \rangle. \tag{6}$$

This function is called the *D-function associated with $f(x)$*. The *D-projection of a point $y \in S$ onto a set $\Omega \subseteq \mathbb{R}^n$* is then defined to be a point $x^* \in \Omega \cap \bar{S}$ for which

$$\min_{z \in \Omega \cap \bar{S}} D(z, y) = D(x^*, y). \tag{7}$$

Referring to the set S as the *zone* of the function f, Censor and Lent [7] collect the conditions under which Bregman's theory applies. The family of all functions that fulfil these conditions is denoted by $\mathcal{B}$ and the functions themselves are called *Bregman functions.*

Bregman showed [4] that D-projections can be employed in a successive manner to solve the convex feasibility problem. Starting from an arbitrary $x^0 \in S$ the iterative step is

$$x^{k+1} = \hat{P}_{Q_{i_k}}(x^k), \tag{8}$$

where $\hat{P}_Q(x)$ stands for the D-projection of x onto Q. No relaxation parameters are present in Bregman's scheme and he proved convergence under cyclic control or the control of largest $D(x, y)$ 'distance', i.e.

$$\max_{i \in I} \min_{x \in Q_i} D(x, x^k) = \min_{x \in Q_{i_k}} D(x, x^k). \tag{9}$$

Obviously, $D(x, y)$ is *not* generally a distance function but the choice $f(x) = \frac{1}{2}\|x\|^2$ with $\Lambda = S = \bar{S} = \mathbb{R}^n$ and $D(x, y) = \frac{1}{2}\|x - y\|^2$ gives back the SOP method of Section 3. The question of incorporating relaxation parameters into (8) without ruining convergence is still open.

8. Concluding remarks

The convex feasibility problem has various important applications and constructive methods for its solution are very desirable. For the case of systems of linear inequalities many of the methods described here actually coincide, but for systems of convex inequalities they present a wealth of iterative algorithms. A development not discussed here is the extension of Cimmino's algorithm [10, 16] to the linear feasibility problem by Censor and Elfving [9], and to the convex feasibility problem [3].

References

[1] S. Agmon, The relaxation method for linear inequalities, Can. J. Math. 6 (1954) 382–392.

[2] R. Aharoni, A. Berman and Y. Censor, An interior points algorithm for the convex feasibility problem, Advances in Applied Mathematics, to appear.

[3] A. Auslender, Optimisation: Méthodes Numériques (Masson, Paris, 1976).

[4] L.M. Bregman, The relaxation method of finding the common point of convex sets and its application to the solution of problems in convex programming, USSR Comput. Math. and Math. Phys. 7 (1967) 200–217.

[5] Y. Censor, D. E. Gustafson, A. Lent and H. Tuy, A new approach to the emission computerized tomography problem: Simultaneous calculation of attenuation and activity coefficients, IEEE Trans. Nucl. Sci. NS-26 (1979) 2775–2779.

[6] Y. Censor and A. Lent, Cyclic subgradient projections, Mathematical Programming 24 (1982) 233–235.

[7] Y. Censor and A. Lent, An iterative row-action method for interval convex programming, J. Optim. Theory Appl. 34 (1981) 321–353.

[8] Y. Censor, Row-action methods for huge and sparse systems and their applications, SIAM Rev. 23 (1981) 444–466

[9] Y. Censor and T. Elfving, New methods for linear inequalities, Linear Algebra Appl. 42 (1982) 199–211.

[9a] Y. Censor, A Cimmino-like algorithm (in preparation).

[10] G. Cimmino, Calcolo approssimato per le soluzioni dei sistemi di equazioni lineari, La Ricerca Scientifica 1938 — XVI, Serie II, Anno IX, Vol. 1, pp. 326–333, Roma.

[11] I.I. Eremin, The relaxation method of solving systems of inequalities with convex functions on the left hand side, Sov. Math. Dokl. 6 (1965) 219–222.

[12] I.I. Eremin, Certain iteration methods in convex programming, Ekon. Math. Method. 2 (1966) 870–886 (Russian); Math. Rev. 43 (1972) #5925.

[13] I.I. Eremin and Vl.D. Mazurov, Iteration method for solving problems of convex programming, Sov. Phys. Dokl. 11 (1967) 757–759.

[14] I.I. Eremin, On systems of inequalities with convex functions in the left sides, Amer. Math. Soc. Transl. (2) 88 (1970) 67–83.

[15] K. Fan, I. Glicksberg and A.J. Hoffman, Systems of inequalities involving convex functions, Proc. Amer. Math. Soc. 8 (1957) 617–622.

[16] N. Gastinel, Linear Numerical Analysis (Hermann, Paris, 1970).

[17] L.G. Gubin, B.T. Polyak and E.V. Raik, The method of projections for finding the common point of convex sets, USSR Comput. Math. and Math. Phys. 7 (1967) 1–24.

[18] A. Lent, Maximum entropy and multiplicative ART, in: R. Shaw, Ed., Image Analysis and Evaluation, SPSE Conference Proceedings (Society of Photographic Scientists and Engineers, Washington, D.C., 1977) pp. 249–257.

[19] A. Lent and Y. Censor, Extensions of Hildreth's row-action method for quadratic programming, SIAM J. Control Optim. 18 (1980) 444–454.

[20] T.S. Motzkin and I.J. Schoenberg, The relaxation method for linear inequalities, Can. J. Math. 6 (1954) 393–404.

[21] W. Oettli, Symmetric duality, and a convergent subgradient method for discrete, linear, constrained approximation problems with arbitrary norms appearing in the objective function and in the constraints, J. Approximation Theory 14 (1975) 43–50.

[22] J.M. Ortega and W.C. Rheinboldt, Iterative Solution of Nonlinear Equations in Several Variables (Academic Press, New York, 1970).

[23] E. Raik, Fejer type methods in Hilbert space, Esti. NSV Tead. Akad. Toimetised Fuus.-Mat. 16 (1967) 286–293 (Russian); Math. Rev. 36 (1968) #2399.

[24] R.T. Rockafellar, Convex Analysis (Princeton University Press, Princeton, N.J., 1970).

Received 29 April 1981

Annals of Discrete Mathematics 20 (1984) 93–101
North-Holland

INEQUALITIES INVOLVING CONVEX SETS AND THEIR CHORDS

G.D. CHAKERIAN

Mathematics Department, University of California, Davis, California 95616, U.S.A.

P.R. GOODEY

Mathematics Department, Royal Holloway College, Egham, Surrey, U.K.

In this paper we give new proofs and generalizations of some results concerning planar convex sets. We work with integrals in which the integrands are determinants. By exploiting the basic properties of determinants we give geometrical interpretations of our integrals. This enables us to give a proof of Minkowski's inequality for convex curves and some generalizations of Holditch's theorem. In the final section we obtain results related to some recent work of Lutwak.

1. Introduction

In this paper we employ a convenient formalism to treat some results in the theory of plane curves. We give a new proof of Minkowski's Inequality for convex curves in the next section, and in Section 3 we derive some generalizations of Holditch's Theorem. The final section deals with related results of Lutwak [6] and Santaló [8, p. 11].

If $\boldsymbol{a} = (a_1, a_2)$ and $\boldsymbol{b} = (b_1, b_2)$ are vectors in the Euclidean plane E^2, we shall let $[\boldsymbol{a}, \boldsymbol{b}]$ denote the 2×2 determinant having $\boldsymbol{a}$ and $\boldsymbol{b}$ as its rows. That is, we define

$$[\boldsymbol{a}, \boldsymbol{b}] := a_1 b_2 - a_2 b_1. \tag{1}$$

In our treatment we shall exploit the basic properties of the determinant, in particular the skew-symmetry, $[\boldsymbol{a}, \boldsymbol{b}] = -[\boldsymbol{b}, \boldsymbol{a}]$, and the linearity in each argument.

Suppose now that C is a continuously differentiable closed curve in E^2, given by the parametric representation

$$\boldsymbol{x} = \boldsymbol{x}(t), \qquad \alpha \leqslant t \leqslant \beta.$$

We associate with C the quantity $I(C)$ defined by

$$I(C) := \tfrac{1}{2} \int [\boldsymbol{x}, \mathrm{d}\boldsymbol{x}] := \tfrac{1}{2} \int_{\alpha}^{\beta} [\boldsymbol{x}(t), \dot{\boldsymbol{x}}(t)] \, \mathrm{d}t. \tag{2}$$

If, in addition, C is a *simple* closed curve, parameterized so that $\boldsymbol{x}(t)$ traverses C once in the counterclockwise sense as t varies from α to β, then Green's Theorem implies that $I(C)$ is equal to $A(D)$, the area of the domain D bounded by C. In general, if $n(z, C)$ is the winding number of C with respect to the point $z \in E^2 - C$, then $I(C)$ is the integral of $n(z, C)$ over all z belonging to $E^2 - C$.

If C_1 and C_2 are continuously differentiable closed curves in E^2 given by parametric representations $\boldsymbol{x}_1 = \boldsymbol{x}_1(t)$ and $\boldsymbol{x}_2 = \boldsymbol{x}_2(t)$, respectively, where each curve is traversed once as the common parameter t varies from α to β, we define a quantity $I(C_1, C_2)$ by

$$I(C_1, C_2) := \tfrac{1}{2} \int [\boldsymbol{x}_1, \mathrm{d}\boldsymbol{x}_2] := \tfrac{1}{2} \int_\alpha^\beta [\boldsymbol{x}_1(t), \dot{\boldsymbol{x}}_2(t)]\,\mathrm{d}t. \tag{3}$$

$I(C_1, C_2)$ is a translation-invariant functional generalizing that defined by (2), since $I(C, C) = I(C)$. Integration by parts, and the skew-symmetry of the determinant, yields:

$$I(C_1, C_2) = I(C_2, C_1). \tag{4}$$

With C_1 and C_2 as above, let λ_1 and λ_2 be any real numbers and let C be the closed curve with parametric representation $\boldsymbol{x}(t) = \lambda_1 \boldsymbol{x}_1(t) + \lambda_2 \boldsymbol{x}_2(t)$, $\alpha \leq t \leq \beta$. The multilinearity of the determinant, together with (4), gives immediately:

$$I(C) = \lambda_1^2 I(C_1) + 2\lambda_1\lambda_2 I(C_1, C_2) + \lambda_2^2 I(C_2). \tag{5}$$

A case of special importance to us is where K_1 and K_2 are compact convex subsets of E^2, whose respective boundary curves C_1 and C_2 are *smooth*. This means that C_1 and C_2 are continuously differentiable and that each tangent line of C_i meets it in only one point, $i = 1, 2$. The latter condition ensures that the supporting function of K_i is continuously differentiable (see Bonnesen and Fenchel [2, p. 26]). With these hypotheses we may parameterize C_i as $\boldsymbol{x}_i = \boldsymbol{x}_i(\phi)$, $0 \leq \phi \leq 2\pi$, where ϕ is the angle between a fixed ray and the tangent line to C_i at $\boldsymbol{x}_i(\phi)$, and where C_i is traversed once in the counterclockwise sense as ϕ varies from 0 to 2π. We have then that $\boldsymbol{x}_i(\phi)$ is continuously differentiable and

$$A(K_1, K_2) = I(C_1, C_2) = \tfrac{1}{2} \int_0^{2\pi} [\boldsymbol{x}_1(\phi), \dot{\boldsymbol{x}}_2(\phi)]\,\mathrm{d}\phi, \tag{6}$$

where $A(K_1, K_2)$ is the mixed area of K_1 and K_2. This expression for the mixed area may be found in Blaschke [1, p. 66]. One way to verify (6) is to observe that if $K = \lambda_1 K_1 + \lambda_2 K_2$, with $\lambda_1, \lambda_2 \geq 0$, then the boundary curve C of K has the parametric representation $\boldsymbol{x}(\phi) = \lambda_1 \boldsymbol{x}_1(\phi) + \lambda_2 \boldsymbol{x}_2(\phi)$, $0 \leq \phi \leq 2\pi$. Since

$$A(K) = \lambda_1^2 A(K_1) + 2\lambda_1\lambda_2 A(K_1, K_2) + \lambda_2^2 A(K_2), \tag{7}$$

we obtain (6) by comparing corresponding terms in (5) and (7), using the fact that $I(C) = A(K)$.

2. Minkowski's Inequality

If K_1 and K_2 are compact plane convex sets, then Minkowski's Inequality states that:

$$A(K_1, K_2)^2 \geq A(K_1)A(K_2), \tag{8}$$

with equality holding if and only if K_1 and K_2 are homothetic.

We shall initially assume that at each point on the boundary of K_1 or K_2 there is a unique supporting line (that is, a tangent). Since both sides of (8) are invariant under translations of K_i and homogeneous of degree 2, it suffices to consider the situation where K_1 is inscribed in K_2 in the sense of Eggleston [5, p. 108]. That is, we shall further assume that $K_1 \subset K_2$ and that either (i) there are three lines, bounding a triangle T enclosing K_2, each of which supports both K_1 and K_2; or (ii) there are two parallel lines, each of which supports both K_1 and K_2.

In what follows we shall let $l_i(\phi)$ denote the tangent line to C_i (the boundary of K_i) making angle ϕ with a fixed reference ray. Without loss of generality we may assume that in case (i) $l_i(\phi_0)$, $l_i(\phi_1)$, and $l_i(\phi_2)$ contain the sides of the triangle T, where $\phi_0 < \phi_1 < \phi_2 < \phi_0 + 2\pi$. The vertices of T we denote by $\boldsymbol{a}_0 = l_i(\phi_0) \cap l_i(\phi_1)$, $\boldsymbol{a}_1 = l_i(\phi_1) \cap l_i(\phi_2)$, and $\boldsymbol{a}_2 = l_i(\phi_2) \cap l_i(\phi_0)$. Now let $\boldsymbol{z} \in C_2$ be on the line $l_2(\phi)$, where $\phi_0 \leq \phi \leq \phi_1$. Then we shall denote by $\boldsymbol{y} = \boldsymbol{y}(\boldsymbol{z})$ that point of $l_1(\phi)$ for which $\boldsymbol{y} - \boldsymbol{z}$ is a vector parallel to a fixed ray in the interior of the angle of T at $\boldsymbol{a}_0$. So we obtain the points $\boldsymbol{y}$ by projecting each point $\boldsymbol{z}$ of this arc of C_2 into the corresponding $l_1(\phi)$ parallel to a single fixed ray through $\boldsymbol{a}_0$. We carry out the analogous projections of those points $\boldsymbol{z} \in C_2$ for which the corresponding values of ϕ lie in $[\phi_1, \phi_2]$, $[\phi_2, \phi_0 + 2\pi]$. In case (ii) we assume that the parallel support lines are $l_i(\phi_0)$ and $l_i(\phi_0 + \pi)$. Here each $\boldsymbol{z} \in C_2 \cap l_2(\phi)$ is projected parallel to $l_i(\phi_0)$ into a point $\boldsymbol{y} \in l_1(\phi)$.

If, in either case, we find that for each $\boldsymbol{z} \in C_2$ the corresponding point $\boldsymbol{y}$ is in C_1, then as $\boldsymbol{z}$ varies over certain arcs of C_2 we see that $\boldsymbol{y} - \boldsymbol{z}$ has a fixed direction. Also, the tangents at $\boldsymbol{z}$ and $\boldsymbol{y}$ are parallel and so the corresponding arcs of C_1 and C_2 must be translates of one another. It follows that they must in fact coincide, since the arcs already coincide at their end-points. So if each $\boldsymbol{y}$ is in C_1 we would have $K_1 = K_2$.

So now we assume that $K_1 \neq K_2$ and consequently we can choose a $\boldsymbol{z}_0 \in C_2 \cap l_2(\phi^*)$ such that the corresponding $\boldsymbol{y}_0$ is in $l_1(\phi^*)$ but not C_1. We let $\theta (\theta > 0)$ be smaller than the two angles between $l_1(\phi^*)$ and the line joining $\boldsymbol{y}_0$ and $\boldsymbol{z}_0$, and choose $\varepsilon > 0$ so that $d(\boldsymbol{y}_0, K_1) > \varepsilon$. Now we can choose convex sets $\tilde{K}_1$ and $\tilde{K}_2$ with smooth boundaries $\tilde{C}_1$ and $\tilde{C}_2$ arbitrarily close to C_1 and C_2, respectively, and with $\tilde{K}_1$ inscribed in $\tilde{K}_2$ as in case (i). We let $\tilde{C}_i$ have the parametric representation $x_i = x_i(\phi)$, $0 \leq \phi \leq 2\pi$, as in the introductory section. If $\boldsymbol{z} =$

$\boldsymbol{x}_2(\phi) \in \tilde{C}_2$ we denote by $\boldsymbol{x}(\phi)$ the point $\boldsymbol{y}$ obtained by projection onto the parallel tangent $L_1(\phi)$ to $\tilde{C}_1$, as described above. Then we see that we may assume that we have chosen our sets $\tilde{K}_1$ and $\tilde{K}_2$ in such a way that for some $\phi^* \in [0, 2\pi]$ we have $d(\boldsymbol{x}(\phi^*), \tilde{K}_1) > \varepsilon$, and the angles between $L_1(\phi^*)$ and the line joining $\boldsymbol{x}(\phi^*)$ and $\boldsymbol{x}_2(\phi^*)$ are both greater than some $\theta > 0$.

We shall denote by C the simple closed curve passing between C_1 and C_2 which is given by the parametric representation $\boldsymbol{x} = \boldsymbol{x}(\phi)$ for $0 \leqslant \phi \leqslant 2\pi$. We let $\boldsymbol{w}(\phi) = \boldsymbol{x}(\phi) - \boldsymbol{x}_2(\phi)$ and assume that the lines $L_1(\phi)$ for $\phi = \phi_0$, ϕ_1 and ϕ_2 contain the sides of the common supporting triangle of K_1 and K_2. Then it is easy to show that:

$$\boldsymbol{w}(\phi) = \frac{H_2(\phi) - H_1(\phi)}{\sin(\phi - \theta_i)}\, \boldsymbol{u}_i \quad \text{for } \phi \in [\phi_i, \phi_{i+1}], \tag{9}$$

where H_1 and H_2 are the support functions of $\tilde{K}_1$ and $\tilde{K}_2$, respectively, and $\boldsymbol{u}_i$ is the unit vector parallel to our direction of projection corresponding to the arc $\phi_i \leqslant \phi \leqslant \phi_{i+1}$, making angle θ_i with some fixed ray. Observing that $H_1(\phi) = H_2(\phi)$ and $\dot{H}_1(\phi) = \dot{H}_2(\phi)$ when ϕ takes the values ϕ_0, ϕ_1, and ϕ_2, we deduce from (9) that $\boldsymbol{w}$ is a continuously differentiable function of ϕ. It follows that C is a continuously differentiable curve.

Observe now that since $\boldsymbol{w}(\phi)$ has constant direction for $\phi_i \leqslant \phi \leqslant \phi_{i+1}$, we have $\dot{\boldsymbol{w}}(\phi)$ parallel to $\boldsymbol{w}(\phi)$ for all ϕ and so

$$[\boldsymbol{w}(\phi), \dot{\boldsymbol{w}}(\phi)] = 0, \qquad 0 \leqslant \phi \leqslant 2\pi. \tag{10}$$

Furthermore, since $\boldsymbol{x}(\phi) - \boldsymbol{x}_1(\phi)$ is parallel to $\dot{\boldsymbol{x}}_2(\phi)$ for all ϕ, the properties of the determinant imply

$$[\boldsymbol{x}_1(\phi), \dot{\boldsymbol{x}}_2(\phi)] = [\boldsymbol{x}(\phi), \dot{\boldsymbol{x}}_2(\phi)], \qquad 0 \leqslant \phi \leqslant 2\pi. \tag{11}$$

We see from (10) that $[\boldsymbol{x} - \boldsymbol{x}_2, \dot{\boldsymbol{x}} - \dot{\boldsymbol{x}}_2] = 0$, and so, using integration by parts and (11), we have:

$$\begin{aligned} I(C) &= \tfrac{1}{2} \int [\boldsymbol{x}, \mathrm{d}\boldsymbol{x}] \\ &= \tfrac{1}{2} \int [\boldsymbol{x}_2, \mathrm{d}\boldsymbol{x}] + \tfrac{1}{2} \int [\boldsymbol{x}, \mathrm{d}\boldsymbol{x}_2] - \tfrac{1}{2} \int [\boldsymbol{x}_2, \mathrm{d}\boldsymbol{x}_2] \\ &= \int [\boldsymbol{x}, \mathrm{d}\boldsymbol{x}_2] - \tfrac{1}{2} \int [\boldsymbol{x}_2, \mathrm{d}\boldsymbol{x}_2] \\ &= \int [\boldsymbol{x}_1, \mathrm{d}\boldsymbol{x}_2] - \tfrac{1}{2} \int [\boldsymbol{x}_2, \mathrm{d}\boldsymbol{x}_2]. \end{aligned} \tag{12}$$

From (6) and (12) we have:

$$I(C) = 2A(\tilde{K}_1, \tilde{K}_2) - A(\tilde{K}_2). \tag{13}$$

Now let P_1 be the half-plane lying on the same side of the line through $\boldsymbol{x}(\phi^*)$ and $\boldsymbol{x}_2(\phi^*)$ as $\boldsymbol{x}_1(\phi^*)$ and let P_2 be the half-plane determined by $L_1(\phi^*)$ and meeting $\tilde{K}_1$. Then if D is the disc of radius ε centred at $\boldsymbol{x}(\phi^*)$ we see that $D \cap P_1 \cap P_2$ is outside $\tilde{K}_1$ and inside the curve C. But C encloses $\tilde{K}_1$ and so

$$I(C) > A(\tilde{K}_1) + \tfrac{1}{2}\varepsilon^2\theta.$$

Hence, from (13) we have:

$$A(\tilde{K}_1, \tilde{K}_2) \geq \tfrac{1}{2}(A(\tilde{K}_1) + A(\tilde{K}_2)) + \tfrac{1}{4}\varepsilon^2\theta.$$

It follows from the continuity of areas and mixed areas that

$$A(K_1, K_2) \geq \tfrac{1}{2}(A(K_1) + A(K_2)) + \tfrac{1}{4}\varepsilon^2\theta. \tag{14}$$

Now if K_1 and K_2 are arbitrary convex sets, with K_1 inscribed in K_2, and if B is the unit disc, we see that $K_1 + B$ is inscribed in $K_2 + B$ and that these sets have unique support lines at all their boundary points. It follows from (14) that either $K_1 + B = K_2 + B$ or

$$A(K_1 + B, K_2 + B) > \tfrac{1}{2}(A(K_1 + B) + A(K_2 + B)).$$

Consequently, we see that if K_1 is inscribed in K_2 we have:

$$A(K_1, K_2) \geq \tfrac{1}{2}(A(K_1) + A(K_2)), \tag{15}$$

with equality if and only if $K_1 = K_2$. So, using the fact that

$$\tfrac{1}{2}(A(K_1) + A(K_2)) \geq \sqrt{A(K_1)A(K_2)},$$

we have Minkowski's Inequality:

$$A(K_1, K_2)^2 \geq A(K_1)A(K_2),$$

with equality if and only if K_1 and K_2 are homothetic. While Minkowski's Inequality is equivalent to the Brunn–Minkowski Inequality, in fact we can also use (15) directly to obtain the Brunn–Minkowski Inequality. For if K_0 and K_1 are arbitrary convex sets, we choose $\phi > 0$ so that ϕK_1 is inscribed in K_0 and let $\theta \in [0, 1]$. Put $\lambda = \theta(\phi - \theta\phi + \theta)^{-1}$, then $\lambda \in [0, 1]$ and

$$(1-\lambda)K_0 + \lambda\phi K_1 = \lambda\phi\theta^{-1}[(1-\theta)K_0 + \theta K_1].$$

Now since ϕK_1 is inscribed in K_0 we can use (15) to see that

$$\begin{aligned} &A[(1-\lambda)K_0 + \lambda\phi K_1] \\ &\quad = (1-\lambda)^2 A(K_0) + 2\lambda(1-\lambda)A(K_0, \phi K_1) + \lambda^2 A(\phi K_1) \\ &\quad \geq (1-\lambda)A(K_0) + \lambda A(\phi K_1), \end{aligned}$$

with equality if and only if $\phi K_1 = K_0$. Consequently,

$$
\begin{aligned}
A^{1/2}[(1-\theta)K_0+\theta K_1] &= \theta(\lambda\phi)^{-1}A^{1/2}[(1-\lambda)K_0+\lambda\phi K_1] \\
&\geqslant \theta(\lambda\phi)^{-1}[(1-\lambda)A^{1/2}(K_0)+\lambda\phi A^{1/2}(K_1)] \\
&= (1-\theta)A^{1/2}(K_0)+\theta A^{1/2}(K_1),
\end{aligned}
$$

which is the Brunn–Minkowski Inequality. It is also easy to see that equality occurs here precisely if either K_0 and K_1 are homothetic or they lie in parallel lines.

3. Holditch's Theorem

Let C_1 be a plane convex curve and suppose a line segment of fixed length σ is turned once through 360° with its end-points moving on C_1. Let C be the curve traced out by a point on the line segment dividing it into segments of fixed lengths σ_1 and σ_2. Then Holditch's Theorem in its simplest form states that the area of the annular region between C and C_1 is $\pi\sigma_1\sigma_2$, independent of the shape of C_1.

A thorough treatment of Holditch's Theorem and several variants may be found in Edwards [4] and Williamson [9]. For recent expositions see Broman [3] and Müller [7]. In equation (19) below we have a generalization similar to that of Elliott (see [9, p. 209]), although his final result was stated differently.

Let C_1 and C_2 be continuously differentiable closed plane curves with parametric representations $\boldsymbol{x}_1=\boldsymbol{x}_1(t)$ and $\boldsymbol{x}_2=\boldsymbol{x}_2(t)$, $\alpha\leqslant t\leqslant\beta$, as in the introductory section. Let $\boldsymbol{z}(t)=\boldsymbol{x}_1(t)-\boldsymbol{x}_2(t)$, $\alpha\leqslant t\leqslant\beta$. If we apply (5) with $\lambda_1=1$ and $\lambda_2=-1$ we obtain:

$$\tfrac{1}{2}\int[\boldsymbol{z},\mathrm{d}\boldsymbol{z}]=I(C_1)-2I(C_1,C_2)+I(C_2). \tag{16}$$

Now let $\lambda_1,\lambda_2\geqslant 0$ be given real numbers with $\lambda_1+\lambda_2=1$, and let C be the closed curve with parametric representation $\boldsymbol{x}(t)=\lambda_1\boldsymbol{x}_1(t)+\lambda_2\boldsymbol{x}_2(t)$, $\alpha\leqslant t\leqslant\beta$. Substituting into (5) the expression for $I(C_1,C_2)$ from (16) and rearranging terms gives:

$$I(C)=\lambda_1 I(C_1)+\lambda_2 I(C_2)-\frac{\lambda_1\lambda_2}{2}\int[\boldsymbol{z},\mathrm{d}\boldsymbol{z}]. \tag{17}$$

The line segment joining corresponding points of C_1 and C_2 has length $\sigma(t)=\|\boldsymbol{x}_1(t)-\boldsymbol{x}_2(t)\|$. Then $\boldsymbol{z}(t)=\sigma(t)\boldsymbol{u}(t)$, where $\|\boldsymbol{u}(t)\|=1$. We may associate with $\boldsymbol{u}(t)$ a differentiable angle function $\theta(t)$ such that $\boldsymbol{u}(t)=(\cos\theta(t),\sin\theta(t))$, $\alpha\leqslant t\leqslant\beta$. Then $[\boldsymbol{u}(t),\dot{\boldsymbol{u}}(t)]=\dot{\theta}(t)$, so

$$\int[\boldsymbol{z},\mathrm{d}\boldsymbol{z}]=\int\sigma^2[\boldsymbol{u},\mathrm{d}\boldsymbol{u}]=\int_\alpha^\beta\sigma^2(t)\dot{\theta}(t)\mathrm{d}t. \tag{18}$$

The point $\boldsymbol{x}(t)$ on the line segment joining $\boldsymbol{x}_1(t)$ to $\boldsymbol{x}_2(t)$ divides it into segments of lengths $\sigma_1(t)=\lambda_1\sigma(t)$ and $\sigma_2(t)=\lambda_2\sigma(t)$. Using (18), we therefore may write (17) in the form:

$$I(C)=\lambda_1 I(C_1)+\lambda_2 I(C_2)-\tfrac{1}{2}\int_{\alpha}^{\beta}\sigma_1(t)\sigma_2(t)\dot{\theta}(t)\,\mathrm{d}t. \tag{19}$$

In the case where C_1, C_2, and C are simple closed curves traversed once in the counterclockwise sense as t varies from α to β, equation (19) gives a relationship among their areas. If in particular $\boldsymbol{x}_1(t)$ and $\boldsymbol{x}_2(t)$ vary over the same path C_1 in such a way that the angle function $\theta(t)$ varies monotonely from 0 to 2π as t varies from α to β, and A_1 is the area enclosed by C_1 and A the area enclosed by C, then (19) implies:

$$A_1-A=\tfrac{1}{2}\int_0^{2\pi}\sigma_1\sigma_2\,\mathrm{d}\theta, \tag{20}$$

which gives Holditch's Theorem in the case where the chord length $\sigma(t)$ and hence both $\sigma_1(t)$ and $\sigma_2(t)$ are constant. One should keep in mind that while the chord rotating with end-points on C_1 is allowed to have variable length, the point tracing out C divides each chord in the same ratio.

Without giving details we mention that the variants of Holditch's Theorem given in [4], such as Leudesdorf's theorem, results involving the locus of points on one lamina sliding over another, and the locus of the centroid of several mass points moving in a plane, can be derived in a straightforward fashion with this formalism.

4. Mean square chord length

Let C be a continuously differentiable convex curve, and suppose that for each θ, $0\leqslant\theta\leqslant 2\pi$, there is given a chord of C with end-points $\boldsymbol{x}_1(\theta)$ and $\boldsymbol{x}_2(\theta)$ making an angle θ with a fixed reference direction. Assume that both $\boldsymbol{x}_1(\theta)$ and $\boldsymbol{x}_2(\theta)$ are continuously differentiable parameterizations of C traversing C in the counterclockwise sense as θ varies from 0 to 2π.

Let $\boldsymbol{w}(\theta)=\boldsymbol{x}_2(\theta)-\boldsymbol{x}_1(\theta)$ and $\sigma(\theta)=\|\boldsymbol{w}(\theta)\|$, the length of the chord in direction θ. Then integration by parts gives:

$$\int[\boldsymbol{x}_1,\mathrm{d}\boldsymbol{w}]=\int[\boldsymbol{w},\mathrm{d}\boldsymbol{x}_1],$$

and we obtain, from $\boldsymbol{x}_2=\boldsymbol{x}_1+\boldsymbol{w}$:

$$\int[\boldsymbol{x}_2,\mathrm{d}\boldsymbol{x}_2]=\int[\boldsymbol{x}_1,\mathrm{d}\boldsymbol{x}_1]+2\int[\boldsymbol{w},\mathrm{d}\boldsymbol{x}_1]+\int[\boldsymbol{w},\mathrm{d}\boldsymbol{w}]. \tag{21}$$

The integrals giving twice the area enclosed by C can be cancelled on both sides of this equation, yielding:

$$\int [\boldsymbol{w}, \mathrm{d}\boldsymbol{x}_1] = -\tfrac{1}{2}\int [\boldsymbol{w}, \mathrm{d}\boldsymbol{w}]. \tag{22}$$

In the same manner as in the previous section we see that $[\boldsymbol{w}, \dot{\boldsymbol{w}}] = \sigma^2\dot{\theta}$. If $\alpha = \alpha(\theta)$ is the angle between $x_1(\theta)$ and $\boldsymbol{w}(\theta)$, we have $[\boldsymbol{w}, \mathrm{d}\boldsymbol{x}_1] = -\sigma \sin\alpha \, \mathrm{d}s$, where $\mathrm{d}s$ is the element of arclength of C at $\boldsymbol{x}_1(\theta)$. Thus (22) becomes:

$$\int_0^{2\pi} \sigma(\theta)\sin\alpha(\theta)\mathrm{d}s(\theta) = \tfrac{1}{2}\int_0^{2\pi} \sigma^2(\theta)\mathrm{d}\theta. \tag{23}$$

As a first application of (23), suppose that the chords have constant length $\sigma(\theta) \equiv \sigma_0$. Then (23) gives:

$$\int_0^{2\pi} \sin\alpha(\theta)\mathrm{d}s(\theta) = \pi\sigma_0. \tag{24}$$

If, in addition, $\alpha(\theta)$ is constant, $\alpha(\theta) \equiv \alpha_0$, we obtain $L(C) = \pi\sigma_0/\sin\alpha_0$, where $L(C)$ is the perimeter of C, a result found in Santaló [8, p. 11].

For another application of (23), note that $\sigma(\theta)\sin\alpha(\theta) \leqslant b(\theta)$, where $b(\theta)$ is the width of C perpendicular to $\dot{\boldsymbol{x}}_1(\theta)$, that is, the distance between the tangent lines of C parallel to $\dot{\boldsymbol{x}}_1(\theta)$. Hence (23) implies:

$$\tfrac{1}{2}\int_0^{2\pi} \sigma^2(\theta)\mathrm{d}\theta \leqslant \int_0^{2\pi} b(\theta)\mathrm{d}s(\theta). \tag{25}$$

If K is the convex set bounded by C, and K^* its reflection through the origin, it is easy to verify that the integral on the right-hand side of (25) is exactly $A(K+K^*)$, the area of the difference set of K. Proceeding in the same way as Lutwak [6], applying the isoperimetric inequality to $K+K^*$ and using the fact that the perimeters satisfy $L(K+K^*) = 2L(K)$, we obtain from (25) the inequality of Lutwak:

$$\left(\frac{1}{2\pi}\int_0^{2\pi} \sigma^2(\theta)\mathrm{d}\theta\right)^{1/2} \leqslant \frac{L(K)}{\pi}. \tag{26}$$

Equality holds here if and only if K has constant width and $\sigma(\theta)$ is the diametral chord of K in direction θ. In [6], Lutwak restricted himself to the perimeter bisectors and area bisectors of K and derived, in addition to (26), several more refined inequalities.

References

[1] W. Blaschke, Vorlesungen über Differential Geometrie, Vol. II (Chelsea Reprint, New York, 1967).
[2] T. Bonnesen and W. Fenchel, Theorie der konvexen Körper (Springer, Berlin, 1934).
[3] A. Broman, Holditch's sats är något djupare ån Holditch trodde år 1858, Normat, Nord. Mat. Tidskr. 3 (1979) 89–100.
[4] J. Edwards, A Treatise on the Integral Calculus, Vol. I (Chelsea Reprint, New York, 1954).
[5] H.G. Eggleston, Convexity (Cambridge, 1958).
[6] E. Lutwak, Isoperimetric inequalities involving bisectors, Bull. London Math. Soc. 12 (1980) 289–295.
[7] H.R. Müller, Zum Satz von Holditch, in: J. Tölke and J.M. Wills, Eds., Contributions to Geometry, Proc. of the Geometry Symposium at Siegen in 1978 (Birkhäuser, Basel, 1979) pp. 330–334.
[8] L.A. Santaló, Integral Geometry and Geometric Probability, Encyclopedia of Mathematics and its Applications, Vol. 1 (Addison-Wesley, Reading, Mass., 1976).
[9] B. Williamson, An Elementary Treatise on the Integral Calculus (Longmans, Green and Co., London, 1920).

Received 16 March 1981

Annals of Discrete Mathematics 20 (1984) 103–114
North-Holland

A SYMMETRICAL ARRANGEMENT OF ELEVEN HEMI-ICOSAHEDRA

H.S.M. COXETER
University of Toronto, Toronto, Ontario M5S 1A1, Canada

To Paul Erdös for his 70th birthday

1. Introduction

When the elements $0, 1, \ldots, 9, t$ of GF[11] are cyclically permuted, the hexad $02678t$ yields a set of 11 hexads. Any two of these hexads share three elements which may be regarded as the vertices of a triangle. Each hexad contains ten such triangles, fitting together like the faces of a non-orientable polyhedron: the *hemi-icosahedron* $\{3,5\}_5$. The 11 hemi-icosahedra fit together like the facets (or cells) of a 'polytope' whose vertex figure is another non-orientable polyhedron: the hemi-dodecahedron $\{5,3\}_5$. This polytope $_5\{3,5,3\}_5$, having 11 vertices, 55 edges, 55 triangular faces, and 11 facets, is self-reciprocal; its 'symmetry group' is PSL(2, 11), of order 660.

By joining the centre of each face to the midpoints of its edges, we obtain the $55+55$ vertices and 165 edges of a trivalent bipartite graph whose automorphism group is PGL(2, 11) of order 1320. By exhibiting the graph as a set of 11 decagons plus $\binom{11}{2}$ edges joining one vertex of each decagon to one vertex of another, we recognize it as a Cayley graph for the metacyclic group $F^{4,2,-1}$ of order 110 with the presentation

$$R^2 = RS^4RS^2RS^{-1} = 1.$$

2. Hadamard matrices

A suitable numbering of the seven points in the finite projective plane PG(2, 2) yields a cyclic block design (seven numbers arranged in seven triads so that every two triads share one number) and a corresponding Hadamard matrix (with rows and columns numbered $0, 1, \ldots, 6, \infty$) [14, p. 312; 6, pp. 117–119; 2, p. 274]:

$$\begin{array}{ccc@{\qquad}cccccccc}
1 & 2 & 4 & - & + & + & - & + & - & - & + \\
2 & 3 & 5 & - & - & + & + & - & + & - & + \\
3 & 4 & 6 & - & - & - & + & + & - & + & + \\
4 & 5 & 0 & + & - & - & - & + & + & - & + \\
5 & 6 & 1 & - & + & - & - & - & + & + & + \\
6 & 0 & 2 & + & - & + & - & - & - & + & + \\
0 & 1 & 3 & + & + & - & + & - & - & - & + \\
 & & & + & + & + & + & + & + & + & +
\end{array}$$

(Here – means -1 and + means 1.) Since 1, 2, and 4 are the quadratic residues modulo 7, the entry in row μ and column ν of the matrix may be expressed as $\mathcal{A}_{\mu\nu}$, where $\mathcal{A}_{\mu\mu} = -1$, $\mathcal{A}_{\mu\infty} = \mathcal{A}_{\infty\nu} = 1$, and otherwise

$$\mathcal{A}_{\mu\nu} = \left(\frac{\nu - \mu}{7}\right)$$

(the Legendre symbol mod 7).

Representing the row

$$(\mathcal{A}_{\mu 0}, \mathcal{A}_{\mu 1}, \ldots, \mathcal{A}_{\mu 6}, \mathcal{A}_{\mu\infty})$$

by the point $\mathcal{A}_\mu$ in Euclidean 8-space having these Cartesian coordinates, we obtain a 7-dimensional simplex whose vertices all belong to the hypercube

$$(\pm 1, \pm 1, \ldots, \pm 1, 1)$$

in the 7-space $x_\infty = 1$. Gauss proved that if p is an odd prime and $\mu_1 \not\equiv \mu_2 \pmod{p}$,

$$\sum_{\nu=1}^{p-1} \left(\frac{\nu}{p}\right) = 0 \text{ and } \sum_{\nu=0}^{p-1} \left(\frac{\nu - \mu_1}{p}\right)\left(\frac{\nu - \mu_2}{p}\right) = \sum_{\nu=0}^{p-1} \left(\frac{(\nu - \mu_1)(\nu - \mu_2)}{p}\right) = -1$$

[1, p. 130]. Applying these results to the case $p = 7$, we verify that all the points $\mathcal{A}_\mu$ are mutually equidistant, so that they form a *regular* simplex α_7 inscribed in the 7-cube γ_7 [4, p. 341].

Analogous considerations, using the prime 3 instead of 7, yield the familiar fact that 'alternate' vertices of an ordinary cube γ_3 belong to a regular tetrahedron α_3.

Using instead the prime 11, we obtain an α_{11} inscribed in γ_{11}, and the block design is based on the quadratic residues modulo 11, which are 1, 3, 4, 5, 9 [15, p. 323; 3, p. 25]. (When rearranged as 1, 3, 9, 5, 4, these are the successive powers of the residue 3.) For our present purposes, Todd's design has to be replaced by the complementary design based on zero and the non-residues 2, 7, 8, 6, t (which consist of the smallest non-residue 2 multiplied by successive powers of 9 or -2), as on the right below:

1	3	9	5	4	0	2	7	8	6	*t*
2	4	*t*	6	5	1	3	8	9	7	0
3	5	0	7	6	2	4	9	*t*	8	1
4	6	1	8	7	3	5	*t*	0	9	2
5	7	2	9	8	4	6	0	1	*t*	3
6	8	3	*t*	9	5	7	1	2	0	4
7	9	4	0	*t*	6	8	2	3	1	5
8	*t*	5	1	0	7	9	3	4	2	6
9	0	6	2	1	8	*t*	4	5	3	7
t	1	7	3	2	9	0	5	6	4	8
0	2	8	4	3	*t*	1	6	7	5	9

(Like Todd, we write t for ten.)

This new design, corresponding to the minuses in the Hadamard matrix, arranges the 11 residues $0, 1, \ldots, 9, t$ in 11 hexads so that every two hexads share three members. Each hexad shares a different triad with each of the remaining ten hexads. This observation picks out, from each hexad, ten of the $\binom{6}{3}$ possible triads. For instance, the ten special triads in $02786t$ are:

$$078,\ 086,\ 06t,\ 0t2,\ 027,\ 267,\ 67t,\ 7t8,\ t82,\ 826.$$

3. The hemi-icosahedron $\{3, 5\}_5$

Regarding each of these ten special triads as a triangle, we observe that the ten triangles fit together, side by side, like the faces of a polyhedron. Since a typical vertex, such as 0, is surrounded by five triangles

$$207, 708, 806, 60t, t02, \tag{1}$$

the 'polyhedron' has a pentagonal *vertex figure*

$$2\quad 7\quad 8\quad 6\quad t,$$

and is thus of type $\{3, 5\}$. It resembles the ordinary icosahedron, except that the number of triangular faces is not 20 but only ten. This reduction from 20 faces to ten can be achieved by identifying all pairs of antipodal points, as in Klein's trick for passing from the sphere to the elliptic plane. One way to reach the antipodes is to travel over the surface of the icosahedron along a *Petrie polygon*: a skew polygon $0\,7\,8\,t\,2\ldots$ in which every two consecutive sides, but no three, belong to a face. The Platonic icosahedron may be denoted by $\{3, 5\}_{10}$, indicating that its Petrie polygon is a skew *decagon* [7, p. 25]. From any vertex we can reach the opposite vertex by proceeding five steps along such a skew decagon. Accordingly, our non-orientable *hemi-icosahedron* with ten faces [11, pp. 112, 117] is denoted by

$$\{3,5\}_5.$$

The 11 hexads can thus be regarded as 11 hemi-icosahedra; see Fig. 1.

Since the symmetry group $[3,5]=[5,3]$ of the icosahedron $\{3,5\}$ (or of the dodecahedron $\{5,3\}$) is $\mathscr{C}_2\times\mathscr{A}_5$, the direct product of the $\mathscr{C}_2$ generated by the central inversion and the icosahedral (rotation) group $\mathscr{A}_5$ [11, pp. 34, 111; 8, p. 35], the symmetry group of the hemi-icosahedron $\{3,5\}_5$ is the simple group $\mathscr{A}_5$ itself. The icosahedral group $\mathscr{A}_5$ thus has the presentation

$$R_\nu^2=(R_1R_2)^3=(R_2R_3)^5=(R_1R_2R_3)^5=1, \qquad R_1\rightleftarrows R_3, \tag{2}$$

conveniently denoted by $G^{3,5,5}$ [5, p. 114] as well as the more familiar presentation

$$A^3=B^5=(AB)^2=1.$$

The hemi-icosahedron is not merely a topological 'map', or a dissection of the elliptic plane into ten triangles; it can be realized in Euclidean 5-space as a regular *skew* polyhedron (cf. [6, p. 80]) whose 6 vertices, 15 edges and 10 faces all belong to a regular simplex α_5. Since any two adjacent faces lie in a 3-flat,

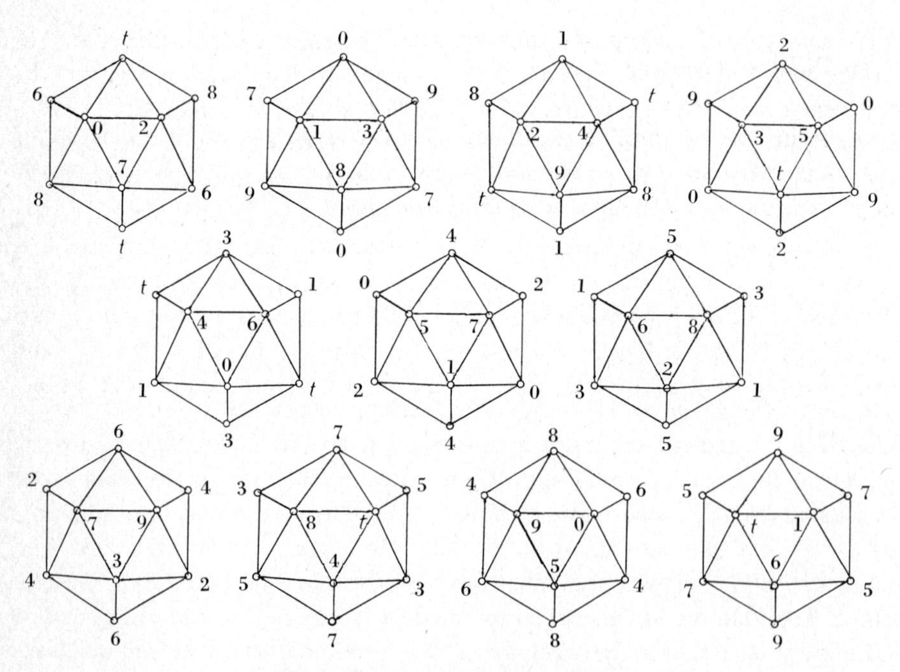

Fig. 1. The 11 hemi-icosahedra $\{3,5\}_5$.

their planes form a dihedral angle arc sec 3, as in the tetrahedron α_3 (unlike the ordinary icosahedron, whose dihedral angle is $\pi - \operatorname{arc\,sin} \frac{2}{3}$).

4. The skew polytope ${}_5\{3,5,3\}_5$

Analogously, our arrangement of 11 hemi-icosahedra with 11 vertices is not merely a topological 'complex' with non-orientable 'cells': it can be realized in Euclidean 10-space as a regular *skew polytope* of type

$$\{3,5,3\},$$

whose 11 vertices, 55 edges and 55 faces belong to a regular simplex α_{10}. The 11 non-orientable cells lie in 11 different 5-flats.

The 11 vertices and $\binom{11}{2} = 55$ edges are *all* the vertices and edges of α_{10}, but the 55 faces are one-third of the $\binom{11}{3}$ faces of α_{10}, namely those derivable from (1) by cyclic permutation (of $0, 1, \ldots, 9, t$).

We have seen that in the hemi-icosahedron 02786*t*, the five triangles surrounding the vertex 0 join this vertex to the sides of the pentagonal vertex figure 2786*t*. Analogously, in the skew polytope, the six hemi-icosahedra surrounding the typical vertex 0 join this vertex to the six pentagons:

2786*t*, 13987, 3452*t*, 95468, 54172, 413*t*6,

which are the vertex figures of those six hemi-icosahedra. (Observe that multiplication by 3 preserves the first pentagon and cyclically permutes the remaining five.) These six pentagons fit together, side by side, like the faces of a 'polyhedron' having, at its vertex 1, the triangular vertex figure 347, so that it is of type $\{5,3\}$. It thus resembles the ordinary dodecahedron, except that the number of pentagonal faces is not 12 but only six. The dodecahedron $\{5,3\}$, like its reciprocal $\{3,5\}$, has a decagonal Petrie polygon (in this case 13954...) along which we can travel five steps to reach the antipodes. Accordingly our *hemi-dodecahedron* with six faces, dual to the hemi-icosahedron, is denoted by $\{5,3\}_5$ [11, p. 117]; see Fig. 2. Finally, the skew polytope, having facet $\{3,5\}_5$ and vertex figure $\{5,3\}_5$, is conveniently denoted by the generalized Schläfli symbol

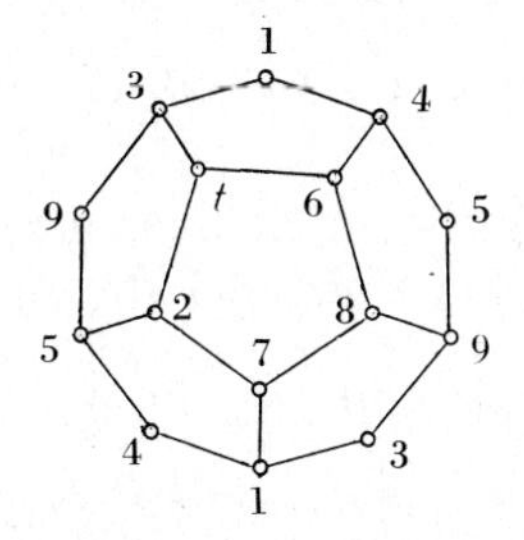

Fig. 2. The hemi-dodecahedron $\{5,3\}_5$.

$${}_5\{3,5,3\}_5,$$

emphasizing the fact that from a topological standpoint, it can be derived from the infinite regular honeycomb $\{3,5,3\}$ in hyperbolic space [6, p. 202; 12, p. 446] by making suitable identifications.

5. A new presentation for PSL(2, 11)

Since the symmetry group of the hyperbolic honeycomb $\{3,5,3\}$ is $[3,5,3]$, with the presentation

$$R_\nu^2 = (R_1R_2)^3 = (R_2R_3)^5 = (R_3R_4)^3 = 1, \qquad R_3 \rightleftarrows R_1 \rightleftarrows R_4 \rightleftarrows R_2, \tag{3}$$

while both $\{3,5\}_5$ and $\{5,3\}_5$ have icosahedral symmetry groups (see (2)), we may expect the symmetry group ${}_5[3,5,3]_5$ of the skew polytope ${}_5\{3,5,3\}_5$ to have the presentation (3) with the extra relations

$$(R_1R_2R_3)^5 = (R_2R_3R_4)^5 = 1. \tag{4}$$

By enumerating the cosets of the icosahedral subgroup generated by R_2, R_3, R_4 in the abstract group so presented, we can soon verify that the number of cosets is 11, so that the group has order $11 \times 60 = 660$, in agreement with the statement of Todd [15, p. 324] that it is the simple group of that order, namely

$$\mathrm{PSL}(2,11).$$

Suitably renumbering the 11 cosets, we can identify them with the vertices of the skew polytope and obtain the representation:

$$R_1 = (0\ 7)(3\ 9)(4\ 5)(6\ t),$$

$$R_2 = (1\ 9)(2\ 6)(4\ 5)(7\ 8),$$

$$R_3 = (2\ 8)(3\ 4)(5\ 9)(6\ t),$$

$$R_4 = (1\ 2)(3\ t)(4\ 5)(6\ 9).$$

By seeing which points are invariant, we find that each generator interchanges two symbol:
the 11 hemi-icosahedra by reflection in their common face: 128 or $03t$ or 017 or 078 (see Fig. 1). Thus, R_1 reverses the edge 07, and the product

$$R_2R_1 = (0\ 7\ 8)(1\ 3\ 9)(2\ t\ 6)$$

rotates the triangle 078, which is invariant for R_4.

It is interesting to observe that the permutations

$$A = (R_1R_2R_4R_3R_2)^2 = (0\ 5\ 7\ 6\ 3)(1\ 9\ 4\ t\ 2),$$

$$B = R_3R_4(R_2R_3R_1)^2R_4R_3 = (8\ 6\ t\ 9\ 3)(7\ 5\ 0\ 1\ 2),$$

$$C = R_2R_3 = (4\ 9\ 1\ 5\ 3)(t\ 6\ 8\ 7\ 2),$$

satisfy the relations

$$A^5 = B^5 = C^5 = (AB)^2 = (BC)^2 = (CA)^2 = (ABC)^2 = 1,$$

which define the group $G^{5,5,5} \cong \mathrm{PSL}(2,11)$ [5, p. 117]. This remark provides a proof that

$$_5[3,5,3]_5 \cong \mathrm{PSL}(2,11), \tag{5}$$

without appealing to Todd.

Abraham Sinkov has found that the relations (4) can be simplified to

$$(R_1R_2R_3)^5 = 1 \tag{6}$$

without changing the group: the period 5 for $R_2R_3R_4$ is a consequence of the relations (3) and (6). He has also provided a direct proof of (5) by exhibiting the generators as linear fractional transformations corresponding to the matrices

$$\begin{bmatrix} -2 & -1 \\ 5 & 2 \end{bmatrix}, \begin{bmatrix} -3 & 3 \\ 4 & 3 \end{bmatrix}, \begin{bmatrix} -1 & 1 \\ -2 & 1 \end{bmatrix}, \begin{bmatrix} 5 & 3 \\ -5 & -5 \end{bmatrix} \pmod{11}.$$

Of course, the matrices themselves, whose squares are

$$\begin{bmatrix} -1 & 0 \\ 0 & -1 \end{bmatrix},$$

generate the group SL(2, 11), for which the presentation

$$R_\nu^2 = (R_1R_2)^3 = (R_2R_3)^5 = (R_3R_4)^3 = (R_3R_1)^2 = (R_1R_4)^2$$
$$= (R_4R_2)^2 = (R_1R_2R_3)^5$$

suffices since it makes R_1 and R_4 behave like i and j in the quaternion group $i^2 = j^2 = (ij)^2$ [8, p. 64].

Since the matrix

$$\begin{bmatrix} -5 & 5 \\ 4 & 5 \end{bmatrix},$$

of determinant -1, transforms

$$R_1,\ R_2,\ R_3,\ R_4 \quad \text{into} \quad R_4,\ R_3,\ R_2,\ R_1,$$

and thus reciprocates the polytope, it follows that the symmetry group of a pair of reciprocal $_5\{3,5,3\}_5$'s is PGL(2, 11). Adopting a symbol analogous to [[3, 4, 3]] [9, p. 323], we thus have:

$$_5[[3,5,3]]_5 \cong \mathrm{PGL}(2,11). \tag{7}$$

6. The Petrie polygon

We have seen that R_1 reverses the edge 07, and R_2R_1 rotates the face 078. Somewhat analogously, the transformation

$$R_3R_2R_1 = (0\ 7\ 8\ t\ 2)(1\ 3\ 4\ 9\ 5)$$

rotates the Petrie polygon 078*t*2 of the hemi-icosahedron 02786*t*, and the 'twist'

$$R_4R_3R_2R_1 = (0\,7\,8\,t\,4\,1)(2\,3)(5\,9\,6) \tag{8}$$

rotates the skew hexagon 078*t*41, which may reasonably be called a *Petrie polygon* of the skew polytope (cf. [7, p. 223]) because 078*t* is part of a Petrie polygon of one hemi-icosahedron, while 78*t*4 is part of a Petrie polygon of another (the ninth in Fig. 1).

Any two adjacent sides of a face are edges belonging to two skew hexagons (for instance, 07 and 78 to 078*t*41 and 078352); thus there are six for each face. But each skew hexagon involves six faces (such as 078, 78*t*, 8*t*4, *t*41, 410, 107). Therefore the total number of Petrie polygons is equal to the number of faces, namely 55, which is also the number of edges. This result is in accordance with (8), where the permutation $R_4R_3R_2R_1$ involves the cycles (2 3)(5 9 6) which belong to one edge 23 and one face 569. We may reasonably call this edge and face *opposites.* It is easy to find the face opposite to a given edge by looking at the faces to which the given edge belongs. (Their number, 3, is the final 3 in the Schläfli, symbol $\{3,5,3\}$.) The three faces to which 23 belongs are 235, 236 and 239 (occurring in the fourth, seventh and ninth hemi-icosahedra); so the face opposite to 23 is 569. Conversely, given a face, we can find the opposite edge by seeing which vertices belong to neither of the hemi-icosahedra which share that face.

The ten edges opposite to the ten faces of one hemi-icosahedron are the edges of the complementary α_4, that is, the joins of all pairs of the five vertices not belonging to the given hemi-icosahedron. The ten faces opposite to the ten edges through one vertex appear as vertex figures of the vertex figure; for instance, the vertex 0 leads to the following ten faces (see Fig. 2):

01	02	03	04	05	06	07	08	09	0*t*
347	57*t*	19*t*	156	249	48*t*	128	679	358	236

7. The perfection of Todd's cyclic design

Looking again at Todd's pentads, we observe that they may be regarded as *cycles* such that each pair of symbols appears as a pair of *adjacent* members of one cycle, *alternate* members of another, in the *same* order; for instance, the pair 07 appears as it stands in the third cycle, and separated by t in the seventh (which is $0t794$). In other words, these cycles (derived by permutation of $0, 1, \ldots, t$ from the cycle of powers of the quadratic residue 3), along with their reversals, form a *perfect cyclic design*, PD(11, 5, 1) in the notation of Mendelsohn [13].

Since each pair now occurs as an *ordered* pair (07 but not 70), the arrangement of 11 cycles has the effect of converting the complete graph $\mathcal{K}_{11}$ into a digraph by the insertion of 55 arrows, that is, of giving every edge of the simplex α_{10} a direction. The ordered pairs

$$02,\ 07,\ 08,\ 06,\ 0t$$

serve to associate the vertex 0 with the cycle of non-residues $2786t$, which appears in Fig. 1 as the vertex figure at that vertex of the first hemi-icosahedron, and in Fig. 2 as the 'middle' face of the hemi-dodecahedron. On the other hand, the ordered pairs

$$10,\ 30,\ 90,\ 50,\ 40$$

involve the cycle of residues 13954, which appears in Fig. 2 as a Petrie polygon of the vertex figure at the vertex 0.

8. The 3-regular trivalent graph with 110 vertices

We saw in an earlier paper [9] that any polytope or honeycomb $\{3, q, 3\}$ (whose facets $\{3, q\}$ come together with 3 at each edge) yields a 3-regular bipartite trivalent graph of girth $2q$ whose vertices are the midpoints of the edges and the centres of the faces. In this manner the hyperbolic honeycomb $\{3, 5, 3\}$ yields an infinite graph of girth 10 and our skew polytope ${}_5\{3, 5, 3\}_5$ yields a graph of girth 10 with $55 + 55$ vertices. Its group of automorphisms, of order $2^3 165 = 1320$, is derived from PSL(2, 11) (with the presentation (3), (6)) by adjoining an involutory element that dualizes the skew polytope, transforming R_1, R_2, R_3, R_4 into R_4, R_3, R_2, R_1. We see from (7) that this enlarged group of order 1320 is PGL(2, 11), and the graph with 110 vertices must be the onc that arosc as the Cayley graph for the metacyclic group $F^{4,2,-1}$ [10, p. 146]. We have thus found a new construction for that known graph, and a new notation for its vertices as *pairs* and *triads* of the 11 symbols $0, 1, \ldots, 9, t$: all the $\binom{11}{2}$ pairs and 55 of the $\binom{11}{3}$ triads, namely those triads not occurring in any of Todd's 11 pentads; see Section 2.

Such a notation, in which each pair occurs in exactly three of the triples, was devised independently by R.M. Foster before the skew polytope was thought of!

He wrote out a complete 'dictionary' translating this notation into that of Coxeter and Frucht. The new notation is far easier to work with. For, as soon as the 55 triads are given (as in Fig. 1) we can see at once which vertices of the graph are joined by edges. The rule is simply that every triad is joined to the three pairs contained in it: the three neighbours of 078 are 07, 08, 78, and the three neighbours of 07 are 017, 027, 078.

9. Eleven disjoint decagons

We have already remarked that the girth of the graph is ten. At each vertex of each hemi-icosahedron, the surrounding edges and faces yield a decagon. A cycle of 11 such decagons can be obtained by using the vertex 0 of the first hemi-icosahedron, vertex 1 of the next, and so on:

02	027	07	078	08	068	06	06*t*	0*t*	02*t*
13	138	18	189	19	179	17	017	01	013
24	249	29	29*t*	2*t*	28*t*	28	128	12	124
35	35*t*	3*t*	03*t*	03	039	39	239	23	235
46	046	04	014	14	14*t*	4*t*	34*t*	34	346
57	157	15	125	25	025	05	045	45	457
68	268	26	236	36	136	16	156	56	568
79	379	37	347	47	247	27	267	67	679
8*t*	48*t*	48	458	58	358	38	378	78	78*t*
09	059	59	569	69	469	49	489	89	089
1*t*	16*t*	6*t*	67*t*	7*t*	57*t*	5*t*	59*t*	9*t*	19*t*

Since these 11 decagons are disjoint, they account for all the 110 vertices of the graph. Accordingly, the graph may be described as *a set of 11 separate decagons along with 55 further edges, each joining one vertex of one decagon to one vertex of another.* For instance, the first decagon is joined to the second by the edge 07 017, to the third by 02*t* 2*t*, to the fourth by 0*t* 03*t*, to the fifth by 06 046, to the sixth by 02 025, to the seventh by 068 68, to the eighth by 027 27, to the ninth by 078 78, to the tenth by 08 089, and to the eleventh by 06*t* 6*t*.

This factorization of the graph, into 11 separate decagons and $\binom{11}{2}$ connecting edges, resembles the structure of the above-mentioned Cayley graph for the metacyclic group $F^{4,2,-1}$ with the presentation

$$R^2 = RS^4RS^2RS^{-1} = 1. \tag{9}$$

The 11 cosets of the subgroup $\mathscr{C}_{10}$ generated by S are represented by the 11 disjoint decagons, and the 55 cosets of the subgroup $\mathscr{C}_2$ generated by R are represented by the 55 connecting edges. In fact, the agreement is precise.

Starting at any vertex, say 01 (in the second decagon), we find that R takes us to 014 (in the fifth), S^4 then takes us (4 steps to the right) to 34t, another R to 3t (in the fourth decagon), S^2 (2 steps to the right) to 03, a third R to 013 (in the second decagon), and finally S^{-1} (1 step to the left) back to the original vertex 01:

	R		S^4		R		S^2		R		S^{-1}	
01		014		34t		3t		03		013		01

The reader may find it amusing to follow through the same pattern starting elsewhere. Various consequences of (9), such as $RS^2RSRS^{-1}RS^{-2} = 1$ [10, p. 147] can be verified in a similar manner:

	R		S^2		R		S		R		S^{-1}		R		S^{-2}	
01		014		14t		1t		16t		16		136		13		01

10. Pairs of opposite vertices of the graph

We saw in Section 6 that the edges and faces of the skew polytope $_5\{3,5,3\}_5$ occur in pairs of *opposites*. The corresponding property of the graph is that every vertex has a unique opposite vertex at graph-theoretic distance 7. (In Foster's notation, the graph has *diameter* 7_1.) From any given vertex there are, of course, three vertices at distance 1, six at distance 2, and twelve at distance 3. The 12 vertices distant 3 from 01 are adjacent (respectively) to the 12 at this same distance from the opposite vertex 347, yielding 12 routes of length 7 from 01 to 347:

01	013	03	039	39	379	37	347
01	013	03	03t	3t	34t	34	347
01	013	13	136	36	346	34	347
01	013	13	138	38	378	37	347
01	014	04	045	45	457	47	347
01	014	04	046	46	346	34	347
01	014	14	14t	4t	34t	34	347
01	014	14	124	24	247	47	347
01	017	07	027	27	247	47	347
01	017	07	078	78	378	37	347
01	017	17	157	57	457	47	347
01	017	17	179	79	379	37	347

Tracing these routes in the list of 11 decagons, we deduce that *any* two opposite vertices of the graph are related by the group element

$$SRS^2RS^2 = SRS^{-2}RSR = S^2RS^{-1}RS^{-1}R = S^3RSRS$$
$$= RS^{-1}RS^2RS^{-1} = RS^{-5}R = RS^5R = RSRSRS^{-2}$$
$$= S^{-1}RS^{-1}RS^{-3} = S^{-1}RSRS^{-1}RS = S^{-2}RS^{-2}RS^{-1} = S^{-3}RS^3,$$

which is the T of Coxeter and Frucht [10, p. 150]. Notice that this involutory element is not central: the opposites of adjacent vertices are never adjacent!

11. Epilogue

After reading a preprint of the preceding ten sections, Branko Grünbaum remarked that he had already discovered the arrangement of 11 hemi-icosahedra as a member of his family of regular 3-*polystromas.* He simply fitted hemi-icosahedra together, three round each edge, until the arrangement closed up. It appears as the climax of his Regularity of Graphs, Complexes and Designs, in *Colloques Internationaux C.N.R.S.*, No. 260 (Orsay, 1976) pp. 191–197.

References

[1] G.E. Andrews, Number Theory (Saunders, Philadelphia, 1971).
[2] W.W.R. Ball and H.S.M. Coxeter, Mathematical Recreations and Essays, 12th edn. (University of Toronto Press, 1974).
[3] R.D. Carmichael, Introduction to the Theory of Groups of Finite Order (Ginn, Boston, Mass., 1937).
[4] H.S.M. Coxeter, Regular compound polytopes in more than four dimensions, J. Math. Phys. (M.I.T.) 12 (1933) 334–345.
[5] H.S.M. Coxeter, The abstract groups $G^{m,n,p}$, Trans. Amer. Math. Soc. 45 (1939) 73–150.
[6] H.S.M. Coxeter, Twelve Geometric Essays (Southern Illinois University Press, 1968).
[7] H.S.M. Coxeter, Regular Polytopes, 3rd edn. (Dover, New York, 1973).
[8] H.S.M. Coxeter, Regular Complex Polytopes (Cambridge University Press, 1974).
[9] H.S.M. Coxeter, The edges and faces of a 4-dimensional polytope, Congressus Numerantium (Winnipeg) 28 (1980) 309–334.
[10] H.S.M. Coxeter and Roberto Frucht, A new trivalent symmetrical graph with 110 vertices, Ann. N.Y. Acad. Sci. 319 (1979) 141–152.
[11] H.S.M. Coxeter and W.O.J. Moser, Generators and Relations for Discrete Groups, 4th edn. (Springer-Verlag, Berlin, 1980).
[12] C.W.L. Garner, Compound honeycombs in hyperbolic space, Proc. Roy. Soc. London A316 (1970) 441–448.
[13] N.S. Mendelsohn, Perfect cyclic designs, Discr. Math. 20 (1977) 63–68.
[14] R.E.A.C. Paley, On orthogonal matrices, J. Math. Phys. (M.I.T.) 12 (1933) 311–320.
[15] J.A. Todd, A combinatorial problem, J. Math. Phys. (M.I.T.) 12 (1933) 321–333.

Received 10 June 1981

Annals of Discrete Mathematics 20 (1984) 115–127
North-Holland

REGULAR INCIDENCE-COMPLEXES AND DIMENSIONALLY UNBOUNDED SEQUENCES OF SUCH, I

L. DANZER
Mathematisches Institut der Universität Dortmund, D 46 Dortmund 50, Fed. Rep. Germany

Dedicated to H. LENZ for his 65th birthday

The notion of a regular polytope is generalized fairly much to what is now called a regular incidence-complex (r.i.c.). The definition is completely in combinatorial terms. All regular d-polytopes and all regular complex d-polytopes are included as well as the projective and other spaces and many well-known configurations. An introductory theory for r.i.c.'s is given together with quite a few examples and also examples of non-existence. Especially, some infinite sequences $(\mathcal{K}_\nu)_{\nu\in\mathbf{N}}$ are described, where $\dim(\mathcal{K}_\nu)=\nu$ and every r.i.c. $\mathcal{K}_\nu$ is a facet of $\mathcal{K}_{\nu+1}$. Furthermore, in part II a construction will be given which associates with every natural number n $(n>1)$ and every non-degenerate r.i.c. $\mathcal{K}$ a non-degenerate r.i.c. $n^{\mathcal{K}}$ where the edges of $n^{\mathcal{K}}$ consist of exactly n vertices while the vertex figures are isomorphic to $\mathcal{K}$.

1

A non-degenerate *incidence-complex* of *dimension* d is defined by the following properties (I1 to I4):

I1 $\left\{\begin{array}{l}(\mathcal{K}, \leqslant) \text{ is a lattice with elements } \emptyset \text{ and } K \text{ such that } F\in\mathcal{K} \text{ implies:} \\ \qquad \emptyset \leqslant F \leqslant K.\end{array}\right.$

I2 $\left\{\begin{array}{l}\text{For all chains } F_0 < F_1 < \cdots < F_i \text{ of members of } \mathcal{K} \text{ the length } i \text{ is} \\ \text{bounded by } d+1\text{; chains of length } d+1 \text{ are called } \textit{flags}\text{. Every chain is} \\ \text{contained in a flag.}\end{array}\right.$

The members of $\mathcal{K}$ are called *faces*. By I2 we may define for every face F

$$\dim(F) := \dim(\langle \emptyset, F\rangle),$$

where $\langle \emptyset, F\rangle$ is the *section-complex* of all faces G with $\emptyset \leqslant G \leqslant F$. Hence, $\dim(\emptyset) = -1$ and $\dim(K) = \dim(\mathcal{K})$. A face F is called a *vertex*, an *edge*, an i-face or a *facet*, iff $\dim(F) = 0$, 1, i, $d-1$, respectively. In general we do not distinguish between F and $\langle \emptyset, F\rangle$. If V is a vertex of $\mathcal{K}$, $\langle V, K\rangle$ is called its *vertex figure*.

I3 $\left\{\begin{array}{l}\text{If } f \text{ and } g \text{ are two different flags of } \mathcal{K} \text{ and } h := f \cap g, \text{ then there are} \\ \text{finitely many flags } f_1 = f, f_2, \ldots, f_n = g, \text{ all containing } h \text{ and such that} \\ f_{\nu+1} \text{ differs from } f_\nu \text{ in exactly one face.}\end{array}\right.$

This means that $\mathcal{K}$ is *connected* (in a sense that is adequate for our purposes).

The final defining property guarantees a certain homogeneity[1] of $\mathcal{K}$:

I4 $\left\{\begin{array}{l}\text{There are cardinal numbers } k_0, k_1, \ldots, k_{d-1}, \text{ not necessarily finite but all} \\ \text{greater than one, such that } F < G \text{ and } \dim(F)+1 = i = \dim(G)-1 \\ \text{implies that the one-dimensional section-complex } \langle F, G\rangle \text{ possesses} \\ \text{exactly } k_i \text{ flags.}\end{array}\right.$

In the case where $\leq$ is a partial order on $\mathcal{K}$, but fails to make $\mathcal{K}$ a lattice, we call $\mathcal{K}$ a *degenerate* incidence-complex, provided $\mathcal{K}$ meets all the other requirements of I1 to I4. If nothing else is said, degeneracy is not excluded. One important property of non-degenerate incidence-complexes is that all the faces may be considered as subsets of the set of vertices of $\mathcal{K}$. And even more:

(1) $\left\{\begin{array}{l}\text{If } i \leq j \text{ every } j\text{-face } F \text{ is uniquely determined by the set of} \\ i\text{-faces } G \text{ with } G \leq F \text{ and vice versa.}\end{array}\right.$

If $\mathcal{K}$ is an incidence-complex and $\leq$ is replaced by $\geq$ while the set of faces is unchanged, we get the *dual* $^*\mathcal{K}$ of $\mathcal{K}$. $^*\mathcal{K}$ of course is an incidence-complex too and is non-degenerate iff $\mathcal{K}$ is. $\mathcal{K}$ is called *self-dual* iff $^*\mathcal{K}$ is isomorphic to $\mathcal{K}$. Such an isomorphism is called a *polarity* iff its square is the identity on $\mathcal{K}$.

2

A *regular* incidence-complex[2] is an incidence-complex $\mathcal{K}$ for which

(R) the group $A(\mathcal{K})$ of automorphisms of $\mathcal{K}$[3] is flag-transitive.

Obviously I1, I2, I3 and (R) imply I4. We say $\mathcal{K}$ is $(m_0, m_1, \ldots, m_{d-1})$-regular iff for every one-dimensional section-complex $\langle F, G\rangle$ with $F < G$ and $\dim(F)+1 = i = \dim(G)-1$ the stabilizer of F and G in $A(\mathcal{K})$ acts m_i-fold transitive on the flags of $\langle F, G\rangle$. Instead of $(m, m, \ldots, m)$-fold regular we simply say *m-fold* regular. By I3 1-fold regularity is equivalent to regularity.

The elementary properties of incidence-complexes and of r.i.c.'s are treated in [11], and some less elementary facts about $A(\mathcal{K})$ are proved in [12]. Therefore it is not necessary to give all proofs in the present paper.

[1] This has nothing to do with the *homogeneity of degree t* to be defined in Section 2.
[2] In the sequel: an r.i.c., or — if the dimension d shall be included in the notation — a d-r.i.c.
[3] Isomorphisms and automorphisms are supposed to preserve order; we write $\cong$.

Two trivial consequences of (R) are:

(2) $\left\{ \begin{array}{l} \text{If } \mathcal{K} \text{ is a non-degenerate } d\text{-r.i.c. and } 0 \leqslant i \leqslant d-1,\ A(\mathcal{K}) \text{ acts faithfully} \\ \text{on the } i\text{-faces of } \mathcal{K} \end{array} \right.$

(cf. (1)); and

(3) $\left\{ \begin{array}{l} \text{if } \mathcal{K} \text{ is an r.i.c., then so is every section-complex } \langle F, G\rangle \text{ of } \mathcal{K}. \text{ The} \\ \text{stabilizer of } F \text{ and } G, \text{ and even the stabilizer of every chain } \emptyset < E_1 < \\ E_2 < \cdots < E_n < F < G < H_1 < H_2 < \cdots < H_m < K \text{ in } A(\mathcal{K}) \text{ is a sub-} \\ \text{group of } A(\langle F, G\rangle).\ \dim(F_1) = \dim(F_2) \text{ and } \dim(G_1) = \dim(G_2) \text{ to-} \\ \text{gether with } F_1 < G_1 \text{ and } F_2 < G_2 \text{ imply:} \\ \langle F_1, G_1\rangle = \langle F_2, G_2\rangle. \end{array} \right.$

Hence, for $F \leqslant G$, $-1 \leqslant i \leqslant j \leqslant d$, and $\dim(F) = i$, $j = \dim(G)$, the number of flags in $\langle F, G\rangle$ does depend only on i, j and $\mathcal{K}$; we denote it by $k_{i,j}(\mathcal{K})$ or briefly by $k_{i,j}$.[4] Obviously, $k_{i,i} = k_{i,i+1} = 1$ and $k_{i-1,i+1} = k_i$ (cf. I 4) for all i. By

$$\begin{pmatrix} k_0, k_1, k_2, \ldots, k_{d-1} \\ k_{-1,2}, k_{0,3}, \ldots, k_{d-3,d} \\ k_{-1,3}, \ldots \qquad . \\ . \qquad\quad . \\ . \qquad . \\ k_{-1,s}, \ldots, k_{d-s-1,d} \end{pmatrix}$$

we denote the family of all isomorphy-types of d-r.i.c.'s, which share the numbers occurring in the scheme. This family will be called a *clan* of *degree s* $(1 \leqslant s \leqslant d)$. A clan of degree d (when the scheme becomes a triangle) we call a *cluster.* In the other extreme case we may also include non-regular incidence-complexes and then talk of the *class* $[k_0, k_1, \ldots, k_{d-1}]$ instead of the corresponding clan of degree 1.

The members of the class [2,2,…,2] are called *polytopes.* Of course, this is a purely combinatorial generalization of convex and other polytopes (cf. Example 2b in Section 4).

$k_{-1,d}$, by definition, is the number of all flags of $\mathcal{K}$; hence $\operatorname{card}(A(\mathcal{K}))$ is a multiple of $k_{-1,d}$. We define the *index* of an r.i.c. $\mathcal{K}$ by

$$\operatorname{ind}(\mathcal{K}) := \frac{\operatorname{card}(A(\mathcal{K}))}{k_{-1,d}} \quad (\text{for finite } k_{-1,d}).$$

In other words, $\operatorname{ind}(\mathcal{K})$ is the order of the stabilizer of a flag of $\mathcal{K}$ in $A(\mathcal{K})$.

A clan of degree s is called *symmetric* iff its scheme is symmetric with respect to its vertical axis; that is, iff

$$k_{i,j} = k_{d-1-j,d-1-i} \quad (\text{for} -1 \leqslant i \leqslant j \leqslant d;\ j - i \leqslant s+1).$$

[4] By definition, $k_{i,j}$ is a cardinal number; it need not be finite.

An r.i.c. is called symmetric iff the cluster it belongs to is symmetric.

Problem 1: Can a symmetric clan of degree 2 contain an asymmetric r.i.c.? (Certainly not among the regular polytopes in the classical sense.)

Problem 2: Must a non-empty symmetric cluster contain at least one self-dual r.i.c.?

Problem 3: Does every self-dual r.i.c. possess a polarity? (Cf. [10, p. 23] or [12].)

I should not be surprised if the answers to these three problems were: no, yes, no, respectively.

Remark 1: Suppose $\mathscr{K}$ is a non-degenerate r.i.c. and there is an abelian subgroup G of $A(\mathscr{K})$, which acts transitively[5] as well on the vertices as on the facets of $\mathscr{K}$. Then $\mathscr{K}$ possesses a polarity and hence is self-dual and a fortiori is symmetric.

Proof: Without loss of generality we may suppose $d := \dim(\mathscr{K}) \geqslant 2$. We consider $\hat{\mathscr{K}} := \{F \mid F \in \mathscr{K} \wedge \dim(F) \in \{-1, 0, d-1, d\}\}$, choose a vertex V_0 and a facet f_0 and define

$$\pi: \quad \begin{aligned} &\hat{\mathscr{K}} \to \hat{\mathscr{K}} \\ &\emptyset \mapsto K \\ &\left.\begin{aligned} &\alpha(V_0) \mapsto \alpha^{-1}(F_0) \\ &\alpha(F_0) \mapsto \alpha^{-1}(V_0) \end{aligned}\right\} \quad \left\{\begin{aligned} &\alpha \text{ running} \\ &\text{through } G \end{aligned}\right. \\ &K \mapsto \emptyset. \end{aligned}$$

Obviously π is bijective and $\pi^2 = id_{\hat{\mathscr{K}}}$. Now suppose $V < F$ for a vertex V and a facet F. By the hypotheses there exist unique β_1 and β_2 in G with

$$V = \beta_1^{-1}(V_0) \quad \text{and} \quad F = \beta_2^{-1}(F_0).$$

It follows that

$$(4) \quad \left\{ \begin{aligned} \pi(F) &= \beta_2(V_0) = (\beta_2 \circ \beta_1)(V) \\ &< (\beta_2 \circ \beta_1)(F) \\ &= (\beta_1 \circ \beta_2)(F) = \beta_1(F_0) \\ &= \pi(V). \end{aligned} \right.$$

[5] Since G is abelian, this implies that G is sharply 1-transitive.

Next we extend π to a mapping of $\mathcal{K}$ onto $\mathcal{K}$ by

$$\pi(E) := \inf\{\pi(V) \mid V \leqslant E \wedge \dim(V) = 0\}$$
$$\text{for } 0 < \dim(E) < d-1.$$

We also consider

$$\nu(E) := \sup\{\pi(F) \mid F > E \wedge \dim(F) = d-1\}.$$

By (4) we have $\nu(E) \leqslant \pi(E)$. But if $\nu(E)$ were less than $\pi(E)$ there would be a vertex V_1 with $V_1 \leqslant \pi(E)$, but $V_1 \not\leqslant \nu(E)$ (cf. (1)). Consequently, again by (4), $\pi(V_1)$ would be above every vertex of E; hence, $\pi(V_1) > E$ and finally $\nu(E) \geqslant \pi(\pi(V_1)) = V_1$. This contradiction shows $\nu = \pi$, and now it is trivial that $\pi^2 = id_\kappa$ and π is a duality. □

A clan of degree s — and also each of its members — is called *homogeneous of degree* t $(t \leqslant s)$ iff

$$k_{i,j} = k_{-1,j-i-1} \qquad (-1 \leqslant i \leqslant j \leqslant d\,; j-i \leqslant t+1);$$

that is to say, iff none of the first t lines of the scheme contains two different numbers. In the case $t = 1$ we simply say *homogeneous* — the term then applying also to non-regular incidence-complexes — and for $\mathcal{K} \in [k_0, k_0, \ldots, k_0]$ define

$$\operatorname{ord}(\mathcal{K}) := k_0 - 1;$$

so the polytopes are of *order* one.

If a cluster is homogeneous of degree d (or, what is the same, of degree $d-1$), we call it *fully homogeneous.*

A d-r.i.c. $\mathcal{K}$ is called *strongly homogeneous of degree* t $(t < d)$ iff all its section-complexes of dimension $i = t$ (and hence for all i less than t) are isomorphic; in the case $t = d-1$ we call $\mathcal{K}$ *totally homogeneous.*

The following applies again also to non-regular incidence-complexes. Since every class $[k_0]$ consists of exactly one r.i.c., strong homogeneity of degree 1 is the same as homogeneity. Especially:

(5) $\left\{\begin{array}{l}\text{every two-dimensional incidence-complex which is}\\ \text{homogeneous is in fact totally homogeneous.}\end{array}\right.$

If any r.i.c. $\mathcal{K}$ is totally homogeneous, its facets are isomorphic to its vertex figures and vice versa.

3

Next we derive some identities and inequalities. Let $\mathcal{K}$ be an r.i.c., $F, G \in \mathcal{K}$, $\dim(F) = i \leqslant r \leqslant \dim(G) = j$ and $F < G$. Under these circumstances we define:

$f^r_{i,j}$ to be the number of r-faces between F and G.

For brevity we put $f^r_{-1,d} =: f^r$ (the number of r-faces of $\mathscr{K}$). Because of the regularity of $\mathscr{K}$, $f^r_{i,j}$ does not depend on F and G (cf. (3)). By definition we have:

$$f^i_{i-1,i+1} = k_{i-1,i+1} = k_i \quad \text{and} \quad f^i_{i,j} = f^j_{i,j} = 1.$$

A count of all flags of $\langle F, G\rangle$ yields:

$$(6) \qquad k_{i,r} f^r_{i,j} k_{r,j} = k_{i,j} \qquad (-1 \leqslant i \leqslant r \leqslant j \leqslant d),$$

with the consequence that $k_{i,j}$ is a common multiple of all products $k_{i,r}k_{r,j}$. As special cases we have:

$$(7) \qquad f^0 k_{0,d} = k_{-1,d} = k_{-1,d-1} f^{d-1}$$

and

(8) $\quad k_{i,j}$ is a common multiple of $k_{i,j-1}$ and $k_{i+1,j}$ $\quad (j \geqslant i+3)$.

For the remainder of this section we assume $\mathscr{K}$ to be non-degenerate.

Let F and G be faces of $\mathscr{K}$ with $F < G$, $\dim(F) = i < r \leqslant s \leqslant j = \dim(G)$. In the case $r > i+1$ we consider an s-face S, two distinct $(r-1)$-faces Q_ν and two r-faces $R_\nu (\nu = 1,2)$ with $F \leqslant Q_\nu \leqslant S \leqslant G$ and $Q_\nu < R_\nu \leqslant G$, but $R_\nu \nleqslant S$. Then $R_1 \neq R_2$ because otherwise we would have $R_1 = \sup\{Q_1, Q_2\}$ and hence $R_1 \leqslant S$. Keeping F, S and G fixed we get:

$$f^r_{i,j} \geqslant f^r_{i,s} + f^{r-1}_{i,s}(f^r_{r-1,j} - f^r_{r-1,s}),$$

and this holds — trivially — also for $r = i+1$. If this inequality for $f^r_{i,j}$ is applied to $f^r_{i,s}$ and $f^{r-1}_{i,s}$, an easy computation shows that the right-hand side increases with s; so the only interesting case is:

$$(9) \qquad f^r_{i,j} \geqslant f^r_{i,j-1} + f^{r-1}_{i,j-1}(f^r_{r-1,j} - f^r_{r-1,j-1}) \quad (\text{for } i+1 < r < j);$$

with its dual counterpart:

$$(10) \qquad f^r_{i,j} \geqslant f^r_{i+1,j} + f^{r+1}_{i+1,j}(f^r_{i,r+1} - f^r_{i+1,r+1}) \quad (\text{for } i < r < j-1).$$

Statements (6), (9) and (10) may be combined in many different ways to obtain lower bounds for the $f^r_{i,j}$ and the $k_{i,j}$. When we put $r := j-1$ in (9) we get $f^{j-1}_{i,j} \geqslant 1 + f^{j-2}_{i,j-1}(k_{j-1} - 1)$ and by induction on j we conclude:[6]

$$(11) \qquad f^{j-1}_{i,j} \geqslant \sum_{\nu=0}^{j-i-1} \left(\prod_{\mu=1}^{\nu} (k_{j-\mu} - 1) \right).$$

By duality we also have:

$$(12) \qquad f^{i+1}_{i,j} \geqslant \sum_{\nu=0}^{j-i-1} \left(\prod_{\mu=1}^{\nu} (k_{i+\mu} - 1) \right).$$

[6] As usual, the empty product is defined to one.

Putting $r:=j-2$ in (9) and applying (12) with $i:=j-3$ we obtain:

$$f_{i,j}^{j-2} \geqslant f_{i,j-1}^{j-2} + f_{i,j-1}^{j-3}(k_{j-2}-1)(k_{j-1}-1),$$

and further, by induction on j:

$$(13) \qquad f_{i,j}^{j-2} \geqslant \sum_{\nu=0}^{j-i-2} \left[\left(\sum_{\mu=1}^{j-i-1-\nu} \prod_{\lambda=1}^{\mu-1} (k_{j-1-\nu-\lambda}-1) \right) \prod_{\mu=1}^{\nu} (k_{j-\mu}-1)(k_{j-\mu-1}-1) \right].$$

Problem 4: Are there similar or better lower bounds for $f_{i,j}^{j-3}$ in terms of the k_ν ($j \geqslant i+6$)?

In order to bound $k_{i,j}$ from below one may use (6) for $r=i+1$, $i+2$, $j-2$, $j-1$ together with (11) and (13) and their dual counterparts. For instance,

$$k_{i,j} = \prod_{\nu=i}^{j-2} f_{\nu,j}^{\nu+1} = \prod_{\nu=i+2}^{j} f_{i,\nu}^{\nu-1}$$

$$= k_{(i+j)/2} \prod_{\nu=1}^{(j-i-2)/2} f_{i-1+\nu,j+1-\nu}^{i+\nu} \cdot f_{i+\nu,j+1-\nu}^{j-\nu} \qquad (j \equiv i \pmod 2),$$

and so on, where all the occurring f's can be estimated by (11) and (12).

Sometimes the results by use of (13) are sharper. In any particular case, when $k_0, k_1, \ldots, k_{d-1}$ are given, we proceed better step by step by applying either (9) or (10); but before taking the next step, first exploit the divisibility-conditions from (6).

In the homogeneous case of order l, where $k_i = l+1$ for every i, all our inequalities reduce to:

$$(14) \qquad f_{i,j}^{i+1}, f_{i,j}^{j-1} \geqslant 1 + l + l^2 + \cdots + l^{j-i-1},$$

$$(15) \qquad f_{i,j}^{i+2}, f_{i,j}^{j-2} \geqslant \frac{(1+l+l^2+\cdots+l^{j-i-2})(1+l+\cdots+l^{j-i-1})}{1+l},$$

and

$$(16) \qquad k_{i,j} \geqslant \prod_{\nu=1}^{j-i-1} (1+l+l^2 \cdots + l^\nu).$$

In the homogeneous case these inequalities are best possible iff l is a prime-power[7] (cf. Examples 1, 3a and 6 in Section 4). In general, the inequalities (9)–(13) are not very sharp (cf. Example 8 in Section 5).

Our estimates can be improved by modifying (9) in the following way. As before, let F, G and S be three faces of the non-degenerate r.i.c. $\mathscr{K}$ ($\dim(F)=i$, $\dim(G)=j$, $\dim(S)=j-1$, $F<S<G$). Now for $i<r<j$:

[7] With the possible exception $j=i+3$. In this case equality in (14) or (15) or (16) is equivalent to the existence of a projective plane of order l whose group of collineations is flag-transitive.

(17) $\quad F \leqslant Q \leqslant S \quad \text{and} \quad \dim(Q) \leqslant r,$

we consider the set $\mathcal{M}(G, S, Q, r)$ of all r-faces R with $Q \leqslant R < G$ and $\inf(R, S) = Q$. Obviously, the set of all r-faces between F and G contains the union of the distinct sets $\mathcal{M}(G, S, Q, r)$, where G, S and r are kept fixed, while Q runs over all faces of $\mathcal{K}$ satisfying (17). The cardinality of $\mathcal{M}(G, S, Q, r)$ — because of the regularity of $\mathcal{K}$ — only depends on i, j, r and the dimension q of Q. Calling this number $g_{i,j}^{q,r}$ we obtain:[8]

$$(18) \quad \begin{cases} r_{i,j}^{r} = \sum_{q=i}^{r} f_{i,j-1}^{q} \, g_{i,j}^{q,r} \\ \qquad \geqslant \sum_{q=i+1}^{r} f_{i,j-1}^{q} \, g_{i,j}^{q,r} \\ \qquad = f_{i,j-1}^{r} + \sum_{q=i+1}^{r-1} f_{i,j-1}^{q} \, g_{i,j}^{q,r}. \end{cases}$$

On the other hand, by the very definition of $g_{i,j}^{q,r}$ we have:

$$(19) \quad \begin{cases} g_{i,j}^{q,r} = f_{q,j}^{r} - \sum_{p=q+1}^{r} f_{q,j-1}^{p} \cdot g_{i,j}^{p,r} \\ \qquad = f_{q,j}^{r} - f_{q,j-1}^{r} - \sum_{p=q+1}^{r-1} f_{q,j-1}^{p} \, g_{i,j}^{p,r} \end{cases}$$

Hence all the g's may inductively be written as polynomials in the f's and the right-hand side of (18) then can be replaced also by a polynomial in the f's, where for the indices of every occurring $f_{\alpha,\gamma}^{\beta}$ the inequalities

$$i \leqslant \alpha \leqslant \gamma \leqslant j, \qquad i+1 \leqslant \beta \leqslant r \quad \text{and} \quad \gamma - \alpha < j - i$$

are valid. In other words, if the scheme of a clan of degree $s-1$ is given and it is known that $\mathcal{K}$ does not degenerate, then the $k_{i-1,i+s}$ can be estimated from below by iterated application of (18) and (19) (and/or of their dual counterparts).

4

Before we go on, some examples may illustrate these ideas.

Example 1: The regular d-simplex $\mathcal{T}^{d}$. It is the only[9] member of the clan

[8] It may happen that $g_{i,j}^{i,r} = 0$, though $g_{i,j}^{q,r} > 0$ for $i+1 \leqslant q \leqslant r$ (cf. Example 3a); it also happens that $g_{i,j}^{i,r} > 0$, but $g_{i,j}^{i+1,r} = 0$ (e.g. $i = -1$, $r = j-1$, $\mathcal{K}$ the j-cube).
Of course, $g_{i,j}^{\alpha,\alpha} = 1$.

[9] The phrase 'the only member of a clan' always means: '... up to isomorphisms'.

$$\begin{pmatrix} 2 & 2 & 2 & \cdot & \cdot & \cdot & 2 & 2 \\ & 6 & 6 & \cdot & \cdot & \cdot & 6 & 6 & \end{pmatrix}$$

of dimension d and we have $k_{i,j} = (j-i)!$. It is non-degenerate, totally homogeneous and hence symmetric. Its section complexes of dimension i all are isomorphic to $\mathcal{T}^i$. This mapping π which maps every vertex V onto the only facet missing v is a polarity (cf. (1)). Finally, we have:[10]

$$A(\mathcal{T}^d) \cong S_{d+1}.$$

As for all polytopes the index equals one, and hence they are sharply 1-fold regular.

Example 2a: The measure-polytope (d-cube) $\mathcal{M}^d$. It is the only member of the cluster

$$\begin{pmatrix} 2 & 2 & 2 & 2 & \cdot & \cdot & \cdot & 2 & 2 \\ & 8 & 6 & 6 & \cdot & \cdot & \cdot & \cdot & 6 \\ & 48 & 24 & \cdot & \cdot & \cdot & \cdot & 24 & \\ & & \cdot & & & & \cdot & & \\ & & & \cdot & & \cdot & & & \\ & & & & 2^d d! & & & & \end{pmatrix}$$

Example 2b: The generalization $\mathcal{M}_m^d$ for $m \geqslant 2$ of $\mathcal{M} = \mathcal{M}_2$ due to SHEPHARD (cf. [3]).[11] Its vertex figures are isomorphic to $\mathcal{T}^{d-1}$, whence $k_{i,j} = (j-i)!$ for $i \geqslant 0$, while $k_{-1,j} = m^j j!$. For $m > 2$ it is no longer a polytope in our sense, though it is a regular complex polytope in the sense of [3] and of [5]. It is non-degenerate. Its facets are isomorphic to $\mathcal{M}_m^{d-1}$. SHEPHARD'S construction is in the unitary geometry over $\mathbb{C}^d$. The unitary isometries which map $\mathcal{M}_m^d$ onto itself form a group isomorphic to $C_m \wr S_d$.[12] In our combinatorial sense, however, we have:[13]

$$A(\mathcal{M}_m^d) \cong S_m \wr S_d,$$

and hence the index equals $((m-1)!)^d$.

Example 2c: The (generalized) cross-polytopes $^*\mathcal{M}_m^d$ ($^*\mathcal{M}^d = {}^*\mathcal{M}_2^d$). They are the duals of the r.i.c.'s of Example 2b.

Of course not only the simplices, cubes and cross-polytopes, but indeed all regular polytopes and honeycombs in the sense of [1] and all regular complex polytopes and honeycombs in the sense of [3] and [5] are r.i.c.'s.

[10] S_n stands for the symmetric group on n letters.
[11] In [5] it is denoted by γ_d^m (p. 118).
[12] C_m denotes the cyclic group of order m, while $\wr$ stands for the wreath-product.
[13] Cf. Example 7b and Section 6.

Example 3a: Let p be a prime, $d, f \in \mathbb{N}$ and $q := p^f$. Then the projective space $\mathrm{PG}(d, \mathrm{GF}(q)) = \mathcal{T}^d(q)$ is at least the only non-degenerate member of the fully homogeneous cluster

$$\begin{pmatrix} q+1, & \cdot & \cdot & & \cdot & \cdot & , & q+1 \\ & t_2, & \cdot & \cdot & \cdot & , & t_2 & \\ & & \cdot & & \cdot & & & \\ & & & \cdot & \cdot & & & \\ & & & \cdot & \cdot & & & \\ & & & & t_d & & & \end{pmatrix},$$

where, independently of d,

$$t_i = \prod_{\nu=2}^{i+1} \frac{q^\nu - 1}{q - 1} \qquad (i = 0, 1, 2, 3, \ldots). \tag{20}$$

It is again totally homogeneous, its faces of dimension i being isomorphic to $\mathcal{T}^i(1)$. There are many polarities. For $d \geq 2$ we have:

$$A(\mathcal{T}^d(q)) \cong \mathrm{P\Gamma}(d+1, \mathrm{GF}(q)),$$

while

$$A(\mathcal{T}^1(q)) \cong S_{q+1}.$$

This shows that for a facet-complex $\mathcal{F}$ of an r.i.c. $\mathcal{K}$, not necessarily $A(\mathcal{F})$ equals the stabilizer of $\mathcal{F}$ in $A(\mathcal{K})$. $\mathcal{T}^d(q)$ is 3-fold regular, and if $d \neq 1$ and $q \neq 4$ it is sharply 3-fold regular.

Cancelling the denominators in (20) and then putting $q := 1$ we get $t_i = (i+1)!$, so justifying our notation. The analogy between the projective spaces of order l and the simplices (of order 1) also becomes evident by the observation, that (20) exactly realizes the right-hand side of (16). (Cf. also Example 3d.)

Example 3b: $\mathrm{PG}(d, L)$ for any natural number d and any field or skew-field L. If L is not a Galois-field, all $k_{i,j}$ are infinite ($j \geq i+2$).

Example 3c: The affine geometry of dimension d over L.

Example 3d: Let q be as before. Consider the non-singular vector space Sy of dimension $2d$ over $\mathrm{GF}(q)$ and the lattice $\mathcal{M}^d(q)$ consisting of $\emptyset$ and of all such subspaces of Sy that contain at least one maximal isotropic subspace of Sy. The latter are considered vertices in $\mathcal{M}^d(q)$, and since they are of dimension d in Sy, the dimension of $\mathcal{M}^d(q)$ becomes d. It turns out to be a non-degenerate r.i.c. with vertex figures obviously isomorphic to $\mathcal{T}^{d-1}(q)$, while the facets are isomorphic to $\mathcal{M}^{d-1}(q)$. So here we have another analogue to $\mathcal{M}^d$. We easily calculate

(21) $$k_{-1,j}(\mathcal{M}^d(q)) = t_{j-1} \prod_{\nu=1}^{j} (q^\nu + 1).$$

For 1 instead of q we obtain, in fact, $k_{-1,j}(\mathcal{M}^d)$.

We have

$$A(\mathcal{M}^d(q)) \cong \mathrm{P}\Gamma Sp(2d, q),$$

and therefore (cf. (21))

$$\mathrm{ind}(\mathcal{M}^d(q)) = f \cdot (q-1)^d \cdot q^{d^2}.$$

Conjecture: For $q \neq 2$, $\mathcal{M}^d(q)$ is sharply 2-fold regular.

For $q = d = 2$ our r.i.c. becomes isomorphic to the CREMONA–RICHMOND 15_3, which is the only member of $(^3{}_{45}{}^3)_{na}$.[14] Hence, it is self-dual because its scheme is symmetric. It also possesses a polarity π as the following representation shows. Here the vertices are identified with the pairs out of $\{1, 2, \ldots, 6\}$. Every row belongs to an edge, E, giving first the vertices of E and then $\pi(E)$:

12	12	12	13	13	13	14	14	14	15	15	15	16	16	16
34	35	36	24	25	26	23	25	26	23	24	26	23	24	25
56	46	45	56	46	45	56	36	35	46	36	34	45	35	34
12	34	56	36	25	14	45	13	26	16	24	35	23	15	46

Obviously $A(\mathcal{M}^2(2)) \cong S_6$ in accordance with $S_6 \cong Sp(4,2) \cong \mathrm{P}Sp(4,2) \cong \mathrm{P}\Gamma Sp(4,2)$.[15] Since S_6 does not contain a subgroup isomorphic to C_{15}, Remark 1 cannot be applied. Since the index is 16, the stabilizer of a flag has order 16. In fact there are even non-identical automorphisms of $\mathcal{M}^2(2)$ with exactly 7 points fixed (corresponding to the isotropic lines in PG(3, GF(2)) which pass through a given point).

The LEVI-graph of $\mathcal{M}^2(2)$ is just TUTTE'S 8-cage (cf. [2, pp. 413, 442]). Closer inspection of the symplectic group in case $2d = 4$ yields:

(22) For every prime-power q the LEVI-graph of $\mathcal{M}^2(q)$ is 5-regular.[16]

Example 3e: Applying the polarity $V \mapsto \pi(V) := V^\perp$ of the symplectic vector space Sy to $\mathcal{M}^d(q)$ we obtain $^*\mathcal{M}^d(q)$, consisting of all isotropic subspaces of Sy with their projective dimension together with $\emptyset$ and, say, $^*M^d(q)$.

[14] We add the subscript *na* ('nicht ausgeartet') to the scheme of a clan in order to describe the subfamily of its non-degenerate members.

[15] $A(\mathcal{M}^2(2))$ may also be interpreted as the group of inner automorphisms of S_6; then π corresponds to the outer automorphism α given by $\alpha((12)(34)(56)) := (12)$, etc. (cf. [2, p. 442]).

[16] A graph is said to be s-regular iff its group of automorphisms acts transitively on the arcs of length s (cf. [2, p. 417]).

Example 4: The MÖBIUS–KANTOR configuration 8_3 (cf. [2, p. 429]).

Example 5: The DESARGUES configuration 10_3 (cf. [2, pp. 434–437]). It is a good example for the following:

Remark 3: If $\mathcal{K}$ is a d-r.i.c. and $-1 \leqslant i < j \leqslant d$, the family of all m-faces of $\mathcal{K}$ with $i < m < j$ together with two new symbols, say $\emptyset_i$ and $K_{i,j}$ form — with respect to the obvious order — a new r.i.c. $\mathcal{K}_{i,j}$ of dimension $j - i - 1$. If $\mathcal{K}$ is non-degenerate, $A(\mathcal{K}_{i,j})$ contains $A(\mathcal{K})$ (cf. (1)). Furthermore:

$$\text{if } \mathcal{K} \left\{ \begin{array}{l} \text{is non-degenerate} \\ \text{is symmetric and} \\ \quad i+j=d-1 \\ \text{is self-dual and} \\ \quad i+j=d-1 \\ \text{possesses a polari-} \\ \text{ty and } i+j=d-1 \end{array} \right\} \text{ then } \mathcal{K}_{i,j} \left\{ \begin{array}{l} \text{is non-degenerate} \\ \text{is symmetric} \\ \\ \text{is self-dual} \\ \\ \text{possesses a} \\ \text{polarity.} \end{array} \right.$$

With this notation $10_3 \simeq \mathcal{T}^4_{0,3}$. Generally for $d \geqslant 2$:

$$\mathcal{T}^{2d}_{d-2,d+1} \in \begin{pmatrix} d+1 & & d+1 \\ & (d+1)\binom{2d+1}{d} & \end{pmatrix}_{\text{na}},$$

with

$$(23) \qquad A(\mathcal{T}^{2d}_{d-2,d+1}) \cong S_{2d+1} \quad \text{and} \quad \operatorname{ind}(\mathcal{T}^{2d}_{d-2,d+1}) = (d!)^2.$$

To prove (23), it is enough to observe that every vertex V of $\mathcal{T}^{2d}$ is uniquely determined by the family of $(d-1)$-faces F with $V \not< F$ and that this family can be described by incidences only between d- and $(d-1)$-faces of $\mathcal{T}^{2d}$.

Even more generally, for $d \geqslant 2$ and $q := p^f$, p a prime, t_i as in (20):

$$\mathcal{T}^{2d}_{d-2,d+1}(q) \in \begin{pmatrix} \dfrac{q^{d+1}-1}{q-1} & & \dfrac{q^{d+1}-1}{q-1} \\ & \dfrac{t_{2d}}{t^2_{d-1}} & \end{pmatrix}_{\text{na}},$$

with

$$A(\mathcal{T}^{2d}_{d-2,d+1}(q)) \geqslant \mathrm{P\Gamma}(2d+1, \mathrm{GF}(q))$$

and index at least

$$f \cdot q^{d(2d+1)} \prod_{\nu=1}^{d} (q^\nu - 1)^2.$$

References

[1] H.S.M. Coxeter, Regular Polytopes (Methuen & Co. Ltd., London, 1948; 3rd edn. 1973).
[2] H.S.M. Coxeter, Self-dual configurations and regular graphs, Bull. Amer. Math. Soc. 56 (1950) 413–455.
[3] G.C. Shephard, Regular complex polytopes, Proc. London Math. Soc. 2 (1952) 82–97.
[4] P. McMullen, Combinatorially regular polytopes, Mathematika 14 (1967) 142–150.
[5] H.S.M. Coxeter, Regular Complex Polytopes (Cambridge University Press, 1974).
[6] H.S.M. Coxeter and G.C. Shephard, Regular 3-complexes with toroidal cells, J. Combinat. Theory B22 (1977) 131–138.
[7] B. Grünbaum, Regular polyhedra — old and new, Aequ. Math. 16 (1977) 1–20.[17]
[8] B. Grünbaum, Regularity of graphs, complexes and designs, in: Coll. Int. C.N.R.S., No. 260 (1977) pp. 191–197.
[9] H.S.M. Coxeter, The Pappus configuration and the self-inscribed octagon, I, II, Proc. Kon. Ned. Ak. A80 (1977) 256–300.
[10] E. Schulte, Reguläre Inzidenzkomplexe, Dissertation, University of Dortmund (1980).
[11] L. Danzer and E. Schulte, Reguläre Inzidenzkomplexe I, revised version of Chapter 1 of [10]. Geom. Dedic. 13 (1982) 295–308.
[12] E. Schulte, Reguläre Inzidenzkomplexe II, revised version of Chapter 2 of [10]. Geom. Dedic. 14 (1983) 33–56.
[13] E. Schulte, Reguläre Inzidenzkomplexe III, revised version of Chapter 3 of [10], Geom. Dedic. 14 (1983) 57–80.
[14] A. Vince, Combinatorial Maps. J. Comb. Theory B, 34 (1983) 1–21.

Received 22 September 1981

[17] Revised version of a lecture given at the LMS-Conference at Durham/England in July 1975.

Annals of Discrete Mathematics 20 (1984) 129–136
North-Holland

SOME OLD AND NEW PROBLEMS IN COMBINATORIAL GEOMETRY

Paul ERDÖS
Hungarian Academy of Sciences, Budapest, Hungary

In this paper several unconnected old and new problems in combinatorial geometry are discussed. The reference list is as complete as possible, including two papers dealing with similar subjects [6, 7]. Many interesting problems and results are in Grünbaum [11], which also has an extensive and useful bibliography and interesting historical remarks.

1

The following problem is attributed to Heilbronn. Let $f(n)$ be the largest number for which there are n points $x_1, \ldots, x_n$ in the unit circle for which the areas of all the triangles $\{x_i, x_j, x_k\}$ are $\geq f(n)$. Determine or estimate $f(n)$ as accurately as possible. Trivially, $f(n) < c/n$ and Heilbronn conjectured

$$\frac{c_1}{n^2} < f(n) < \frac{c_2}{n^2}. \tag{1}$$

I observed that $f(n) > c_1/n^2$ holds. The first non-trivial upper bound,

$$f(n) < \frac{c}{n(\log \log n)^{1/2}}, \tag{2}$$

was proved by Roth [16], his result was improved by Schmidt [18] and later Roth proved $f(n) < n^{-11/10}$ [17]. Very recently Komlós, Pintz and Szemerédi disproved the conjecture of Heilbronn. In fact they proved

$$\frac{c_1 \log n}{n^2} < f(n) < \frac{c_2}{n^{8/7}}. \tag{3}$$

Their proof of the lower bound in (3) is extremely noteworthy and uses a very interesting new combinatorial idea. They seem to believe that the lower bound in (3) gives the correct order of magnitude of $f(n)$.

This problem led Szemeredi and Erdös to formulate the following conjecture. Let $z_1, \ldots, z_n$ be n points in the unit circle. Denote by $d(z_i, z_j)$ the distance between z_i and z_j. Put $D(z_i, \ldots, z_j) = \min_{1 \leq i \leq j \leq n} d(z_i, z_j)$. Denote by $\alpha(z_1, \ldots, z_n)$ the smallest angle determined by the n points. It is well known and easy to see that

$$\max \alpha(z_1,\ldots,z_n)=\frac{\pi}{n}$$

and for a certain c:

$$\max D(z_1,\ldots,z_n)=(c+o(1))n^{-1/2}.$$

Now we conjecture that

$$\alpha(z_1,\ldots,z_n)D(z_1,\ldots,z_n)=o\left(\frac{1}{n^{3/2}}\right), \tag{4}$$

and perhaps it is less than c/n^2. The regular polygon shows that if true, this is the best possible.

Corrádi, Hajnal and Erdös noticed more than 20 years ago that the following simple question seems to present some difficulties. Let $z_1,\ldots,z_n$ be n points not all on a line. Is it true that they determine a positive angle which is $\leq \pi/n$? We have not even been able to show that it is less than c/n for some absolute constant c.

Schmidt [18] proved that one can find n points in the unit circle so that the area of the least convex domain determined by any four of them is always greater than $c/n^{3/2}$. He also observed that it seems very difficult to show that among any of these n points there always are four of them so that the area of the convex domain determined by them is $o(1/n)$. Indeed, this beautiful problem is still open.

As far as I know the following question has not yet been investigated. Let $f_3(n)$ be the largest number for which there are n points in the unit sphere so that all the triangles determined by them have an area $\geq f_3(n)$. Determine or estimate $f_3(n)$ as accurately as possible. Trivially, $f_3(n)<(c/n)^{2/3}$ and I believe one can show that $f_3(n)>c/n$. The first problem would be to prove $f_3(n)=o(1/n^{2/3})$.

2

Straus, Purdy and Erdös [19] proved that if $x_1,\ldots,x_n$ are any n points in the plane not all on a line, then they always determine two triangles of non-zero area so that the ratio of their areas is not less than $[(n+1)/2]$, and it is easy to see that $[(n+1)/2]$ is the best possible. We further show that equality is only possible if the points are all situated on two parallel lines.

Several interesting generalizations are possible, e.g. for quadrilaterals instead of triangles and for the points being in higher dimensions. We also thought of the following possible sharpening of Roth's Theorem. Let $x_1,\ldots,x_n$ be n points in the unit circle, with at most $o(n^{1/2})$ of them on a line; then they always determine a triangle of non-zero area which is $o(1/n)$. The lattice points show immediately that the condition $o(n^{1/2})$ cannot be relaxed. Many further conjectures of this type can be stated but so far we have no results.

3

Let $x_1, \ldots, x_n$ be n points in the plane. We join every pair of them to obtain the lines $L_1, \ldots, L_m$. It is well known that if $m > 1$, then $m \geq n$. The possible values of $m \geq n$ are fairly well known. For $m < cn^{3/2}$ Kelly and Moser [13] have fairly complete results, and answering a question of Grünbaum Erdös [4] proved that every $cn^{3/2} < m < \binom{n}{2}$ can occur, except $\binom{n}{2} - 1$ and $\binom{n}{2} - 3$.

Denote by u_i the number of points on the line L_i, $u_1 \geq u_2 \geq \cdots \geq u_m = 2$. Very little is known about the possible choices of $\{u_1, \ldots, u_n\}$. Trivially

$$\sum \binom{u_i}{2} = \binom{n}{2}$$

and by the well-known Gallai–Sylvester Theorem $u_m = 2$ (unless $m = 1$, $u_1 = n$). Erdös conjectured that the number of possible choices of $\{u_1, \ldots, u_m\}$ is less than $\exp c \cdot n^{1/2}$. It is easy to see that this conjecture if true is best possible, apart from the value of c. Purdy and Erdös conjectured that

$$\sum_{u_i > cn^{1/2}} \binom{u_i}{3} < c' n^{3/2},$$

and perhaps the number of $u_i > n^{\alpha}$ is less than $C \cdot n^{1-\alpha}$, but so far no progress has been made with these conjectures.

Denote by $\alpha_n(k)$ the largest integer for which there are $\alpha_n(k)$ lines, with $u_i = k$, and by $\beta_n(k)$ the largest integer for which there are $\beta_n(k)$ lines, with $u_i = k$, under the assumption that $u_j = 0$ for all $u_j > k$ (i.e. there are no lines containing more than k points). Clearly $\alpha_n(k) \geq \beta_n(k)$.

As far as I know Sylvester was the first to investigate $\alpha_n(3)$. He proved

$$\alpha_n(3) > \frac{n^2}{6} - cn. \tag{5}$$

His results were improved in a recent paper by Burr, Grünbaum and Sloane but the exact value of $\alpha_n(3)$ is not yet known. It seems certain that $\alpha_n(3) = \beta_n(3)$.

Croft and Erdös observed that the example of the lattice points shows that for every k, $\alpha_n(k) > C_k n^2$. Trivially, $\alpha_n(k) < n(n-1)/k(k-1)$ for every $k > 2$ and we conjectured that

$$\lim_{n \to \infty} \frac{\alpha_n(k)}{\binom{n}{2}} < \frac{1}{\binom{k}{2}}. \tag{6}$$

Expression (6) should not be difficult to prove, but we have not yet done so. It would be of interest to determine the value of the limit in (6). Croft and Erdös further made the following much sharper conjecture. For every $\varepsilon > 0$ there is a $k_0(\varepsilon)$ and an $n = \eta(\varepsilon)$ so that

$$\sum_{k_0 < k < \eta n} \alpha_n(k) \binom{k}{2} < \varepsilon \binom{n}{2}. \tag{7}$$

In other words, the number of pairs of points (x_i, x_j) for which the line (x_i, x_j) has at least k_0 and fewer than ηn points is less than $\varepsilon \binom{n}{2}$.

I conjectured many years ago that for every $k > 3$:

$$\lim_{n\to\infty} \beta_n(k)/n = \infty, \qquad \lim_{n\to\infty} \beta_n(k)/n^2 = 0. \tag{8}$$

The first conjecture of (8) was proved by Karteszi who in fact proved that for every k:

$$\beta_n(k) > c_k n \log n,$$

and this was strengthened by Grünbaum to:

$$\beta_n(k) > c_k n^{1+1/k}, \tag{9}$$

which, as far as I know, is the strongest inequality known at present and is perhaps the best possible.

The second conjecture of (8) is still open and I offer 2000 shekels or 100 dollars to anyone who can prove or disprove it.

As stated previously, Gallai proved that unless all our points are on a line there is at least one ordinary line, i.e. a line with $|L_i| = 2$, and Dirac proved that there are at least three such lines. De Bruijn and Erdös conjectured that the number of ordinary lines tends to infinity. This was proved by Motzkin [15] who conjectured that the number of ordinary lines is, for $n > n_0$, at least $n/2$, and remarked that if true, this is the best possible. Kelly and Moser [13] proved that the number of ordinary lines is $\geq 3n/7$.

An old conjecture of Dirac states that if $x_1, \ldots, x_n$ are not all on a line and if one joins every two of them, then at least one of the points is incident to $n/2 - c$ lines, where c is an absolute constant. It is not even known that there is an $\varepsilon > 0$, so that there is a point incident to more than εn lines.

An old conjecture of mine states that if at most $n - k$ of the points $x_1, \ldots, x_n$ are on a line, then they determine at least ckn distinct lines. I offer (100 dollars or 2000 shekels) for a proof or disproof. If the conjecture holds, it would be of some interest to determine the largest value of c for which it holds. The results of Kelly and Moser [13] imply the conjecture for $k < c_1 n^{1/2}$.

To end this section I mention the following lovely conjecture of Kupitz, which is several years old but I only learned it at this conference. Let $z_1, \ldots, z_n$ be any n points in the plane, then there always is a line l going through at least two of our points so that the difference between the number of points on the two sides of L is at most one. It is not even known that this difference can be made less than a bound independent of n.

4

An old problem of E. Klein (Mrs. Szekeres) states: Let $H(n)$ be the smallest integer so that every set of $H(n)$ points in the plane with no three on a line contains the vertices of a convex n-gon. Klein proved $H(4)=5$, Turán and Makai proved $H(5)=9$, and Erdös and Szekeres [5, 8, 9] proved

$$2^{n-2}+1 \leqslant H(n) \leqslant \binom{2n-4}{n-2}. \tag{10}$$

We conjectured $H(n)=2^{n-2}+1$, thus $H(6) \doteq 17$; but this is not yet known.

Recently, I found the following interesting modification of this problem. Let M_n be the smallest integer so that every set of M_n points in the plane no three on a line always contains the vertices of a convex n-gon which contains none of the other points in its interior. Trivially, $M_4=5$, Ehrenfeucht proved that M_5 is finite and Harborth [12] proved that $M_5=10$. It is not yet known if M_6 exists.

Denote by $f_k(n)$ the largest integer for which every set of n points $x_1, \ldots, x_n$ no three on a line contains $f_k(n)$ convex k-gons. It is easy to see that for every $k \geqslant 4$

$$\lim_{n \to \infty} f_k(n) \Big/ \binom{n}{k} = c_k, \qquad 0 < c_k < 1.$$

It would be of some interest to determine the value of c_k. Guy conjectured more than 20 years ago that $c_4 = \frac{3}{8}$.

Denote by $F(n)$ the largest integer so that there are always at least $F(n)$ convex subsets of the x_i. It is not hard to deduce from (10) that:

$$n^{c_1 \log n} < F(n) < n^{c_2 \log n}. \tag{11}$$

It would be of interest to improve (11). No doubt $\lim_{n \to \infty} \log F(n)/(\log n)^2$ exists and I would like to know its value. To obtain an exact formula for $F(n)$ may be hopeless, but perhaps there is some chance for an asymptotic formula.

5. Miscellaneous problems

Denote by $\alpha_r(n)$ the largest angle not exceeding π so that every set of n

points in r-dimensional space determines an angle greater than $\alpha_r(n)$. Erdös and Szekeres proved that α_2:

$$\alpha_2(2^n) = \pi\left(1 - \frac{1}{n}\right). \tag{12}$$

The problem remains to determine the smallest $m_2(n)$ for which $\alpha_2(m_2(n)) = \pi(1-1/n)$. We could not disprove that for large n, $m_2(n) = 2^{n-1}+1$. Very likely $m_2(n) < 2^n$, but we only could prove that every set of $2^n - 1$ points in the plane determines an angle $\geqslant \pi(1-1/n)$, or for every $\varepsilon > 0$, $\alpha_2(2^n - 1) > \pi(1 - 1/n - \varepsilon)$.

Very much less is known about $\alpha_r(n)$ for $r > 2$. I conjectured and Danzer and Grünbaum [3] proved that every set of $2^n + 1$ points in n-dimensional space determines an angle $> \pi/2$, but the value of $\alpha_n(2^n+1)$ is not yet known and may be difficult to determine. I asked: determine the smallest $h(n)$ for which $h(n)$ points in n-dimensional space always determine an angle $\geqslant \pi/2$. Trivially, $h(2) = 4$ and Croft proved $h(3) = 6$. His proof is not simple; as far as I know $h(4)$ is not yet known, and it is not at all impossible that $h(n)$ grows linearly, but I cannot exclude the possibility that $h(n)$ grows exponentially.

Denote by $f_r(n)$ the largest integer for which there are $f_r(n)$ points in r-dimensional space which determine n distinct distances. Trivially, $f_r(1) = r+1$. As far as I know Coxeter first asked for the determination of $f_r(2)$. The sharpest current upper bound is due to Blokhuis [1]; he proved $f_2(r) \leqslant \frac{1}{2}(r+1)(r+2)$.

Is it true that there is an absolute constant c so that 2^n points in n-dimensional space determine more than cn distances? The n-dimensional cube determines $n+1$ distinct distances.

Denote by $f_r(n)$ the largest integer so that among any n points in r-dimensional space one can always select $f_r(n)$ of them so that all their distances should be different. I conjectured that for $r=1$, $f_1(n) = (1+o(1))n^{1/2}$, and that perhaps the minimum is assumed if the points are equidistant on the line. Komlós, Sulyok and Szemeredi [14] proved that $f_1(n) > cn^{1/2}$ for some absolute constant $c < 1$. In fact, they obtained a very much more general theorem which is very interesting in itself. $f_r(n) > n^{\varepsilon_r}$ is not hard to prove, but the value of

$$\lim_{n\to\infty} \log f_r(n)/\log n = c_r$$

is not known. Perhaps $c_r = 1/(r+1)$. All these problems become much easier for infinite sets. I proved more than thirty years ago (without using the continuum hypothesis) that if S is an infinite subset of r-dimensional space and the power of S is m, then S has a subset S' of power m so that all the distances of the points determined by the points of S' are different.

Let $x_1, \ldots, x_n$ be n (distinct) points in the plane. Denote by $d_1 > d_2 > \cdots > d_m$

the distances determined by these points and assume that the distance d_i occurs u_i times (i.e. there are u_i pairs (x_u, x_v) with $d(x_u, x_v) = u_i$). I have several problems and results on the largest possible value of $\max u_i$ and on the smallest possible value of m. I refer for a detailed account of these problems to my papers quoted earlier [6, 7]. Here I discuss related but slightly different problems. We evidently have

$$\sum_{i=1}^{m} u_i = \binom{n}{2} .$$

I conjectured that for $n > 4$ the set $u_1, \ldots, u_m$ cannot be a permutation of $1, 2, \ldots, n-1$ unless the x_i's are equidistant points on a line or a circle. For $n = 4$ this is clearly possible. Let x_1, x_2, x_3, be the vertices of an isosceles triangle and x_4 the centre of its circumscribed circle. Pomerance found an example which shows that my conjecture fails for $n = 5$. Let x_1, x_2, x_3 be the vertices of an equilateral triangle, x_4 the centre of the triangle, and x_5 one of the intersections of the circumbscribed circle of x_1, x_2, x_3 and the perpendicular bisector of the segment (x_3, x_4). Nevertheless I am fairly sure that my conjecture holds for sufficiently large n, perhaps for all $n > 5$. A recent communication from L. Berkes, a Hungarian high school student in Kecskemét, shows that my conjecture also fails for $n = 6$. More generally, one could ask: How many distinct values can the u_i's take? Clearly, at most $n-1$, but perhaps for large n this is possible only if the x_i's are equidistant points on a line.

Not much is known about the possible values of $\{u_1, \ldots, u_m\}$. My previous conjecture can be restated as follows. Assume that the u_i's are all distinct. Then for $n \geq 7$, m cannot be $n-1$ (not all the x_i's on a line or circle). If my conjecture is true, one could ask what is the largest possible value of m if we assume that the u_i's are all distinct. Observe that for the regular $(2k+1)$-gon $m = k$ and all the u_i's are $2k+1$. Clearly, many questions could be asked about the possible values of the u_i's. By an old result of Pannwitz the diameter of $x_1, \ldots, x_n$ can occur at most n times, thus $\min u_i \leq n$ equality holds when n is odd and the x_i's form a regular polygon. Many problems and conjectures can be formulated: e.g. let $x_1, \ldots, x_n$ be n points and assume all the u_i's are equal. For which values is this possible? Denote this value by t_n, then $t_n = 1$ is of course possible and $t_n = n$ is possible if and only if n is odd. Also by Pannwitz's result, $t_n \leq n$. What values are possible for t_n? Clearly,

$$t_n \,\Big|\, \binom{n}{2} .$$

References

[1] A. Blokhuis, A new upper bound for the cardinality of 2-distance sets in Euclidean space, this volume.

[2] S.A. Burr, B. Grünbaum and N.J.A. Sloane, The orchard problem, Geometriae Dedicata 2 (1974) 397–424.

[3] L. Danzer and B. Grünhaum, Über zwei Probleme bezüglich konvexer Körper von P. Erdös und V. Klee, Math. Z. 9 (1962) 90–99.

[4] P. Erdös, On a problem of Grünbaum, Can. Math. Bull. 15 (1972) 23–25.

[5] P. Erdös, The Art of Counting: Selected Writings (MIT Press, 1973).

[6] P. Erdös, On some problems of elementary and combinatorial geometry, Ann. Math. Ser. IV, V 103 (1975) 99–108.

[7] P. Erdös, Some combinatorial problems in geometry, in: Geometry and Differential Geometry, Lecture Notes in Mathematics 792 (1979) pp. 46–53.

[8] P. Erdös and G. Szekeres, A combinatorial problem in geometry, Compos. Math. 2 (1935) 463–470.

[9] P. Erdös and G. Szekeres, On some extremum problems in elementary geometry, Ann. Univ. Sci. Budapest Sect. Math. 3–4 (1961) 313–320.

[10] B. Grünbaum, New views on some old questions of combinatorial geometry, Coll. Intern. Sulla Teoria Comb., Rome, Vol. 1, Math. Rev. 57 10605 (1973).

[11] B. Grünbaum, Arrangements and Spreads, Conference Board on Mathematical Sciences, Amer. Math. Soc. No. 10.

[12] H. Harborth, Konvexe Funfecke in ebenen Punktmengen, Elem. Math. 33 (1978) 116–118.

[13] L.M. Kelly and W. Moser, On the number of ordinary lines determined by n points, Can. J. Math. 10 (1978) 210–219.

[14] J. Komlós, M. Sulyok and E. Szemerédi, Linear problems in combinatorial number theory, Acta Math. Acad. Sci. Hungary 26 (1975) 113–121.

[15] T.S. Motzkin, The lines and planes connecting the points of a finite set, Trans. Amer. Math. Soc. 70 (1951) 451–464.

[16] K.F. Roth, On a problem of Heilbronn, J. London Math. Soc. 26 (1951) 198–204.

[17] K.F. Roth, On a problem of Heilbronn, II and III, J. London Math. Soc. 25 (1972) 193–212.

[18] W. Schmidt, On a problem of Heilbronn, J. London Math. Soc. 4 (2) (1971/72) 545–550.

[19] E. Straus, G. Purdy and P. Erdös, Can. J. Math. (to appear).

Received March 1981

Annals of Discrete Mathematics 20 (1984) 137–146
North-Holland

PSEUDO-BOOLEAN FUNCTIONS AND THEIR GRAPHS

Aviezri S. FRAENKEL
Department of Applied Mathematics, The Weizmann Institute of Science, Rehovot, Israel

Peter L. HAMMER
Rutgers University, RUTCOR, Hill Center, New Brunswick, NJ, U.S.A.

With any form of a pseudo-Boolean function f, two graphs are associated: the *conflict* graph $C(f)$ and the *disjunctive-normal-form* graph $D(f)$. Various connections between these graphs and between them and $\max(f)$ and $\min(f)$ are given, including one between $\min(f)$ and the independence number $\alpha(D(f))$, which improves a previous result. Finally, a characterization of $D(f)$ graphs is given.

1. Introduction

A pseudo-Boolean function is a mapping f: $\{0,1\}^n \to$ reals (here $\{0,1\}^n$ is the n-fold cartesian product over $\{0,1\}$ for some fixed positive integer n). Since the mapping is into a finite subset of the reals, it is clear that every pseudo-Boolean function can be written in the form of a polynomial:

$$f = c + \sum_{i=1}^{m} a_i \cdot (y_{i1} \wedge \cdots \wedge y_{i,n_i}) \qquad (n_i \geq 1,\ 1 \leq i \leq m),$$

with real constant c and real coefficients a_i, literals $y_{ij} \in \{0,1\}$, where each literal y_{ij} is either a Boolean variable x_k or its complement $\bar{x}_k = 1 - x_k$, $\wedge$ denotes conjunction, and the symbol $\cdot$, which is often omitted, denotes ordinary multiplication (over the reals). We assume throughout that every conjunct contains each literal at most once, and that if it contains a literal, it does not contain its complement. The *Boolean support* of f is defined to be the DNF (= disjunctive normal form) function:

$$\varphi = \bigvee_{i=1}^{m} (y_{i1} \wedge \cdots \wedge y_{i,n_i}).$$

It is clear that (by using $\bar{x}_i = 1 - x_i$) every pseudo-Boolean function f can be written with positive (negative) coefficients only (except possibly for the constant term), in which case f is called a *posiform* (*negaform*). Let

$$f_1 = c_1 + \sum_{i=1}^{m_1} a_i (y_{i1} \wedge \cdots \wedge y_{i,n_i}),$$

$$f_2 = c_2 + \sum_{i=1}^{m_2} b_i (z_{i1} \wedge \cdots \wedge z_{i,k_i})$$

be two pseudo-Boolean functions in the same variables. We write $f_1 = f_2$ if f_1 and f_2 have the same values for every assignment of truth-values to the variables, and $f_1 \equiv f_2$ if $c_1 = c_2$, $m_1 = m_2$, $n_i = k_i$, $y_{i1} = z_{i1}, \ldots, y_{i,n_i} = z_{i,k_i}$ and $a_i = b_i$ $(1 \leqslant i \leqslant m_1)$. Thus, $f_1 \equiv f_2 \Rightarrow f_1 = f_2$.

Write

$$f = c + \sum_{i=1}^{m} a_i t_i, \quad \text{where } t_i = y_{i1} \wedge \cdots \wedge y_{i,n_i}.$$

Then

$$f = c + \sum_{i=1}^{m} a_i (1 - \bar{t}_i), \qquad \bar{t}_i = \bar{y}_{i1} \vee \cdots \vee \bar{y}_{i,n_i},$$

where $\vee$ denotes disjunction. Now $y_1 \vee y_2 = y_1 + y_2 - y_1 \cdot y_2 = y_1 + \bar{y}_1 \cdot y_2 = y_1 + (\bar{y}_1 \wedge y_2)$, and, by induction,

$$y_1 \vee \cdots \vee y_n = \sum_{i=1}^{n} (\bar{y}_1 \wedge \cdots \wedge \bar{y}_{i-1} \wedge y_i).$$

Therefore

$$\bar{t}_i = \sum_{j=1}^{n_i} (y_{i1} \wedge \cdots \wedge y_{i,j-1} \wedge \bar{y}_{ij}),$$

and so

$$f = c + \sum_{i=1}^{m} a_i \left(1 - \sum_{j=1}^{n_i} s_{ij}\right), \quad \text{where } s_{ij} \equiv y_{i1} \wedge \cdots \wedge y_{i,j-1} \wedge \bar{y}_{ij}.$$

Thus:

$$f = g, \quad \text{where } g = a - \sum_{i=1}^{m} a_i \sum_{j=1}^{n_i} s_{ij}, \qquad a = c + \sum_{i=1}^{m} a_i. \tag{1}$$

Note that f is a posiform of a pseudo-Boolean function F if and only if g is a negaform of the same function F.

To appreciate the multitude of the different forms a pseudo-Boolean function can assume, note that each of the $n_i!$ permutations of the literals in each conjunct $y_{i1} \wedge \cdots \wedge y_{i,n_i}$ of f induces its own form of the function g given by (1). Also, f can be written in many forms. For example, write

$$g = a - \sum_{i=1}^{m} a_i \sum_{j=1}^{n_i} (1 - \bar{s}_{ij}) = a - \sum_{i=1}^{m} a_i n_i + \sum_{i=1}^{m} a_i \sum_{j=1}^{n_i} \bar{s}_{ij},$$

where

$$\bar{s}_{ij} = \bar{y}_{i1} \vee \cdots \vee \bar{y}_{i,j-1} \vee y_{ij} = y_{i1} \wedge \cdots \wedge y_{ij} + \sum_{k=1}^{j-1} y_{i1} \wedge \cdots \wedge y_{i,k-1} \wedge \bar{y}_{ik}.$$

Thus, if we put

$$b = a - \sum_{i=1}^{m} a_i n_i,$$

then if $f = c + \sum_{i=1}^{m} a_i t_i$ is a posiform of F, also

$$f_1 = b + \sum_{i=1}^{m} a_i \sum_{j=1}^{n_i} \bar{s}_{ij}$$

is a posiform of F, and generally $f_1 \not\equiv f$, since f_1 has generally more terms than f.

Problem. Define a natural canonical posiform and negaform for each pseudo-Boolean function.

With every pseudo-Boolean function form $f = c + \sum_{i=1}^{m} a_i t_i$ and its associated Boolean support $\varphi = \bigvee_{i=1}^{m} t_i$, associate two graphs:

(i) A *conflict graph* $C(f) = C(\varphi) = (V, E)$, where $V = \bigcup_{i=1}^{m} t_i$, and $(t_i, t_j) \in E$ if and only if there is a 'conflict' between t_i and t_j, that is, some x_k appears in one and $\bar{x}_k$ in the other. With vertex t_i of $C(f)$ associate the *weight* a_i. The graph $C(\varphi)$ is unweighted.

(ii) A DNF graph $D(f) = D(\varphi)$ defined as follows. Each *occurrence* of a literal is a vertex and no other vertices are allowed. Two vertices y_1 and y_2 are joined by an edge if and only if they correspond to nonconflicting literals in distinct conjuncts t_i, that is, if and only if $y_1 \in t_i \Rightarrow y_2 \not\in t_i$ and $y_1 \neq \bar{y}_2$. With all occurrences of a literal in a conjunct t_i of $D(f)$, associate the weight a_i. The graph $D(\varphi)$ is unweighted.

In Section 2 we give a simple connection between max(f), min(f) and the maximum weighted independent sets of $C(f)$ and $C(g)$. In Section 3 we exhibit a connection between $D(f)$ and the complement $\bar{C}(g)$ of $C(g)$ for every g of the form (1), and prove the equivalence between min(f) = 0 and the independence number $\alpha(\bar{D}(f)) = \alpha(C(g)) = m$, where m is the number of conjuncts of f, and $\bar{D}$ is the complement of D. In the final section, Section 4, we give a complete characterization of $\bar{D}(f)$ graphs.

2. Connection with conflict graph

Theorem 1. (i) *If f is a posiform with constant term c and $C(f)$ its conflict graph, then* $\max f = c$ + *maximum weighted independent set of $C(f)$.* (ii) *If g is given by*

(1), *then* $\min f = a +$ *minimum weighted independent set of* $C(g)$, *where* $C(g)$, *which has negative weights, is the conflict graph of* g.

Proof. Obvious. □

Note. Since the decision problem INDEP SET is NP-complete, the decision problems $\max f$ and $\min f$ are a fortiori NP-complete. The former seems to be NP-complete even when the graph C is a tree. However, it is polynomial if C is bipartite. Even the problem of finding a maximum weighted independent set in a weighted bipartite graph is polynomial (see Berenguer and Diaz [1]).

Problems.

(i) Characterize the forms f corresponding to special types of conflict graphs such as trees, bipartite graphs, etc.

(ii) Characterize properties of f in terms of properties of $C(f)$ in addition to $\max(f)$ and $\min(f)$.

(iii) Given the conflict graph of a posiform $f = c + \sum_{i=1}^{m} a_i\,(y_{i1} \wedge \cdots \wedge y_{i,n_i})$, what can be said about the structure of the conflict graph of g, where g is given by (1)? Of course, $C(f)$ has m vertices and $C(g)$ has $\sum_{i=1}^{m} n_i$ vertices and all n_i vertices of weight a_i form a clique of size n_i $(1 \leqslant i \leqslant m)$ in $C(g)$. We shall say something about a related problem in the next section.

Example. Let $f = a_1(x \wedge \bar{y} \wedge z) + a_2(\bar{x} \wedge \bar{w}) + a_3(y \wedge w \wedge \bar{z})$. Then

$$g = a_1 + a_2 + a_3 - a_1(\overset{1}{\bar{x}} + \overset{2}{(x \wedge y)} + \overset{3}{(x \wedge \bar{y} \wedge \bar{z})}) - a_2(\overset{4}{x} + \overset{5}{(\bar{x} \wedge w)}).$$

$$- a_3(\overset{6}{\bar{y}} + \overset{7}{(y \wedge \bar{w})} + \overset{8}{(y \wedge w \wedge z)})$$

(see Fig. 1).

The independent sets of $C(f)$ are $\{a_1\}$, $\{a_2\}$, $\{a_3\}$ and $\emptyset$. Those of $C(g)$ are $\{1,5,6\}$, $\{1,5,8\}$, $\{1,7\}$, $\{2,4,7\}$, $\{2,4,8\}$, $\{3,4,6\}$ and their subsets. Hence, $\max f = \max\{a_1, a_2, a_3\}$ and $\min f = 0$.

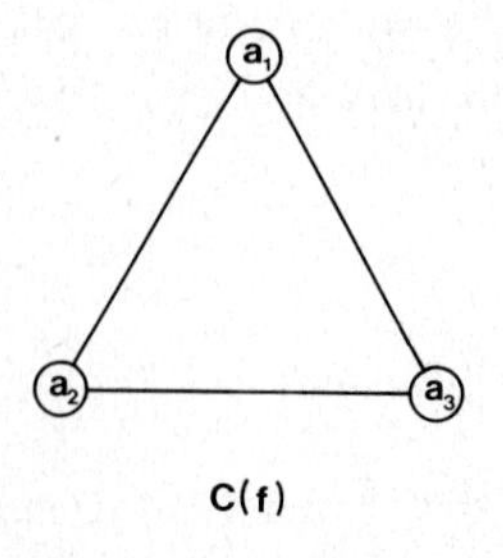

C(f)

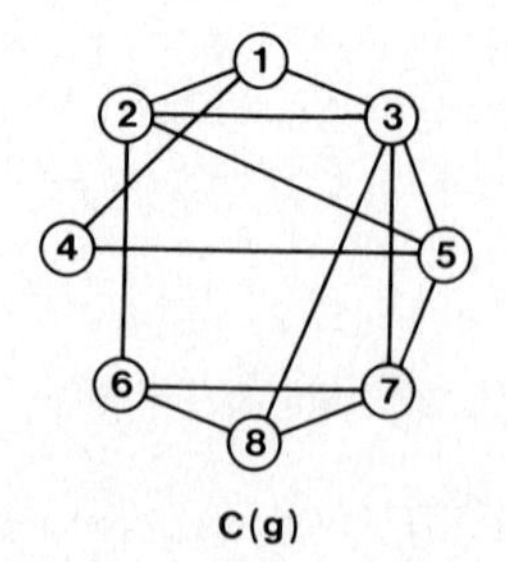

C(g)

Fig. 1.

3. Connection with DNF graph

Lemma 1 (*see Cook* [2] *in proof of Theorem* 2). *A Boolean support φ is a nontautology if and only if $D(\varphi)$ has a clique of size m.*

Proof. Obvious. □

Definition. Let $G=(V,E)$ be an undirected graph. Then
(i) $\bar{G}=(V,\bar{E})$ is the *complement* of G, where $e \in \bar{E}$ if and only if $e \notin E$.
(ii) $G_1=(V,E_1)$ is a *partial graph* of G if $E_1 \subseteq E$.

Theorem 2. *Suppose that f is a posiform and g is any form given by* (1). *Then $\bar{C}(g)$ is isomorphic to a partial graph of $D(f)$. Moreover, the union of $\bar{C}(g)$ over all forms g induced by f by taking the different permutations of literals in each conjunct of f, is isomorphic to $D(f)$.*

Proof. Recall that if

$$f=c+\sum_{i=1}^{m} a_i(y_{i1}\wedge\cdots\wedge y_{i,n_i}),$$

then

$$g=a-\sum_{i=1}^{m} a_i \sum_{j=1}^{n_i} y_{i1}\wedge\cdots\wedge y_{i,j-1}\wedge \bar{y}_{ij}.$$

Note that $D(f)$ and $\bar{C}(g)$ have $\sum_{i=1}^{m} n_i$ vertices each. Define a vertex correspondence between the two graphs by

$$y_{i1}\wedge\cdots\wedge y_{i,j-1}\wedge\bar{y}_{ij}\in V(\bar{C}(g)) \leftrightarrow y_{ij}\in V(D(f)).$$

Suppose $e=(u_1,u_2)\in E(\bar{C}(g))$. Then $u_1=a_i(y_{i1}\wedge\cdots\wedge y_{i,j-1}\wedge\bar{y}_{ij})$, $u_2=a_k(y_{k1}\wedge\cdots\wedge y_{k,l-1}\wedge\bar{y}_{kl})$, where $i\neq k$ and no literal is the complement of any other. Now u_1 and u_2 map into y_{ij} and y_{kl} under the above correspondence. Moreover, $(y_{ij},y_{kl})\in E(D(f))$, since y_{ij} and y_{kl} are nonconflicting literals in distinct conjuncts, and so $\bar{C}(g)$ is isomorphic to a partial graph of $D(f)$. Now assume that $(y_{ij},y_{kl})\in E(D(f))$. Then y_{ij} and y_{kl} are nonconflicting literals in distinct conjuncts of f. There is some permutation of literals in f such that under the above correspondence $u_1\leftrightarrow y_{ij}$ and $u_2\leftrightarrow y_{kl}$. For such a form of g we clearly have $(u_1,u_2)\in E(\bar{C}(g))$. □

Examples.
(i) For the example considered above, see Fig. 2.

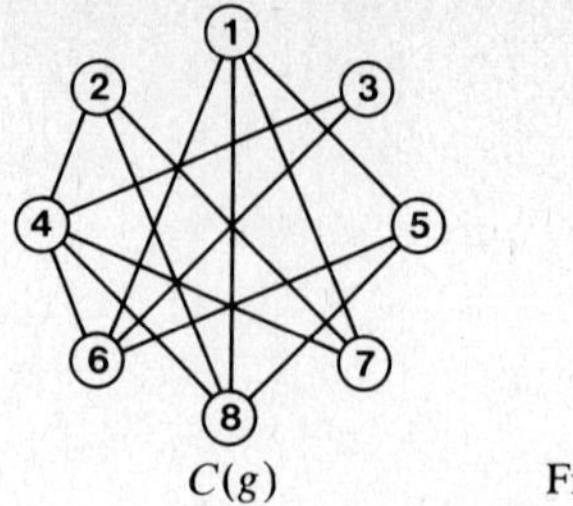

C(g)

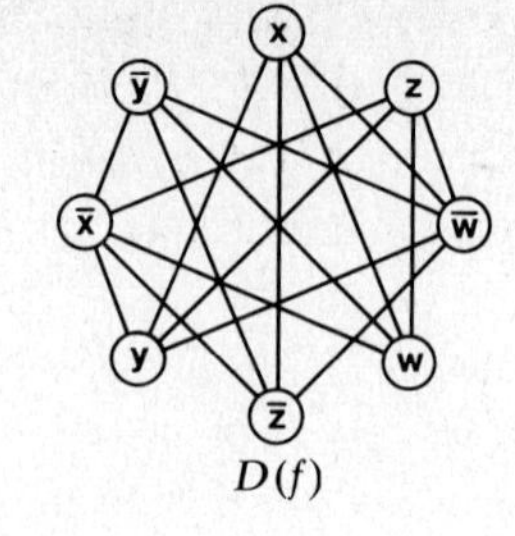

D(f)

Fig. 2.

(ii) Let

$$f = a_1(\overset{1}{\bar{x}} \wedge \overset{2}{y}) + a_2(\overset{3}{x} \wedge \overset{4}{y}),$$

$$g_1 = a_1 + a_2 - a_1(\overset{1}{x} + (\bar{x} \overset{2}{\wedge} \bar{y})) - a_2(\overset{3}{\bar{x}} + (x \overset{4}{\wedge} \bar{y})),$$

$$g_2 = a_1 + a_2 - a_1(\overset{2}{\bar{y}} + (y \overset{1}{\wedge} x)) - a_2(\overset{3}{\bar{x}} + (x \overset{4}{\wedge} \bar{y})).$$

The relevant graphs appear in Fig. 3, and it is seen that the union of $\bar{C}(g_1)$ and $\bar{C}(g_2)$ is isomorphic to $D(f)$.

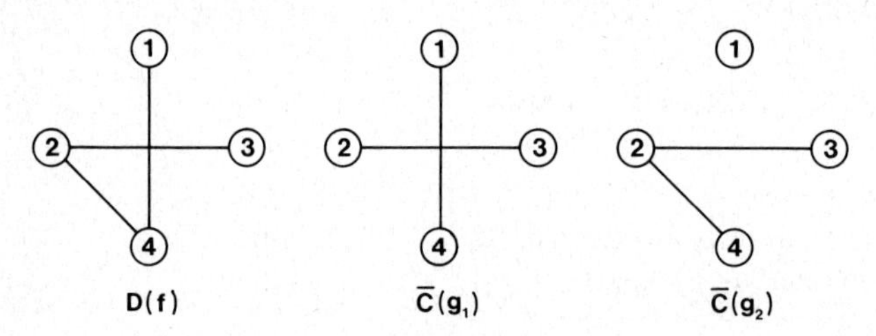

Fig. 3.

(iii) Let

$$f = a_1(\overset{1}{\bar{x}} \wedge \overset{2}{y}) + a_2(\overset{3}{x} \wedge \overset{4}{\bar{y}}),$$

$$g = a_1 + a_2 - a_1(\overset{1}{x} + (\bar{x} \overset{2}{\wedge} \bar{y})) - a_2(\overset{3}{\bar{x}} + (x \overset{4}{\wedge} y)).$$

Here already $\bar{C}(g)$ is isomorphic to $D(f)$ (see Fig. 4).

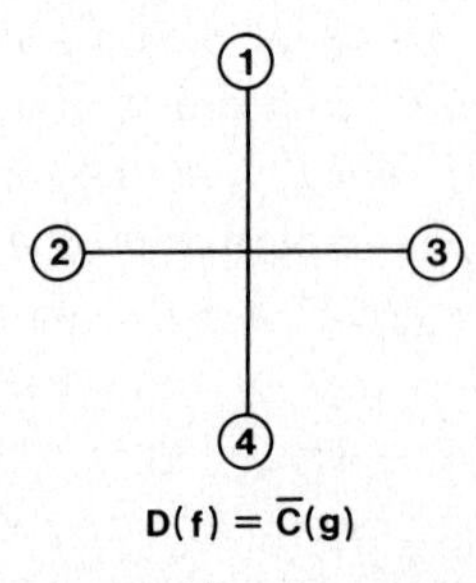

D(f) = C̄(g)

Fig. 4.

Problems. For what forms f is it true that $\bar{C}(g)$ is isomorphic to $D(f)$:

(i) for all g of the form (1);

(ii) for some g of the form (1);

(iii) for no g of the form (1)?

Theorem 3. *Suppose that* $f = c + \sum_{i=1}^{m} a_i (y_{i1} \wedge \cdots \wedge y_{i,n_i})$ *is a posiform with Boolean support* $\varphi = \bigvee_{i=1}^{m} (y_{i1} \wedge \cdots \wedge y_{i,n_i})$. *Then the following conditions are equivalent*:

(i) $\min f = c$,

(ii) $\alpha(C(g)) = m$, *where* g *is any form given by* (1),

(iii) $\alpha(\bar{D}(f)) = m$,

(iv) φ *is a nontautology.*

Proof.

(i) $\Rightarrow$ (ii). By Theorem 1, $\min f = c$ implies $-\sum_{i=1}^{m} a_i =$ minimum weighted independent set S of $C(g)$. In $C(g)$ the vertices s_{ij} $(1 \leqslant j \leqslant n_i)$ of weight $-a_i$ form a clique of size n_i $(1 \leqslant i \leqslant m)$. Hence, S has at most one member in $\{s_{ij}\}_{j=1}^{n_i}$ and so $|S| \leqslant m$. Since the weight of S is $-\sum_{i=1}^{m} a_i$, exactly one vertex from each $\{s_{ij}\}_{i=1}^{n_i}$ is in S, which thus has size m.

(ii) $\Rightarrow$ (iii). By Theorem 2, $\bar{D}(f)$ is isomorphic to a partial graph of $C(g)$, hence $\alpha(\bar{D}(f)) \geqslant m$. Since the occurrences of the literals in every conjunct of f form a clique in $\bar{D}(f)$, $\alpha(\bar{D}(f)) \leqslant m$, hence $\alpha(\bar{D}(f)) = m$.

(iii) $\Rightarrow$ (iv). Follows directly from Cook's Lemma.

(iv) $\Rightarrow$ (i). The truth values for which $\varphi = 0$ make $f = c$. □

Theorem 2 shows that Theorem 3 is an improvement of Cook's Lemma, in the sense that φ is a nontautology if and only if any one of a collection of *partial* graphs of $D(f)$ has a clique of size m.

4. A characterization of $\bar{D}(\varphi)$ graphs

Given any finite graph $G = (V, E)$, we can construct a Boolean support function φ such that $C(\varphi)$ is isomorphic to G. Cover G with a collection of complete bipartite graphs $\{B_i = (V_0^i \cup V_1^i, E^i)$: $i = 1, \ldots, k\}$ and assign the variable x_i to every vertex in V_0^i and $\bar{x}_i$ to every vertex in V_1^i $(1 \leqslant i \leqslant k)$. If the collection $\{B_i\}$ is sufficiently large, then the conjunction of all literals associated with a vertex gives a conjunct of φ which is distinct from all other conjuncts, and φ is the disjunction of all these conjuncts. However, not every graph is a $D(\varphi)$ graph, as is easily seen. We characterize $D(\varphi)$ graphs — actually $\bar{D}(\varphi)$ graphs, for convenience — in Theorem 4 below. First we shall prove jointly with Bruno Simeone Lemma 2 below.

Definition. A *k-path* ($k \geq 3$) in a graph $G=(V,E)$ is a path of the form $(u_1,\dots,u_k)$, $u_i \in V$ $(1 \leq i \leq k)$, $(u_i,u_{i+1}) \in E$ $(1 \leq i < k)$, such that there are no other edges among the u_i, that is $(u_i,u_j) \in E \Rightarrow j = i+1$.

Lemma 2. *A connected graph $G=(V,E)$ is a complete bipartite graph if and only if*

(i) *G contains no triangle, and*

(ii) *G contains no 4-path.*

Proof. Let $G=(V,E)$ be a graph satisfying (i) and (ii). Let b be any vertex, V_1 the set of all vertices adjacent to b, and V_2 the set of all vertices x for which there is a vertex $y \in V_1$ such that $(x,y) \in E$.

Suppose that $c \in V_1 \cap V_2$. Since $c \in V_1$, we have $(b,c) \in E$. Since $c \in V_2$, there exists $d \in V_1$ such that d is adjacent to both b and c. Hence (b,c,d) is a triangle. Therefore $V_1 \cap V_2 = \emptyset$.

Suppose that $c \in V$, $c \notin V_1 \cup V_2$. Then $c \neq b$, since $b \in V_2$ (if $|V|>1$, as we may assume). Since G is connected, there is a shortest path $(b,d,g,e,\dots,c)$ between b and c, where $d \in V_1$, $g \in V_2$ and e is a vertex (possibly c) distinct from d and g. Since G contains no triangle and no 4-path, $(b,e) \in E$, hence there is a shorter path between b and c, a contradiction. Thus, $V_1 \cup V_2 = V$.

Let $c \in V_1$, $d \in V_2$, and $d \neq b$. Then $(c,b) \in E$ and there exists $g \in V_1$ such that $(b,g),(g,d) \in E$. If $g=c$, then $(c,d) \in E$. Otherwise, since there is no triangle, (c,b,g,d) is a 4-path unless $(c,d) \in E$. Thus, every $c \in V_1$ and $d \in V_2$ are adjacent. If $g,e \in V_i$ ($i=1$ or 2) and $(g,e) \in E$, then for every $h \in V_{3-i}$, (g,e,h) is a triangle, a contradiction.

Conversely, if $G=(V,E)$ is a complete bipartite graph, it cannot contain a triangle. If (u_1,u_2,u_3,u_4) is a path in G, then clearly $(u_1,u_4) \in E$, hence G contains no 4-path. □

Theorem 4. *A graph $G=(V,E)$ is isomorphic to a $\bar{D}(\varphi)$ graph if and only if E can be partitioned into a set E_r of red edges and a set E_y of yellow edges ($E_r \cup E_y = E$, $E_r \cap E_y = \emptyset$), such that*

(i) *there exists no red 3-path,*

(ii) *every triangle is red, and*

(iii) *there exists no yellow 4-path.*

Moreover, if G has the property that every $u \in V$ either belongs to a triangle or is isolated, then the partitioning is unique if it exists. In this case there is, up to a renaming of literals, a unique form φ such that $\bar{D}(\varphi)$ is isomorphic to G.

Proof. Suppose that G is a graph for which (i), (ii) and (iii) hold. Let C be a red connected component, and b and c two vertices in C. Then there is a red path

$b=1,\ldots,m=c$. If $1 \leqslant i < j \leqslant m$, then $(i,j) \in E_r$ by (i). Thus, every red component is a clique, and so the red partial graph is partitioned into vertex-disjoint red cliques.

Let B be any yellow connected component. By (ii), (iii) and Lemma 2, B is a complete bipartite graph. Suppose that B contains two vertices, b and c, in some red clique C. Let $b=1,\ldots,m=c$ be a shortest yellow path. Then $m \geqslant 3$. Hence, by (ii), $m > 3$. By (iii), $m \leqslant 3$. This contradiction shows that B intersects each red clique in at most one vertex.

Let $V(B) = V_0 \cup V_1$ with $V_0 \cap V_1 = \emptyset$ such that every edge of $E(B)$ is between a vertex in V_0 and one in V_1. Allowing the extreme case $V_0 = \emptyset$ (or $V_1 = \emptyset$) we can assume that the entire vertex set V of G is covered by the $V(G)$'s. Assign the literal x_B to all vertices in V_0 and $\bar{x}_B$ to all vertices in V_1. Do this for every yellow connected component and define the conjuncts of φ to be the conjunctions of all literals in each red clique. Then each literal appears at most once in any red clique, and if it appears in a red clique C, its complement does not appear in C. If φ is defined as the disjunction of all these conjuncts, where any isolated vertex represents a conjunct consisting of a single literal distinct from all others, then clearly $\bar{D}(\varphi)$ is isomorphic to G.

Conversely, assume that G is isomorphic to $\bar{D}(\varphi)$ for some DNF-function φ. For every conjunct C of φ, color the edges between literals in C red, and the other edges adjacent to these literals yellow. Then (i)–(iii) are clearly satisfied.

Let a be any vertex belonging to a triangle $\Delta(a,b,c)$. Suppose there are two partitionings of G into E_r and E_y satisfying (i)–(iii). Let a be in a red clique Q_i in one of these partitionings and in a red clique Q_j in the other. By (ii), $\Delta(a,b,c) \subseteq Q_i \cap Q_j$. If there is a vertex d outside the triangle with $d \in Q_i$, then, since Q_i is a clique, $\Delta(a,b,d) \subseteq Q_i$ and by (ii) also $\Delta(a,b,d) \subseteq Q_j$, so $d \in Q_j$. Thus, $Q_i \subseteq Q_j$ and, by symmetry, $Q_j \subseteq Q_i$; hence, $Q_i = Q_j$. This implies that the partitioning into red cliques is unique if every vertex belongs to a triangle or is isolated. Hence, the partitioning of E into E_r and E_y is unique if it exists.

Suppose there exist Boolean supports φ and ψ such that both $\bar{D}(\varphi)$ and $\bar{D}(\psi)$ are isomorphic to G. For every conjunct of C of φ, color the edges between literal occurrences of C in $\bar{D}(\varphi)$ red, and the other edges adjacent to these literals yellow. This constitutes a partitioning satisfying (i)–(iii). Doing the same for $\bar{D}(\psi)$, another partitioning satisfying (i)–(iii) is obtained. Thus, if the partitioning is unique, then $\varphi \equiv \psi$ up to a renaming of variables. □

Note. Let c be a vertex not belonging to a triangle and not isolated. Then there is a vertex d adjacent to c. If, for example, (c,d) is a connected component in G, then there is a partitioning in which $(c,d) \in E_r$ (then c and d represent different nonconflicting literals) and one in which $(c,d) \in E_y$ (then c represents a literal and d its complement), and so in this case there is uniqueness neither in the partitioning of G satisfying (i)–(iii), nor in the Boolean support φ for which $\bar{D}(\varphi)$ is isomorphic to G.

References

[1] X. Berenguer and J. Diaz, The weighted Sperner's set problem, in: P. Dembiński, Ed., Proceedings, 9th Symposium Mathematical Foundations of Computer Science, Rydzyna, Poland, Lecture Notes in Computer Science, Vol. 88 (1980) pp. 137–141.

[2] S. Cook, The complexity of theorem-proving procedures, 3rd Annual ACM Symp. on Theory of Computing, Assoc. for Comput. Mach., New York (1971) pp. 151–158.

Received 29 March 1981

Annals of Discrete Mathematics 20 (1984) 147–150
North-Holland

VALUATIONS OF TREES, POLYNOMIAL EQUATIONS AND POLYTOPES

H. FRIED*
Department of Mathematics and Computer Science, Bar-Ilan University, Ramat Gan, Israel

All graphs to be discussed in this note are finite simple graphs without loops or multiple edges (mainly trees).

A *valuation* of a finite graph Γ is an injective map f from the vertex set $v(\Gamma)$ of Γ into the integers; if $v \in v(\Gamma)$, $f(v) \in Z$.

Given a valuation f of the graph Γ, let $e = (v, w)$, $v, w \in v(\Gamma)$ be an edge of Γ. If $f(v) < f(w)$, we call v the lower end of e and w its upper end. One may regard each edge as oriented from its lower end to its upper one, and the resulting *coboundary map* on $e(\Gamma)$ maps each edge onto some positive integer; we designate this map by f as well:

$$e = (v, w), \qquad f(v) < f(w) \Rightarrow f(e) = f(w) - f(v).$$

On a *bipartite* graph a valuation f will be called monotone if the induced orientation is consistent with the colour partition; that is, if all the lower ends of the edges are one colour class, and the upper ends, another.

A valuation f on a graph Γ having m edges and no more than $m+1$ vertices is a *β-valuation* [3] (or a 'graceful numbering' [1]), if

(i) the range of f on $v(\Gamma)$ is contained in the interval $[0, m]$, and

(ii) the range of f on $e(\Gamma)$ coincides with the interval $[1, m]$.

A monotone valuation on a bipartite graph which is also a β-valuation will be called a monotone β-valuation.

Given a tree T on $n+1$ vertices, let us designate these vertices in some order by $v_0, v_1, \ldots, v_n$, and the n edges by $e_1, e_2, \ldots, e_n$. A tree being a bipartite graph, $v(T)$ is partitioned into two classes, R (red) and B (blue), say, each edge having one end in R and one in B. If f is a monotone β-valuation of T, write x_i for $f(v_i)$ and y_k for $f(e_k)$. We may suppose each edge oriented from *red* to *blue*, so that

$$e_k \in e(\Gamma), \qquad e_k = (v_i, v_j), \qquad v_i \in R,\ v_j \in B \Rightarrow y_k = x_j - x_i. \tag{1}$$

Therefore, f will be a monotone β-valuation of T if, and only if,

* Submitted and presented at the Conference by S. Schreiber

(α) the sequence $\langle x_0 \cdots x_n \rangle$ is a permutation of the sequence $\langle 0, 1, \ldots, n \rangle$, and
(β) the sequence $\langle y_1 \cdots y_n \rangle$ is a permutation of the sequence $\langle 1, 2, \ldots, n \rangle$.

Thus, to find a monotone β-valuation of T, it is necessary and sufficient to find a sequence $\langle x_i \rangle$ satisfying (α) and, after the substitution (1), also (β).

Notice that when the condition of monotonicity is dropped, (β) still has to be satisfied, but (1) will then be replaced by

$$y_k = |x_j - x_i|, \tag{1'}$$

or by

$$y_k^2 = (x_j - x_i)^2. \tag{1''}$$

We now give two interpretations of (α) and (β). Let $\alpha_0 < \alpha_1 < \cdots < \alpha_m$ be real, $\sum \alpha_i = \boldsymbol{A}$, and designate the sum of the first $\boldsymbol{r}$ elements of this sequence by $\boldsymbol{b}_r$ and the sum of the last $\boldsymbol{r}$ by $\boldsymbol{c}_r$; thus, $\boldsymbol{b}_1 = \alpha_0$, $\boldsymbol{c}_1 = \alpha_m$, $\boldsymbol{b}_{m+1} = \boldsymbol{c}_{m+1} = A$.

Let $C(\alpha_0, \alpha_1, \ldots, \alpha_m)$ be the convex hull of the $(m+1)!$ points in affine (or Euclidean) $\boldsymbol{m}+1$-space, whose coordinates are $\alpha_0, \alpha_1, \ldots, \alpha_m$ in some order. It is readily checked that this is an $\boldsymbol{m}$-dimensional polytope, given by the equality

$$\sum_{i=0}^{m} x_i = A \tag{2}$$

and, for each choice of $\boldsymbol{r}$ out of $\boldsymbol{m}+1$ coordinates, by the inequalities

$$\boldsymbol{b}_r \leqslant x_{i,1} + x_{i,2} + \cdots + x_{i,r} \leqslant \boldsymbol{c}_r, \tag{3}$$

half of which are redundant because of (2). Indeed, since $\boldsymbol{b}_r + \boldsymbol{c}_{m+1-r} = A$, the left inequality for $\boldsymbol{r}$ implies the right one for $\boldsymbol{m}+1-\boldsymbol{r}$ and vice versa. Moreover, for each $\boldsymbol{r}$ there are $\binom{m+1}{r}$ inequalities of type (3) and (considering only the left-hand inequalities, say), it is trivial to show that a set of $\boldsymbol{m}+1$ numbers satisfying (2) and (3), with one equality for each set of $\binom{m+1}{r}$ conditions, is indeed a permutation of $\langle \alpha_0, \alpha_1, \ldots, \alpha_m \rangle$. Thus, we obtain:

Lemma 1. *A set of $\boldsymbol{m}+1$ numbers is a permutation of $\langle \alpha_0, \alpha_1, \ldots, \alpha_n \rangle$ if and only if it represents a vertex of the polytope given by* (2) *and* (3).

Corollary 1. *The sequence $\langle x_0, x_1, \ldots, x_n \rangle$ satisfies condition (α) if and only if the point $(x_0, x_1, \ldots, x_n)$ is a vertex of the polytope P_1 given by*

$$\sum_{i=0}^{n} x_i = \binom{n+1}{2} \tag{4}$$

and, for $1 \leqslant r \leqslant n$, the inequalities

$$\binom{r}{2} \leqslant x_{i,1} + x_{i,2} + \cdots + x_{i,r} \leqslant \binom{n+1}{2} - \binom{n+1-r}{2} = r \cdot n - \binom{r}{2}. \tag{5}$$

As remarked above, this is an n-dimensional polytope and, moreover, because of (4), one of the two inequalities in (5) becomes redundant.

Corollary 2. *The sequence* $\langle y_1, y_2, \ldots, y_n \rangle$ *satisfies condition* (β) *if and only if the point* $(y_1, y_2, \ldots, y_n)$ *is a vertex of the polytope* P_2 *given by*

$$\sum_{k=1}^{n} y_k = \binom{n+1}{2} \tag{6}$$

and, for $1 \leq r \leq n-1$, *the inequalities*

$$\binom{r+1}{2} \leq y_{k,1} + y_{k,2} + \cdots + y_{k,r} \leq \binom{n+1}{2} - \binom{n+1-r}{2} = r \cdot n - \binom{r}{2}. \tag{7}$$

In the variables y_k, P_2 is an $n-1$-dimensional polytope. Substituting for (1), it becomes a *convex cylinder set* of n dimensions in $n+1$-dimensional affine x-space; this follows readily from the fact that the l.h.s. of (1) is unaffected by replacing each x_i by $x_i + h$, for any real h. Moreover, if the $n+1$-dimensional x-space is taken to be *Euclidean*, equation (6), after substituting for (1), will represent a hyperplane perpendicular to (4). Thus, by intersecting P_2 with (4) to obtain an $n-1$-dimensional polytope P_2', no new vertices occur. However, and this may be shown by numerical examples, intersecting P_1 by (6) gives an $n-1$-dimensional polytope P_1', *not* all of whose vertices are also vertices of P_1; and conceivably — though I know of no such example — some of these *accessory* vertices may be vertices of P_2'. Thus, we may formulate our result as follows:

Proposition 1. *Let the edges of the tree* T *be related to the vertices by* (1); *then* T *has a monotone* β*-valuation if and only if the* n*-dimensional affine polytope* P_1 *given by* (4) *and* (5) *has* (*in* (6)) *a vertex in common with the* $n-1$*-dimensional polytope* P_2' *given by* (4), (6) *and* (7).

Our second interpretation requires no preparation.

Let t be an auxiliary variable; then, from (α) and (β), one has directly:

$$\prod_{i=0}^{n} (t - x_i) = \prod_{i=0}^{n} (t - i), \tag{8}$$

and

$$\prod_{k=1}^{n} (t - y_k) = \prod_{k=1}^{n} (t - k). \tag{9}$$

Comparison of the coefficients of powers of t between both sides of (8) gives $n+1$ polynomial equations in the $n+1$ unknowns x_i; substituting for (1) in the l.h.s. of (9) gives n more equations. Therefore

Proposition 2. *Let the* $\boldsymbol{n}$ *edges of the tree T be related to the vertices by* (1); *then T has a monotone* β*-valuation if and only if the* $2\boldsymbol{n}+1$ *polynomial equations derived from* (8) *and, after substituting* (1), *from* (9), *have at least one common solution* $x_0, x_1, \ldots, x_n$.

If one does not require monotonicity, the corresponding values y_k and x_i will still be connected by (1″). Then we may replace (9) by

$$\prod_{k=1}^{n} (t - y_k^2) = \prod_{k=1}^{n} (t - k^2), \tag{9''}$$

and modify the statement to read:

Proposition 2″. *Let the* $\boldsymbol{n}$ *edges of the tree T be related to the vertices by* (1″); *then T has a* β*-valuation if and only if the* $2\boldsymbol{n}+1$ *polynomial equations derived from* (8) *and, after substituting* (1″), *from* (9″), *have at least one common solution* $x_0, x_1, \ldots, x_n$.

A system of $2\boldsymbol{n}+1$ polynomial equations in $\boldsymbol{n}+1$ unknowns is highly overdetermined and a priori it would seem very unlikely that a solution should be found for any given tree, even for moderate values of $\boldsymbol{n}$. However, Rosa [3] has verified that all trees on 15 edges or less have (general) β-valuations, and the present author has found monotone β-valuations for all trees on 10 edges or less. Solutions of remarkable generality have recently appeared in [2]. In the absence of a reasonable algorithm for dealing with such systems of equations it is hard to look for a counterexample, and thus the question of whether every tree has a β-valuation or a monotone β-valuation remains open.

References

[1] S.W. Golomb, How to number a graph, in: Graph Theory and Computing (Academic Press, New York, 1972) pp. 23–27.
[2] K.M. Koh, D.G. Rogers and T. Tan, Products of graceful trees, Discr. Math. 31 (1980) 279–292.
[3] A. Rosa, On certain valuations of the vertices of a graph, in: Theory of Graphs, Proc. International Symposium, Rome, 1966 (Gordon & Breach, New York, 1967) pp. 349–355.

Received March 1981

Annals of Discrete Mathematics 20 (1984) 151–159
North-Holland

SPECTRA OF TREES

C.D. GODSIL

Institut für Mathematik und Angewandte Geometrie, Montanuniversität Leoben, A-8700, Leoben, Austria

The paper surveys some little-known results and presents some new results concerning the characteristic polynomials and adjacency matrices of trees and forests. The new results include an explicit expression for the entries of the adjoint matrix of $xI - A$, when A is the adjacency matrix of a forest, and bounds on the value of the smallest positive eigenvalue of such a matrix.

1. Introduction

Let G be a graph with vertex set $V(G) = \{1, \ldots, n\}$. The *adjacency matrix* $A(G) = (a_{ij})$ of G is defined by setting $a_{ij} = 1$ when i and j are adjacent, and setting $a_{ij} = 0$ otherwise. (In particular, $a_{ii} = 0$.) The *characteristic polynomial* $\varphi(G, x)$ is $\det(xI - A(G))$, where I is the identity matrix. We will sometimes abbreviate $\varphi(G, x)$ to $\varphi(G)$ and $A(G)$ to A.

In this paper we consider the relation between the graph-theoretic properties of G and the algebraic properties $A(G)$ and $\varphi(G)$, under the additional assumption that G is a forest. The general problem, when G is an arbitrary graph, has been studied extensively (a convenient reference is [1]).

Since forests are a special class of graphs one would hope to find that their adjacency matrices and characteristic polynomials manifest some unusual behaviour. Until now there has been little evidence of this. Here we hope to go some way toward overcoming this shortage by presenting some new results (see Theorem 4, Lemma 2, Corollary 3 and Theorem 7) and by surveying some results which, although 'known', do not seem to be common knowledge. (These latter results come from a paper by Heilmann and Lieb [6] on an apparently unrelated topic.)

2. Basic properties

In this section we present some of the basic properties of the characteristic polynomial of a forest. None of these results is new.

Definition 1. Let F be a graph on n vertices. A *k-matching* in G is a set of k vertex disjoint edges. The number of k-matchings in F will be written as $m(F, k)$ and we assume that $m(F, 0) = 1$. A matching M with the property that every vertex in F is incident with an edge in M is called a *1-factor.* The *matchings polynomial* of F (sometimes also called the matching polynomial) is

$$\mu(F, x) = \sum_{k=0}^{[n/2]} (-1)^k m(F, k) x^{n-2k}.$$

Our first theorem has a complicated history. It is a special case of Theorem 1.2 in [1]. For the proof and the history we refer the reader to that source.

Theorem 1. *Let F be a forest. Then $\varphi(F, x) = \mu(F, x)$.*

One immediate consequence of Theorem 1 is that we have an explicit expression for the coefficients in $\varphi(F, x)$. A second consequence is:

Corollary 1. *Let F be a forest. Then $A(F)$ is invertible if and only if F has a 1-factor.*

Proof. By the definition of $\mu(F, x)$ and Theorem 1 we see that $|\det(A)|$ is the number of 1-factors. □

We will say more about the inverse of $A(F)$ (when it exists) in Section 5. We must also point out that the converse of Theorem 1 is true — if $\varphi(F, x) = \mu(F, x)$, then F is a forest. This is given as Corollary 4.2 in [5].

The matchings polynomial of a graph has many interesting properties. For a survey see [5]. The paper by Heilmann and Lieb [6] is also an important source of information.

Our next result provides relations between the characteristic polynomial of a forest and the characteristic polynomials of its subforests. We require some preliminary definitions first, however.

Definition 2. If G is a graph and $S \subseteq V(G)$, then $G \setminus S$ is obtained by deleting the vertices in S. If $e = [1, 2]$ is an edge in G, then $G \setminus e$ is obtained by deleting e, but not the vertices 1 or 2, from G. If $i, j \in V(G)$ then we write $i \sim j$ to denote that i and j are adjacent.

Lemma 1. *Let F be a forest. Then*

(a) *if F is the disjoint union of forests F_1 and F_2, $\varphi(F) = \varphi(F_1)\varphi(F_2)$,*

(b) $d\varphi(F)/dx = \sum_{i \in V(F)} \varphi(F \setminus i)$,

(c) *if* $e=[1,2]$ *is an edge in* F, $\varphi(F)=\varphi(F\setminus e)-\varphi(F\setminus\{1,2\})$,
(d) *if* $1\in V(F)$, $\varphi(F)=x\varphi(F\setminus 1)-\Sigma_{i\sim 1}\varphi(F\setminus\{1i\})$.

Proof. For (a) see Theorem 2.4 in [1] (where 'disjoint union' is rendered as 'direct sum'). For (b) see Theorem 2.14 in [1]. Claim (c) follows from Theorem 2.12 in [1]. Finally (d) can be derived by repeated applications of (c). It is also equivalent to equation (4.1) in [6].

It should be clear from our proof that Lemma 1 is a summary of standard results! We note that (a) and (b) of Lemma 1 are valid for arbitrary graphs, that (c) holds for graphs such that e lies on no circuit, and (d) holds for graphs where 1 lies on no circuit.

It is easy to find examples to show that the characteristic polynomial of a forest does not, in general, identify it. (For examples see the tables in [1]). However, a stronger statement is true:

Theorem 2 (Schwenk [8]). *Let* p_n *denote the proportion of trees on* n *vertices which are determined up to isomorphism by their characteristic polynomials. Then* $p_n\to 0$ *as* $n\to\infty$.

3. The zeros of $\varphi(F,x)$.

One interesting problem in algebraic graph theory is to provide bounds on graph-theoretic parameters of a given graph G in terms of the eigenvalues of $A(G)$. In this section we state what is known when F is a forest. We remind the reader that $A(G)$ is symmetric and so the zeros of $\varphi(G)$ are always real, for any graph G.

Definition 3. We use $\Delta=\Delta(G)$ to denote the maximum degree of the graph G and $\hat{d}(G)$ to denote the average degree. We denote the largest eigenvalue of $A(G)$ by $\lambda_1(G)$.

Theorem 3. *Let* F *be a forest with* $\Delta>1$. *Then*
(a) $\lambda_1(F)>\hat{d}(F)$, *and*
(b) $\sqrt{\Delta}\leqslant\lambda_1(F)<2\sqrt{\Delta-1}$.

Proof. Part (a) follows from Theorem 3.8 of [1], on observing that since $\Delta>1$, F cannot be regular. The lower bound in (b) follows from Exercise 11.14(a) in [7] or from Exercise 2 on p. 112 of [1].

To prove the upper bound in (b) is valid we need a construction. We define the *rooted Bethe tree* $B(d, n)$ as follows. $B(d, 1)$ is a single vertex, which is simultaneously the root. $B(d, n)$ consists of a vertex v_n which is joined by edges to the roots of each of $d-1$ copies of $B(d, n-1)$. The vertex v_n is the root of $B(d, n)$. By the comment at the end of Section 3 in [6], $\lambda_1(B(d, n)) < 2\sqrt{d-1}$.

Our proof is completed by noting that if F has maximal degree Δ, then F is a subforest of $B(\Delta, n)$ when n is large enough. So, by Exercise 11.13 of [7], for example, $\lambda_1(F) \leqslant \lambda_1(B(\Delta, n))$. □

The lower bounds in Theorem 3 are valid for all graphs. However, for arbitrary graphs G we have only $\lambda_1(G) \leqslant \Delta(G)$, with equality occurring only if G is regular (for a proof see Exercise 11.14(a) in [7] again). The proof of the upper bound in Theorem 3(b) was suggested to the author by Ivan Gutman. The bound itself is due to Heilmann and Lieb (see Theorem 4.3 in [5]).

Since a forest is a bipartite graph, its eigenvalues are always symmetric about the origin. This is easy to see using Theorem 1 and the definition of $\mu(G, x)$. Alternatively, we refer to Exercise 11.19 in [7].

Virtually nothing is known about the eigenvalues of $A(F)$, other than $\lambda_1(F)$. Some information is available when $A(F)$ is invertible. We will summarize this in Section 5. The only remaining information we know of comes from the next result.

Theorem 4. *Let G be a graph with minimum vertex degree δ. Then $\varphi(G, x)$ has a zero θ with $\theta^2 \leqslant \delta$.*

Proof. Since $A = A(G)$ is symmetric it has $n = |V(G)|$ orthogonal eigenvectors $z_1, \ldots, z_n$. Let Z_i be the $n \times n$ matrix $z_i z_i^t$. We assume each z_i has unit length. It follows from this and their orthogonality that $Z_1 + Z_2 + \cdots + Z_n$ is the identity matrix.

Let λ_i be the eigenvalue associated with z_i.

Thus, $AZ_i = \lambda Z_i$ and

$$A^m = A^m \sum_{i=1}^{m} Z_i = \sum_{i=1}^{n} \lambda_i^n Z_i. \tag{1}$$

We may assume that $1 \in V(G)$ has degree δ. Then $(A^2)_{11}$ equals δ. Set $c_i = (Z_i)_{11}$. Taking $m = 0$ and $m = 2$ in turn in (1) yields:

$$1 = \sum_{i=1}^{n} c_i, \qquad \delta = \sum_{i=1}^{n} c_i \lambda_i^2. \tag{2}$$

Now, by definition, the diagonal entries of Z_i are the squares of the corresponding entries of the vector z_i. Thus, $c_i \geqslant 0$ and so, by (2), δ is a convex combination of the numbers λ_i^2. Accordingly, for some index j we must have $\lambda_j^2 \leqslant \delta$, as required. □

Corollary 2. *Let f be a forest. Then $\varphi(F, x)$ has a zero in the interval* $[-1,1]$.

If G is connected, then, with more care, it is possible to show that equality holds in Theorem 4 only when $\delta = 1$ and G is the complete graph on two vertices.

4. Christoffel–Darboux formulae

In Section 6 of [6] Heilmann and Lieb derive a number of identities for the matchings polynomial of a graph. These identities can be viewed as analogues of the Christoffel–Darboux formulae in the theory of orthogonal polynomials (for details of this see any reference book on the subject).

Our next result is a summary of the results of Section 6 in [6], insofar as they apply to trees. We use the notation P_{ij} to denote a path in a graph joining vertex i to j. In a tree there is a unique path joining any two vertices, so in this case P_{ij} is determined by i and j. We represent by $T \setminus P_{ij}$ the graph obtained when all vertices in P_{ij} (including i and j) are deleted from T.

Theorem 5 (Heilmann and Lieb [6]). *Let T be a tree and suppose* $V(T) = \{1, 2, \ldots, n\}$. *Then*

(a) $$\varphi(T \setminus i, x)\varphi(T, y) - \varphi(T \setminus i, y)\varphi(T, x) = (x - y) \sum_{j=1}^{n} \varphi(T \setminus P_{ij}, x)\varphi(T \setminus P_{ij}, y),$$

(b) $$\varphi(T \setminus i, x)\varphi'(T, x) - \varphi'(T \setminus i, x)\varphi(T, x) = \sum_{j=1}^{n} \varphi(T \setminus P_{ij}, x)^2,$$

(c) $$\varphi(T \setminus i, x)\varphi(T \setminus j, x) - \varphi(T \setminus \{ij\}, x)\varphi(T, x) = \varphi(T \setminus P_{ij}, x)^2.$$

Proof. Equations (a), (b) and (c) are special cases of equations (6.4), (6.5) and (6.6), respectively, in [6]. □

Using a different proof technique we can obtain analogous identities for arbitrary graphs. We will present this in a future paper [4].

It is known that if G is a graph and $v \in V(G)$, then the zeros of $\varphi(G \setminus v)$ *interlace* those of $\varphi(G)$. That is, between any two zeros of $\varphi(G)$ lies a zero of $\varphi(G \setminus v)$. (See, for example, Theorem 0.10 in [1].) One way of stating this result is by saying that $\varphi(G \setminus v)/\varphi(G)$ has only simple poles.

For trees this last statement can be strengthened:

Lemma 2. *Let P be a path in the tree T. Then $\varphi(T \setminus P, x)/\varphi(T, x)$ has only simple poles.*

Proof. Summing equation (b) of Theorem 5 over all vertices in $V(T)$ and then applying equation (b) of Lemma 1 we obtain:

$$\varphi'(T,x)^2 - \varphi''(T,x)\varphi(T,x) = \sum_{i,j} \varphi(T \setminus P_{ij}, x)^2. \tag{3}$$

Suppose θ is a zero of $\varphi(T)$ with multiplicity $m > 1$. Then it is zero of the LHS of (3) with multiplicity at least $2m-2$. Since the RHS is a sum of squares and θ is real, we conclude that, for any i and j, θ is a zero with multiplicity $2m-2$ of $\varphi(T \setminus P_{ij})^2$. Hence, θ has multiplicity at least $m-1$ in $\varphi(T \setminus P_{ij})$, whence the lemma follows. □

Lemma 2 is false for arbitrary graphs. For example, take G to be a complete graph with $n \geqslant 3$ vertices and P a Hamiltonian path in G. Then $\varphi(G \setminus T) = 1$, while $\varphi(G)$ has -1 as a zero with multiplicity $n-1$.

5. The adjoint matrix of $(xI - A(F))$

In this section we present an explicit expression for the adjoint matrix of $xI - A$, when $A = A(F)$ is the adjacency matrix of a forest F. We use this to obtain some information about the eigenvectors of F. Our next result could also be derived from (1) in [2], but we give a direct proof.

We use B^* to denote the adjoint matrix of a given matrix B.

Theorem 6. *Let F be a forest. Then the ij-entry of $(xI - A(F))^*$ is $\varphi(F \setminus P_{ij}, x)$.*

Proof. Let $A = A(F)$ and let $B = (b_{ij})$ be defined by setting $b_{ij} = \varphi(F \setminus P_{ij}, x)$. To prove the theorem it will suffice to show that $(xI - A)B = \varphi(F, x)I$.

The ith diagonal entry of $(xI - A)B$ is

$$x\varphi(F \setminus P_{ii}, x) - \sum_{j \neq i} a_{ij}\varphi(F \setminus P_{ij}, x). \tag{4}$$

Since $a_{ij} = 0$ unless i and j are adjacent, and since P_{ij} is just the edge $\{i, j\}$, when i and j are adjacent, it follows from Lemma 1(d) that the polynomial in (4) is just $\varphi(F, x)$.

If $i \neq j$, then the ij-entry of $(xI - A)B$ is

$$x\varphi(F \setminus P_{ij}) = \sum_{i \neq k} a_{ik}\varphi(F \setminus P_{kj})$$

$$= x\varphi(F \setminus P_{ij}) - \sum_{k \sim i} \varphi(F \setminus P_{kj}). \tag{5}$$

Let $\bar{k}$ denote the vertex adjacent to i in the path P_{ij}. Note that if $k \sim i$ and $k \neq \bar{k}$, then $P_{\bar{k}j}$ is a subpath of P_{kj}.

Hence we have

$$\sum_{k \sim i} \varphi(F \setminus P_k) = \varphi(F \setminus P_{\bar{k}j}) + \sum_{\substack{k \sim i \\ k \neq \bar{k}}} \varphi(F \setminus G_{\bar{k}j}) \setminus \{ik\}). \tag{6}$$

By Lemma 1(d) the RHS of (6) is just $x\varphi(F \setminus P_{ij})$, whence the RHS of (5) is zero.

Consequently, $(xI - A)B = \varphi(F, x)I$ and so $B = (xI - A)^*$, as required. □

Corollary 3. *Let T be a tree with $V(T) = \{1, \ldots, n\}$. Let $z_T(x)$ be the vector with* ith *component equal to $\varphi(T \setminus P_{1i}, x)$. Then if λ is a zero of $\varphi(T, x)$, $z_T(\lambda)$ is an eigenvector of T with eigenvalue λ.*

Proof. The vector $z_T(\lambda)$ is just the first column of $(\lambda I - A(T))^*$. Since $(xI - A)(xI - A)^* = \varphi(T, x)I$, it follows that when λ is a zero of $\varphi(T)$, $(\lambda I - A)z_T(\lambda) = 0$. Hence, $z_T(\lambda)$ is an eigenvector for $A(T)$ with eigenvalue λ.

□

Definition 4. Let T be a tree with n vertices and let $z = (z_i)$ be an n-dimensional vector. Suppose $e = \{i, j\}$ is an edge in T. We say there is a *sign-change* in z on the edge e if $z_i z_j < 0$. The *number of sign-changes* in z is the number of edges in T for which a sign-change occurs in z.

Theorem 7. *Let T be a tree and suppose θ is a zero of $\varphi(T)$ such that no component of $z = z_T(\theta)$ is zero. Then the number of sign-changes in z is equal to the number of zeros of $\varphi(T)$ greater than θ.*

Proof. We begin by proving that the numbers of sign-changes in $z_T(\theta)$ equals the number of zeros of $\varphi(T \setminus 1)$ greater than θ. We will prove this by induction on $n = V(T)$. The result can be verified for $n = 2$ by direct calculation.

Assume $n > 2$. If G is a graph we denote by $\nu(G)$ the number of zeros of $\varphi(G)$ greater than θ. Choose a vertex in T adjacent to 1. We assume this vertex is 2. Let U be the component of $T \setminus 1$ containing 2 and set $U' = (T \setminus 1) \setminus U$.

Let y be the vector consisting of the components of z corresponding to the vertices in U. It is straightforward to show that $y = \varphi(U', x)z_U(\theta)$, whence, by our induction hypothesis, the number of sign-changes in y equals the number of zeros of $\varphi(U \setminus 2)$ greater than θ.

We next determine whether a sign-change occurs in z on the edge $\{1, 2\}$. Let z_i be the ith component of z. We have $\varphi(T \setminus 1) = \varphi(U')\varphi(U)$ and so:

$$\text{sign}(z_1) = (-1)^{\nu(T\setminus 1)} = (-1)^{\nu(U')+\nu(U)},$$

$$\text{sign}(z_2) = (-1)^{\nu(T\setminus 1)} = (-1)^{\nu(U')+\nu(U\setminus 2)}.$$

Hence, the sign of $z_1 z_2$ is $(-1)^{\nu(U)-\nu(U)-\nu(U\setminus 2)}$. The zeros of $U\setminus 2$ interlace those of U, therefore $\nu(U)-\nu(U\setminus 2)\in\{0,1\}$. In other words, $z_1 z_2 < 0$ if and only if $\nu(U)-\nu(U\setminus 2)=1$.

Thus, we have that the number of sign-changes in z on edges in U or on $\{1,2\}$ equals $\nu(U)$. Since the component U was chosen arbitrarily and since $\nu(T\setminus 1)$ is just the sum of the numbers $\nu(U)$, where U is a component of $T\setminus 1$, it follows that the number of sign-changes in z equals $\nu(T\setminus 1)$, as claimed.

To complete the proof we must show that $\nu(T\setminus 1)=\nu(T)$. We have assumed that no $z_i = 0$. Therefore $\varphi(T\setminus 1, \theta)\neq 0$, since this is just z_1. Accordingly, θ is not a zero of $\varphi(T\setminus 1)$. Since the zeros of $\varphi(T\setminus 1)$ interlace those of $\varphi(T)$, we conclude that $\nu(T\setminus 1)=\nu(T)$. □

When T is a path its adjacency matrix $A(T)$ can be assumed to be tridiagonal. In this case it is known that the polynomials $\varphi(T\setminus P_{1i}, x)$ form a Sturm sequence and the conclusion of Theorem 7 is a known property of such sequences.

The case when θ is the second largest zero of $\varphi(T,x)$ was first studied by A. Neumaier. He verified that Theorem 7 holds in this instance. (His proof was communicated privately to the author.) When θ is the largest zero of $\varphi(T)$, Theorem 7 is a consequence of the Perron–Frobenius theorem.

Finally, we discuss some further consequences of Theorem 6. We mentioned above that $A(F)$ is invertible if and only if F has a 1-factor. Now the subgraph of a graph G formed by taking the union of two distinct 1-factors contains at least one circuit. Hence, a forest has at most a single 1-factor. By Theorem 1 we see then that if $A(F)$ is invertible, then $\det(A)=\pm 1$.

In particular, the inverse of $A(F)$ will be an integral matrix. But applying our observation that $\det(A)^2 = 1$ to Theorem 6 yields the conclusion that the entries of $A(F)^{-1}$ lie in the set $\{0, 1, -1\}$. This fact was first pointed out in [2].

If we ignore the signs of the elements of $A(F)^{-1}$, then we may view the resulting non-negative matrix as the adjacency matrix of a graph, which we will denote by F^+. It is shown in [2] that F^+ contains F as a subgraph. However, we have been able to show that if F is a forest such that $A(F)$ is invertible, then there is a diagonal matrix Λ, with diagonal entries equal to ± 1, such that $\Lambda^{-1}A(F)^{-1}\Lambda$ is a non-negative matrix. The proof of this will appear in [3].

In other words, the matrices $A(F)^{-1}$ and $A(F^+)$ are similar. Hence, $\lambda_1(F^+)$ is the reciprocal of the smallest positive zero of $\varphi(F)$. Denote this eigenvalue by τ. Then since F is a subgraph of F^+ we have $\lambda_1(F)\leq 1/\tau$ or, conversely, $\tau\leq 1/\lambda_1(F)$. This provides the only known reasonable upper bound on τ. We have also been able to show (in [3]) that if P_n is the path on $n = |V(F)|$ vertices and $A(F)$ is invertible, then $\tau(F) \geqslant \tau(P_n)$. Again, this is the only known lower bound (better than zero) for $\tau(F)$.

References

[1] D.M. Cvetković, M. Doob and H. Sachs, Spectra of Graphs (Academic Press, New York, 1980).

[2] D.M. Cvetković, I. Gutman and S.K. Simić, On self pseudo-inverse graphs. Univ. Beograd, Publ. Elektrotehn. Fak. Ser. Math. Fiz. 602–633 (1978) 111–117.

[3] C.D. Godsil, Inverses of trees, Combinatorica, to appear.

[4] C.D. Godsil, Walk generating functions, Christoffel–Darboux identities and the adjacency matrix of a graph (in preparation).

[5] C.D. Godsil and I. Gutman, On the theory of the matching polynomial, J. Graph Theory 5 (1981) 137–144.

[6] O.J. Heilmann and E.H. Lieb, Theory of Monomer–Dimer systems, Commun. Math. Phys. 25 (1972) 190–232.

[7] L. Lovász, Combinatorial Problems and Exercises (North-Holland, Amsterdam, 1979).

[8] A.J. Schwenk, Almost all trees are cospectral, in: New Directions in the Theory of Graphs, Proc. Third Ann Arbor Conference, Michigan (Academic Press, New York, 1973).

Received March 1981

Annals of Discrete Mathematics 20 (1984) 161–163
North-Holland

THE VALENCE-FUNCTIONAL

Peter GRITZMANN
Universität Siegen, Siegen, Fed. Rep. Germany

1. Introduction

Properties involving embeddings and colourings of graphs on two-dimensional manifolds have been studied for a long time. There are, however, a number of interesting problems similar to those considered for convex polytopes, investigation of which started just a few years ago (compare [1]).

For example, a theorem of Steinitz shows that any map on the sphere which is 3-connected is isomorphic to the graph of all edges of some convex 3-polytope. If a map is drawn on some other two-dimensional manifold, very little is known about when a polyhedral structure isomorphic to it exists. It has in general not even been possible to determine the region of f-vectors of polyhedral realizations of such manifolds, i.e. the vectors of possible numbers of vertices, edges and facets.

In this paper we consider a special functional — the valence-functional — which turns out to be an applicable tool in connection with such problems. The upper and lower bounds established for this functional enable us to solve completely the problem concerning the f-vectors of polyhedral realizations of the torus.

2. Bounds for the valence-functional

Let g be a non-negative integer and let (M, g) denote the closed orientable two-dimensional manifold of genus g. *K-realizations* $\mathcal{P}$ of (M, g) are geometric cell-complexes in R^3 with convex facets such that set $(\mathcal{P})$ — the union of all faces of $\mathcal{P}$ — is homeomorphic to M.

For vertices v of $\mathcal{P}$ let $\mathrm{val}(v, \mathcal{P})$ denote the number of edges of $\mathcal{P}$ being incident with v. Then we have $\mathrm{val}(v, \mathcal{P}) \geq 3$ for every vertex v of $\mathcal{P}$ and we define the valence-value $v(\mathcal{P})$ of $\mathcal{P}$ to be the weighted vertex-sum $\Sigma(\mathrm{val}(v, \mathcal{P}) - 3)$ taken over all vertices of $\mathcal{P}$.

Since we want to obtain results which characterize the genus g rather than some special K-realizations of (M, g), it is natural to define the valence-functional to be the minimum of $v(\mathcal{P})$ taken over all K-realizations $\mathcal{P}$ of (M, g), i.e. $V(g) := \min[v(\mathcal{P}) \mid \mathcal{P}$ is a K-realization of $(M, g)]$. It is easily shown that the sphere is characterized by the property $V(g) = 0$, and one may ask if $V(g)$ is closely related to g in general, too.

The following theorem, proved in [3], gives an affirmative answer.

Theorem. *Let g be an integer with $g \geq 1$. Then* $2g + 1 \leq V(g) \leq 3g + 3$.

One has to be careful in interpreting this result. The theorem does not show that the number of vertices which are more than 3-valent increases with increasing g. Surprisingly, this is not even true. In fact, slight modification of an example given in [4] proves the existence of K-realizations of (M, g), all but six vertices of which are 3-valent.

For applications, the lower bounds are especially important because they exclude the existence of several types of K-realizations of (M, g) which are combinatorially possible.

In general the given lower bounds for $V(g)$ are not the best possible. By means of extensive geometrical studies it is shown in [4] that the lower bounds can be replaced by 6.

Theorem. *Let g be an integer with $g \geq 1$. Then* $6 \leq V(g)$.

This is an improvement for $g = 1, 2$ and yields the exact value of $V(1)$.

Corollary. $V(1) = 6$.

These inequalities allow several interesting applications. The valence-functional, which is in a certain sense a combinatorial translation of the geometrical fact that all facets of considered realizations of (M, g) are convex, is related to polyhedral colouring problems (compare [1]), to geometrical problems of the Eberhard-type, etc.

Now we want to sketch the relevance of the valence-functional for the problem of determining the f-vectors of k-realizations of (M, g) by giving the solution of this problem for the torus.

3. f-vectors of K-realizations of the torus

For a torus $\mathcal{T}$ let $f_0(\mathcal{T})$, $f_1(\mathcal{T})$, $f_2(\mathcal{T})$ denote the number of vertices, edges and facets of $\mathcal{T}$, respectively. It is easy to show that there are the following combinatorial restrictions for the f-vectors:

$$f_0(\mathcal{T}) - f_1(\mathcal{T}) + f_2(\mathcal{T}) = 0$$

$$f_2(\mathcal{T}) \geq \tfrac{1}{2} f_0(\mathcal{T}), \qquad f_0(\mathcal{T}) \geq \tfrac{1}{2} f_2(\mathcal{T})$$

$$f_2(\mathcal{T}) \geq \tfrac{1}{2} f_0(\mathcal{T})(11 - f_0(\mathcal{T})), \qquad f_0(\mathcal{T}) \geq \tfrac{1}{2} f_2(\mathcal{T})(11 - f_2(\mathcal{T})).$$

Let R be the set of all points with integer coordinates lying in the region determined by these relations. Which of the points of R occur as f-vectors of K-realizations of the torus? In the case of the sphere every combinatorially possible f-vector is, in the sense of k-realizations, geometrically possible, too.

The torus behaves quite differently. Since $f_2(\mathcal{T}) \geq \frac{1}{2} f_0(\mathcal{T}) + \frac{1}{2} V(1)$ we get the 'geometrical' restriction $f_2(\mathcal{T}) \geq \frac{1}{2} f_0(\mathcal{T}) + 3$.

Let R_K denote the points of R which fulfill this restriction as well. Then we have the following theorem stated in [2] (compare also [5]).

Theorem. (f_0, f_1, f_2) *is the* f*-vector of a* K*-realization of the torus if and only if* $(f_0, f_1, f_2) \in R_K$.

It is shown in [2] that the points of $R \setminus R_K$ are geometrically realizable as well if one weakens the assumptions and allows the facets to be non-convex polygons.

References

[1] D. Barnette, Polyhedral maps on 2-manifolds (to appear).

[2] P. Gritzmann, Der Bereich der f-Vektoren für polyedrische Realisierungen des Torus, Mitt. Math. Ges. Hamburg, XI, 3, to appear.

[3] P. Gritzmann, Upper and lower bounds of the valence functional, Israel J. Math. 43 (1982) 237–243.

[4] P. Gritzmann, Polyedrische Realisierungen geschlossener 2-dimensionaler Mannigfaltigkeiten im $\mathbb{R}^3$, Dissertation, Siegen (1980).

[5] J. Simutis, Geometric realisations of toroidal maps, PhD. Thesis, University of California, Davis (1977).

Received 9 June 1981

Annals of Discrete Mathematics 20 (1984) 165–175
North-Holland

MATROIDS WITH WEIGHTED BASES AND FEYNMAN INTEGRALS

P. HELL*

Departments of Computing Science and Mathematics, Simon Fraser University, Burnaby, British Columbia, Canada

E.R. SPEER**

Department of Mathematics, Rutgers University, New Brunswick, New Jersey, U.S.A.

We ask when it is possible to weight the bases B of a matroid E by non-negative real numbers w_B, so that the sum of w_B over all bases is 1 and, for each element $e \in E$, the sum of w_B over all bases B containing e is bounded below and above by specified real numbers λ_e, μ_e. In other words, given a matroid E with the rank function r and the set of bases $\mathcal{B}$, and given real vectors $(\lambda_e)_{e\in E}$ and $(\mu_e)_{e\in E}$, we wish to determine if the linear constraints

$$\begin{aligned} &\sum_{B\in\mathcal{B}} w_B = 1, \\ &\lambda_e \leqslant \sum_{B\ni e} w_B \leqslant \mu_e \quad \text{(for each } e \in E), \qquad (1)\\ &w_B \geqslant 0 \quad \text{(for each } B \in \mathcal{B}), \end{aligned}$$

are feasible. We shall find the following necessary and sufficient conditions for the feasibility of (1):

$$\begin{aligned} &\lambda_e \leqslant \mu_e \quad \text{(for each } e \in E), \\ &\sum_{e\in A} \lambda_e \leqslant r(A) \quad \text{(for each } A \subseteq E), \qquad (2)\\ &\sum_{e\notin A} \mu_e \geqslant r(E) - r(A) \quad \text{(for each } A \subseteq E). \end{aligned}$$

We found it interesting to compare two different proofs of this fact. One uses only the duality theorem of linear programming and the greedy algorithm for finding an optimal base; the other depends directly on some deeper results of matroid theory. The question was motivated by a problem of uniform estimation

* With support from the President's Research Grant, Simon Fraser University.
** Partially supported by the National Science Foundation, grant PHY-80-03298.

of certain Feynman integrals [10]; we shall describe briefly this application of our theorem. We shall also consider the special case $\lambda_e = \mu_e = \lambda$ (for each $e \in E$), especially in the context of graphic matroids, and make an observation relating edge-connectivity and edge-density.

Theorem. *The constraints* (1) *are feasible if and only if* (2) *holds.*

It is easy to see that the conditions (2) are necessary for the feasibility of (1). Indeed, if $(w_B)_{B\in\mathcal{B}}$ is a solution of (1), then clearly $\lambda_e \leqslant \mu_e$ for each $e \in E$, and for each $A \subseteq E$,

$$\sum_{e\in A} \lambda_e \leqslant \sum_{e\in A} \sum_{B\ni e} w_B = \sum_{B\in\mathcal{B}} |A \cap B| w_B \leqslant r(A) \sum_{B\in\mathcal{B}} w_B = r(A),$$

$$\sum_{e\notin A} \mu_e \geqslant \sum_{e\notin A} \sum_{B\ni e} w_B = \sum_{B\in\mathcal{B}} |B \setminus A| w_B = \sum_{B\in\mathcal{B}} (|B| - |A \cap B|) w_B$$

$$\geqslant r(E) - r(A).$$

Lemma 1. *The constraints* (1) *are feasible if and only if for every pair of vectors* $(x_e)_{e\in E}$ *and* $(y_e)_{e\in E}$, *with non-negative coordinates,*

$$\sum_{e\in E} (\lambda_e x_e - \mu_e y_e) \leqslant \max_{B\in\mathcal{B}} \sum_{e\in B} (x_e - y_e). \tag{3}$$

Proof. This can be easily seen using the Farkas Lemma [3] or its equivalents [7], and amounts to an application of the duality theorem of linear programming. For example, according to [4, Theorem 2.8], the constraints

$$\sum_{B\in\mathcal{B}} w_B \leqslant 1, \qquad \sum_{B\in\mathcal{S}} - w_B \leqslant -1,$$

$$\sum_{B\ni e} w_B \leqslant \mu_e, \qquad \sum_{B\ni e} - w_B \leqslant -\lambda_e \quad \text{(for each } e \in E\text{)},$$

$$w_B \geqslant 0 \qquad\qquad \text{(for each } B \in \mathcal{B}\text{)}$$

(equivalent to (1)) are feasible if and only if the constraints

$$\sum_{e\in B} (y_e - x_e) + v - t \geqslant 0 \quad \text{(for each } B \in \mathcal{B}\text{)},$$

$$\sum_{e\in E} (\mu_e y_e - \lambda_e x_e) + v - t < 0,$$

$$v \geqslant 0, \qquad t \geqslant 0,$$

$$x_e \geqslant 0, \qquad y_e \geqslant 0 \qquad \text{(for each } e \in E\text{)},$$

are not. Setting $u = v - t$, we obtain the condition that for any vectors $(x_e)_{e\in E}$ and $(y_e)_{e\in E}$, with non-negative coordinates,

$$\sum_{e\in B} (x_e - y_e) \leqslant u, \text{ for all } B \in \mathcal{B}, \text{ implies } \sum_{e\in E} (\lambda_e x_e - \mu_e y_e) \leqslant u,$$

which is equivalent to (3).

Given vectors $(x_e)_{e\in E}$ and $(y_e)_{e\in E}$, we write $z_e = x_e - y_e$. The quantity $\max_{B\in\mathcal{B}} \sum_{e\in B} z_e$, which occurs in (3), can be found by applying the greedy algorithm [8, p. 277], which for our purposes may be described as follows. Let $E = \{e_1, e_2, \ldots, e_m\}$ be an enumeration of the elements of E such that $z_i = z_{e_i}$ satisfy $z_1 \geqslant z_2 \geqslant \cdots \geqslant z_m$. Let $A_0 = \emptyset$, $A_k = \{e_1, e_2, \ldots, e_k\}$ for $1 \leqslant k \leqslant m$, and $B^* = \{e_k : 1 \leqslant k \leqslant m,\ r(A_k) = r(A_{k-1}) + 1\}$. Then B^* is a base of E and $\sum_{e\in B^*} z_e = \max_{B\in\mathcal{B}} \sum_{e\in B} z_e$. (In fact, we really need only the fact that B^* is a base.)

To complete the first proof of the theorem, it remains to show that (2) implies that for any vectors $(x_e)_{e\in E}$ and $(y_e)_{e\in E}$ of non-negative numbers,

$$\sum_{e\in E} (\lambda_e x_e - \mu_e y_e) \leqslant \sum_{e\in B^*} z_e.$$

Assume that $z_1 \geqslant \cdots \geqslant z_n \geqslant 0 \geqslant z_{n+1} \geqslant \cdots \geqslant z_m$ and write $\lambda_i = \lambda_{e_i}$, $\mu_i = \mu_{e_i}$. Then

$$\sum_{k=1}^{n} \left[(z_k - z_{k+1}) \sum_{i=1}^{k} \lambda_i \right] \leqslant \sum_{k=1}^{n} (z_k - z_{k+1}) r(A_k)$$

(where in this inequality only we set $z_{n+1} = 0$), whence

$$\sum_{k=1}^{n} \lambda_k z_k \leqslant \sum_{k=1}^{n} z_k (r(A_k) - r(A_{k-1})) = \sum_{e\in B^*\cap A_n} z_e.$$

Similarly,

$$\sum_{k=n+1}^{m} \left[(z_{k-1} - z_k) \sum_{i=k}^{m} \mu_i \right] \geqslant \sum_{k=n+1}^{m} (z_{k-1} - z_k)(r(E) - r(A_{k-1}))$$

(where again, in this inequality only, we set $z_n = 0$), and so

$$\sum_{k=n+1}^{m} \mu_k z_k \leqslant \sum_{k=n+1}^{m} z_k (r(A_k) - r(A_{k-1})) = \sum_{e\in B^*\setminus A_n} z_e.$$

Thus,

$$\begin{aligned}\sum_{e\in E} (\lambda_e x_e - \mu_e y_e) &= \sum_{k=1}^{n} [\lambda_k z_k + (\lambda_k - \mu_k) y_k] + \sum_{k=n+1}^{m} [\mu_k z_k + (\lambda_k - \mu_k) x_k] \\ &\leqslant \sum_{k=1}^{n} \lambda_k z_k + \sum_{k=n+1}^{m} \mu_k z_k \leqslant \sum_{e\in B^*} z_e,\end{aligned}$$

as required.

For the second proof of the theorem we observe that, without loss of generality, the vectors $(\lambda_e)_{e\in E}$ and $(\mu_e)_{e\in E}$ may be assumed to have rational coordinates. Indeed, if real vectors $(\lambda_e)_{e\in E}$ and $(\mu_e)_{e\in E}$ satisfy (2), then so do any rational vectors $(\lambda_e^i)_{e\in E}$ and $(\mu_e^i)_{e\in E}$, where $\lambda_e^i \leqslant \lambda_e$ and $\mu_e \leqslant \mu_e^i$ for each $e \in E$; hence, for any rational sequences $\{\lambda_e^i\}_{i=1}^{\infty}$ and $\{\mu_e^i\}_{i=1}^{\infty}$ (for each $e \in E$), with $\lambda_e^i \leqslant \lambda_e$, $\mu_e \leqslant \mu_e^i$, $\lim_{i\to\infty} \lambda_e^i = \lambda_e$ and $\lim_{i\to\infty} \mu_e^i = \mu_e$ (for each $e \in E$), the sets

$$S_i = \left\{ (w_B)_{B\in\mathscr{B}} : \sum_{B\in\mathscr{B}} w_B = 1,\ \lambda_e^i \leqslant \sum_{B\ni e} w_B \leqslant \mu_e^i (e \in E),\ w_B \geqslant 0\ (B \in \mathscr{B}) \right\}$$

are nonempty compact subsets of $\mathbb{R}^{\mathscr{B}}$. Therefore $\bigcap_{i=1}^{\infty} S_i \neq \emptyset$, and (1) is feasible.

Let k, l_e $(e \in E)$ and m_e $(e \in E)$ be integers such that, for all $e \in E$, $\lambda_e = l_e/k$ and $\mu_e = m_e/k$. It remains to show that

$$\begin{aligned} &l_e \leqslant m_e && \text{(for each } e \in E\text{)},\\ &\sum_{e\in A} l_e \leqslant kr(A) && \text{(for each } A \subseteq E\text{)}, \qquad (4)\\ &\sum_{e\notin A} m_e \geqslant k(r(E) - r(A)) && \text{(for each } A \subseteq E\text{)}, \end{aligned}$$

imply that

$$\begin{aligned} &\sum_{B\in\mathscr{B}} w_B = k, &&\\ &l_e \leqslant \sum_{B\ni e} w_B \leqslant m_e && \text{(for each } e \in E\text{)}, \qquad (5)\\ &w_B \geqslant 0 && \text{(for each } B \in \mathscr{B}\text{)}, \end{aligned}$$

is feasible.

Consider the sets $\tilde{E} = \{(e, i): e \in E, 1 \leqslant i \leqslant m_e\}$ and $\tilde{\mathscr{B}}$ defined to consist of all subsets X of $\tilde{E}$ satisfying:

(a) for each $e \in E$ there is at most one i such that $(e, i) \in X$, and

(b) $\{e: (e, i) \in X$ for some $i\}$ is a base of E.

It is easy to verify that $\tilde{\mathscr{B}}$ is the set of bases of a matroid on $\tilde{E}$; we shall refer to that matroid as $\tilde{E}$ and its rank function shall be denoted by $\tilde{r}$. Finally, let $T = \{(e, i): e \in E, 1 \leqslant i \leqslant l_e\}$.

Lemma 2. (5) *has an integer solution if and only if* $\tilde{E}$ *admits* k *disjoint bases whose union contains* T.

Proof. Let $X_1, X_2, \ldots, X_k$ be disjoint bases of $\tilde{E}$ whose union contains T. Let $B_j = \{e : (e, i) \in X_j$ for some $i\}$, $1 \leqslant j \leqslant k$; by (b) each B_j is a base of $\tilde{E}$, but the B_j's are not necessarily disjoint, or even distinct. Let, for $B \in \mathscr{B}$, w_B be the number of j's such that $B_j = B$. Then evidently $\sum_{B\in\mathcal{B}} w_B = k$; moreover, for any $e \in E$, $\sum_{B\ni e} w_B \leqslant m_e$ because the X_j's are disjoint; finally, for each $e \in E$, $\sum_{B\ni e} w_B \geqslant l_e$ because $\bigcup_{j=1}^{k} X_j$ contains T. Hence, w_B, $B \in \mathcal{B}$, are integers satisfying (5). Conversely, given integers w_B, $B \in \mathcal{B}$, satisfying (5), it is easy to construct k disjoint bases of $\tilde{E}$ whose union contains T.

Lemma 3. *There exist k disjoint bases of $\tilde{E}$ whose union contains T if and only if there exist k disjoint bases of $\tilde{E}$ and there exist k bases of $\tilde{E}$ whose union contains T.*

Proof. Lemma 3 holds for an arbitrary matroid $\tilde{E}$ and any subset T. Accordingly, in this proof no use shall be made of the specific form of $\tilde{E}$ and T. We first construct a new matroid $\tilde{E}^{(k)}$ on $\tilde{E}$, whose independent sets are precisely the sets that can be written as unions of k sets, independent in $\tilde{E}$. (In other words, $X \subseteq \tilde{E}$ is independent in $\tilde{E}^{(k)}$ if and only if $X = I_1 \cup I_2 \cup \cdots \cup I_k$, with each I_j independent in $\tilde{E}$.) It follows from [9] that $\tilde{E}^{(k)}$ is indeed a matroid. Assume that $\tilde{E}$ has k disjoint bases $X_1, \ldots, X_k$. Then $\bigcup_{i=1}^{k} X_i$ is an independent set in $\tilde{E}^{(k)}$ and hence the rank of $\tilde{E}^{(k)}$ is $k\tilde{r}(\tilde{E})$. In other words, every base of $\tilde{E}^{(k)}$ is the union of k disjoint bases of $\tilde{E}$. If at the same time $\tilde{E}$ admits k bases $Y_1, Y_2, \ldots, Y_k$ such that $T \subseteq \bigcup_{i=1}^{k} Y_i$, then T is also an independent set in $\tilde{E}^{(k)}$. Therefore T can be extended to a base of $\tilde{E}^{(k)}$; i.e. some k disjoint bases of $\tilde{E}$ contain T in their union. The converse is obvious.

Our second proof is now completed by an application of the covering and packing theorems of Edmonds [2]. They imply that $\tilde{E}$ has k disjoint bases if and only if $|\tilde{E} \setminus A| \geqslant k(\tilde{r}(\tilde{E}) - \tilde{r}(A))$ for all $A \subseteq \tilde{E}$; this is easily seen to be equivalent to $\sum_{e\notin A} m_e \geqslant k(r(E) - r(A))$ for all $A \subseteq E$. They also imply that T can be covered by k bases (or, equivalently, independent sets) of $\tilde{E}$ if and only if $|A| \leqslant k\tilde{r}(A)$ for all $A \subseteq \tilde{E}$; this is easily seen to be equivalent to $\sum_{e\in A} l_e \leqslant kr(A)$ for all $A \subseteq E$. Hence, conditions (4) are, by Lemmas 2 and 3, sufficient to ensure that (5) admits an integer solution.

Corollary 1. *There is an integer solution to* (5) *if and only if* (4) *holds.*

Proof. The necessity of (4) follows from the Theorem. Letting $\lambda_e = l_e/k$ and $\mu_e = m_e/k$, feasibility of (5) implies feasibility of (1) and thus (2) holds, which is equivalent to (4).

There is an obvious interpretation of an integer solution to (5) as a set of k bases (with possible repetition) with the property that each $e \in E$ belongs to at least l_e and at most m_e of these bases.

Corollary 2.

(a) *The constraints*

$$\sum_{B\in\mathcal{B}} w_B = 1,$$

$$\sum_{B\ni e} w_B \geqslant \lambda_e \quad \text{(for each } e\in E\text{)},$$

$$w_B \geqslant 0 \quad \text{(for each } B\in\mathcal{B}\text{)},$$

are feasible if and only if

$$\sum_{e\in A} \lambda_e \leqslant r(A) \quad \text{(for each } A\subseteq E\text{)}.$$

(b) *The constraints*

$$\sum_{B\in\mathcal{B}} w_B = 1,$$

$$\sum_{B\ni e} w_B \leqslant \mu_e \quad \text{(for each } e\in E\text{)},$$

$$w_B \geqslant 0 \quad \text{(for each } B\in\mathcal{B}\text{)},$$

are feasible if and only if

$$\sum_{e\notin A} \mu_e \geqslant r(E)-r(A) \quad \text{(for each } A\subseteq E\text{)}.$$

Proof. These results follow from our theorem by choosing, in (a), $\mu_e = 1$ for all $e\in E$, and in (b), $\lambda_e = 0$ for all $e\in E$. We remark that (a) and (b) are related by duality, that is, (a) for E is equivalent to (b) for the matroid dual to E, and vice versa. We also remark that similar one-sided integral versions of (5) may be deduced from Corollary 1 by taking $m_e = k$ (respectively $l_e = 0$) for all $e\in E$.

Corollary 2(b) for graphic matroids has been applied in [10] to estimate the magnitude of the *Feynman amplitudes* which are associated with (Feynman) graphs in perturbative quantum field theory. The essential idea may be described in the context of an arbitrary matroid as follows. To the matroid E and positive real numbers a and $(b_e)_{e\in E}$ we associate the amplitude

$$I_E(a,b) = \int_0^\infty \cdots \int_0^\infty U(\alpha)^{-a} \prod_{e\in E} \{\Gamma(b_e)^{-1}\alpha_e^{b_e-1}\exp(-\alpha_e)\mathrm{d}\alpha_e\}, \tag{6}$$

where

$$U(\alpha) = \sum_{B\in\mathcal{B}} \prod_{e\in B} \alpha_e \quad\text{and}\quad \Gamma(x) = \int_0^\infty y^{x-1}\exp(-y)\mathrm{d}y \qquad (x>0).$$

Note that $I_E(0,b)\equiv 1$; on the other hand, if (b_e) is fixed and a is sufficiently large, the integral diverges. In fact, the condition for convergence is known to be

$$c(A) \equiv \sum_{e \in A} b_e - a[r(E) - r(E \setminus A)] > 0, \tag{7}$$

for each $A \subseteq E$.

The sufficiency of (7) for convergence follows from the argument below. We briefly sketch a standard argument (see, for example, [11]) showing necessity also. Consider the integral (6) taken over a sector $\alpha_{e_1} \leqslant \cdots \leqslant \alpha_{e_m}$ and make the variable change $\alpha_{e_k} = \prod_{j=k}^{m} t_j$, $k = 1, \ldots, m$, with the Jacobian $\prod_{j=1}^{m} t_j^{j-1}$. Let $A_j = \{e_1, \ldots, e_j\}$ and

$$B^* = \{e_k : 1 \leqslant k \leqslant m,\ r(A_k) = r(A_{k-1}) + 1\},$$

as in the first proof of the theorem. Now

$$U(\alpha(t)) = \sum_{B \in \mathscr{B}} \prod_{j=1}^{m} t_j^{|A_j \cap B|} = \prod_{j=1}^{m} t_j^{[r(E) - r(E \setminus A_j)]} V(t),$$

where, since $|A_j \cap B| \geqslant r(E) - r(E \setminus A_j)$, $V(t)$ is a polynomial. Moreover,

$$U(\alpha(t)) \geqslant \prod_{e \in B^*} \alpha_e(t) = \prod_{j=1}^{m} t_j^{[r(E) - r(E \setminus A_j)]},$$

so $V(t) \geqslant 1$, and, since $V(t)$ is independent of t_m, it is uniformly bounded above on the compact set $\{0 \leqslant t_j \leqslant 1 : j = 1, \ldots, m-1\}$. Thus, the integral over this sector becomes

$$\prod_{e \in E} \Gamma(b_e)^{-1} \int_0^\infty \mathrm{d}t_m \int_0^1 \mathrm{d}t_{m-1} \cdots \int_0^1 \mathrm{d}t_1 \prod_{j=1}^{m} t_j^{c(A_j) - 1} V(t)^{-a} \exp\left[- \sum_{e \in E} \alpha_e(t) \right], \tag{8}$$

and the upper and lower bounds on V imply that (8) converges if and only if $c(A_j) > 0$, $j = 1, \ldots, m$.

The problem in [10] is to give an estimate for I_E of the special form

$$I_E(a, b) \leqslant K^{|E|}, \tag{9}$$

where the constant K does not depend on E. Such a uniform estimate will not of course hold in the entire convergence region specified by (7), but only on compact subregions; here we specify such a subregion by a parameter $\varepsilon > 0$ and by the condition

$$\sum_{e \in A} b_e - a[r(E) - r(E \setminus A)] \geqslant \varepsilon |A| \quad (\text{for all } A \subseteq E) \tag{10}$$

(the specification in [10] is slightly different). Then we have

Corollary 3. *For all matroids E and non-negative parameters a, $(b_e)_{e \in E}$ satisfying* (10), *there is a constant K depending only on ε such that* (9) *holds.*

Proof. By Corollary 2(b) there are non-negative weights $(w_B)_{B\in\mathscr{B}}$ with $\sum_{\mathscr{B}} w_B = 1$ and

$$\sum_{B\ni e} w_B \leqslant a^{-1}(b_e - \varepsilon).$$

From the standard inequality between arithmetic and geometric means,

$$U(\alpha) \geqslant \sum_{B\in\mathscr{B}} w_B \prod_{e\in B} \alpha_e$$

$$\geqslant \prod_{B\in\mathscr{B}} \left(\prod_{e\in B} \alpha_e\right)^{w_B} = \prod_{e\in E} \alpha_e^{(\sum_{B\ni e} w_B)}.$$

Thus,

$$I_\varepsilon(a,b) \leqslant \prod_{e\in E} \Gamma(b_e)^{-1} \int_0^\infty \alpha_e^{d_e-1} \exp(-\alpha_e)\,\mathrm{d}\alpha_e = \prod_{e\in E} \Gamma(d_e)/\Gamma(b_e),$$

where $d_e = b_e - a\sum_{B\ni e} w_B$ satisfies $b_e \geqslant d_e \geqslant \varepsilon$. Since for $t>0$, $\Gamma(t)$ is convex with a minimum at $t=1$, we have:

$$\Gamma(d_e)/\Gamma(b_e) \leqslant \begin{cases} \Gamma(\varepsilon)/\Gamma(1) = \Gamma(\varepsilon), & \text{if } d_e \leqslant 1, \\ 1 \leqslant \Gamma(\varepsilon), & \text{if } d_e \geqslant 1. \end{cases}$$

Thus (9) holds with $K = \Gamma(\varepsilon)$.

Next we explore the Theorem in the special case $\lambda_e = \mu_e$, $e \in E$. It implies that the system of linear equations

$$\sum_{B\in\mathscr{B}} w_B = 1,$$

$$\sum_{B\ni e} w_B = \lambda_e \quad \text{(for each } e \in E\text{)},$$

admits a non-negative solution w_B, $B \in \mathscr{B}$, if and only if

$$r(E) - r(E\setminus A) \leqslant \sum_{e\in A} \lambda_e \leqslant r(A) \quad \text{(for each } A \subseteq E\text{)}. \tag{11}$$

Note that for $A = E$, (11) implies that $\sum_{e\in E} \lambda_e = r(E)$. In particular, when each $\lambda_e = \mu_e = \lambda$ $(e \in E)$, (11) implies that $\lambda = r(E)/|E|$. In fact, if $\sum_{B\in\mathscr{B}} w_B = x$ and $\sum_{B\ni e} w_B = y$ (for each $e \in E$), then either one of x, y determines the other, since

$$y|E| = \sum_{e\in E}\sum_{B\ni e} w_B = \sum_{B\in\mathscr{B}}\sum_{e\in B} w_B = r(E)\sum_{B\in\mathscr{B}} w_B = xr(E).$$

Hence, if $y = r(E)/|E|$, then $x = 1$, and the feasibility of (1) with all $\lambda_e = \mu_e = r(E)/|E|$ $(e \in E)$, is equivalent to the feasibility of

$$\sum_{B \ni e} w_B = \frac{r(E)}{|E|} \quad \text{(for each } e \in E\text{)},$$

$$w_B \geq 0 \qquad \text{(for each } B \in \mathcal{B}\text{)},$$

or of

$$\sum_{B \ni e} w_B = 1 \qquad \text{(for each } e \in E\text{)}, \tag{12}$$

$$w_B \geq 0 \qquad \text{(for each } B \in \mathcal{B}\text{)}.$$

The system (12) is particularly interesting. We are trying to weight the bases of E by non-negative reals so that the sum of weights of all bases containing each fixed element e is 1. Matroids in which this is possible, i.e. in which (12) is feasible, shall be called *baseable.*

Corollary 4. *A matroid E is baseable if and only if any of the following conditions holds*:

(a) $\dfrac{|A|}{r(A)} \leq \dfrac{|E|}{r(E)}$ *(for each* $A \subseteq E$*),*

(b) $\dfrac{|K|}{r(K)} \geq \dfrac{|E|}{r(E)}$ *(for each contraction K of E),*

(c) *There exist bases* $B_1, B_2, \ldots, B_{|E|}$ *of E (not necessarily distinct), such that each* $e \in E$ *belongs to precisely* $r(E)$ *bases among* $B_1, B_2, \ldots, B_{|E|}$.

(d) (12) *is solvable by integer multiples of* $1/r(E)$.

Proof. The feasibility of (12) is equivalent to (11) with each $\lambda_e = r(E)/|E|$, i.e. to

$$r(E) - r(E \setminus A) \leq \frac{r(E)}{|E|}|A| \leq r(A) \quad \text{(for each } A \subseteq E\text{)}.$$

Since a contraction K of E has $r(K) = r(E) - r(E \setminus K)$ [13, p. 63], E is baseable if and only if (a) and (b) hold. On the other hand, (a) and (b) are algebraically equivalent, so either is equivalent to the baseability of E. Condition (c) is equivalent to the integer feasibility of

$$\sum_{B \ni \mathcal{B}} w_B = |E|,$$

$$\sum_{B \ni e} w_B = r(E) \quad \text{(for each } e \in E\text{)}, \tag{13}$$

$$w_B \geq 0 \qquad \text{(for each } B \in \mathcal{B}\text{)}.$$

(Given $B_1, B_2, \ldots, B_{|E|}$, let w_B equal the number of times B occurs among $B_1, B_2, \ldots, B_{|E|}$; given w_B, take each B w_B times to form $B_1, B_2, \ldots, B_{|E|}$.) Since we have already observed that the first equation in (13) is redundant, (c) and (d) are equivalent. Moreover, according to Corollary 1, (13) has an integer solution if and only if (a) and (b) hold.

We define a graph to be *arboreal* if its cycle matroid is baseable. The *density* of a connected graph $G = (V, E)$ is the ratio $|E|/(|V| - 1)$. Using condition (a) of Corollary 4 and the fact that $(|E_1| + |E_2|)/((|V_1| - C_1) + (|V_2| - C_2))$ lies between $|E_1|/(|V_1| - C_1)$ and $|E_2|/(|V_2| - C_2)$, we can see that G is arboreal if and only if all components of G are arboreal and have the same density. Therefore we shall confine our remarks to connected arboreal graphs. There is the obvious interpretation of Corollary 4, where 'base' is replaced by 'spanning tree', 'matroid E' by 'connected graph $G = (V, E)$', and 'baseable' by 'arboreal'. It is easy to see that (a) and (b) can be weakened to

(a′) *No connected subgraph of G has greater density than G;*
(b′) *No contraction of G has smaller density than G.*

Corollary 4(a) has led us to wonder when can (12) be solved by a constant vector $(w_B)_{B \in \mathscr{B}}$ (i.e. when can all w_B be chosen the same)? In [12] we asked for connected graphs having the property that each edge lies in the same number of spanning trees. Mendelsohn suggested that such graphs be called *equiarboreal.* Equiarboreal graphs are necessarily arboreal. Obviously trees and edge-transitive graphs are equiarboreal. Also, equiarboreal graphs with the same density can be attached together at a vertex to form another equiarboreal graph. Godsil [5] has shown that any colour class in an association scheme is equiarboreal. This means that any distance regular graph [1], and in particular any strongly regular graph, is equiarboreal. Even more generally, he defined a graph to be *1-homogeneous* if, for every l, the number of closed walks of length l through a vertex v is the same for every v, and the number of closed walks of length l through an edge vw is the same for every vw. Godsil gave an elegant proof showing that 1-homogeneous graphs are equiarboreal, and observed that colour classes of association schemes are 1-homogeneous, and that 1-homogeneous graphs are closed under conjunction. Thus there is a wide supply of equiarboreal graphs.

Godsil also noticed that the edge-connectivity of a connected equiarboreal graph is at least as large as its density [5], and was led to a number of interesting results and conjectures about the edge-connectivity of highly regular graphs [5, 6]. He later extended his observation on edge-connectivity to all regular arboreal graphs [6]. In our final corollary, we show how Corollary 4(c) can be used to extend Godsil's proof to all arboreal graphs:

Corollary 5. *The edge-connectivity of a connected arboreal graph is at least as large as its density.*

Proof. By Corollary 4(c), a connected arboreal graph $G=(V,E)$ admits spanning trees $T_1, T_2, \ldots, T_{|E|}$ with the property that each edge of G belongs to precisely $|V|-1$ of them. Consider a set of c edges whose removal disconnects G. Each spanning tree of G must contain one of the c edges, and hence

$$|E| \leqslant c(|V|-1).$$

Therefore the edge-connectivity of G is at least $|E|/(|V|-1)$.

It may also be fruitful to study 'equibaseable' matroids, i.e. matroids in which each element belongs to the same number of bases.

References

[1] N. Biggs, Algebraic Graph Theory (Cambridge University Press, 1974).
[2] J. Edmonds, Lehman's switching game and a theorem of Tutte and Nash-Williams, J. Res. Nat. Bur. Stand. 69B (1965) 73–77.
[3] J. Farkas, Über die Theorie der einfachen Ungleichungen, J. Reine Angew. Math. 124 (1902) 1–24.
[4] D. Gale, The Theory of Linear Economic Models (McGraw-Hill, New York, 1958).
[5] C.D. Godsil, Equiarboreal graphs, Combinatorica 1 (1981) 163–167.
[6] C.D. Godsil, personal communciation.
[7] H.W. Kuhn, Solvability and consistency for linear equations and inequalities, Amer. Math. Monthly 63 (1956) 217–232.
[8] E.L. Lawler, Combinatorial Optimization (Holt, Rinehart and Winston, 1976).
[9] C. St. J.A. Nash-Williams, An application of matroids to graph theory, in: Theory of Graphs, International Symposium Rome (Dunod, Paris, 1966) pp. 263–265.
[10] V. Rivasseau and E. Speer, The Borel transform in Euclidean φ^4_ν. Local existence for Re $\nu < 4$, Commun. Math. Phys. 72 (1980) 293–302.
[11] E.R. Speer, Generalized Feynman Amplitudes (Princeton University Press, 1969).
[12] Vancouver Problem Book, N.S.E.R.C. Summer Research Workshop in Combinatorics, Simon Fraser University (1979).
[13] D.J.A. Welsh, Matroid Theory (Academic Press, 1976).

Received 31 May 1981

Annals of Discrete Mathematics 20 (1984) 177–182
North-Holland

BALANCED PAIRS

Franz HERING
University of Dortmund, Dortmund, Fed. Rep. Germany

1. Introduction

In block designs with correlated observations it is sometimes useful and possible to balance not only the assignment of treatment units to the blocks, but also to balance the correlation structure of the design. That is to say, we assume that the blocks carry the structure of a graph in such a way that two observations are correlated if and only if they are observed on plots of the same blocks which are neighbours on that graph. Then we want to balance this neighbourly-relation as follows:

(i) *Every two-set of different treatments appears equally often as an edge in the graphs of the blocks.*

(ii) *Any two different block graphs have the same number of common edges.*

Some balanced incomplete block designs, in which on each block a graph with the above balancing properties is defined, have been investigated in [3]. The present paper is a continuation of [3]. It contains new results concerning the small cases, obtained by a computer as well as a new infinite series of such designs.

2. Cyclically balanced pairs of small order

Definition 1. Let M denote a set of m elements and W denote a set of m pairwise conjugate permutations on M. Then (M, W) is a *balanced pair* if it has the following properties:

(i) *For every two different permutations $\pi, \sigma \in W$ there are exactly two different elements $i, j \in M$ such that*

$$(j = \pi(i) \text{ or } i = \pi(j)) \quad \textit{and} \quad (j = \sigma(i) \text{ or } i = \sigma(j)). \qquad *$$

(ii) *For every two different elements $i, j \in M$ there are exactly two different permutations π, σ such that $(*)$ holds.*

Then m is the *order* of (M, W).

If (M, W) is a balanced pair, then from Proposition 1, p. 203 of [3] or by a simple counting argument, it follows that every $\pi \in W$ fixes exactly one different $i \in M$. We write π_i for the $\pi \in W$ fixing $i \in M$, so $(M, W) = (M, \{\pi_i : i \in M\})$.

Definition 2. A balanced pair $(M, \{\pi_i : i \in M\})$ is *cyclical* if π_i acts cyclically on $M \setminus \{i\}$ for every $i \in M$.

Suppose $(M, \{\pi_i : i \in M\})$ is a balanced pair of order m. We regard the trivial symmetric balanced incomplete block design with parameters $(m, m-1, m-2)$ having as treatments the elements of M and as blocks the sets $M \setminus \{i\} : i \in M$. On $M \setminus \{i\}$ we define a graph as follows: $u, v \in M$ are joined by an edge, if and only if $\pi_i(u) = v$ or $\pi_i(v) = u$. Then, by definition, this set of block graphs has the balancing properties listed in the introduction.

In this section we present a survey of cyclically balanced pairs of small order. The results have been mainly obtained by a computer program.

An *automorphism* of a balanced pair (M, W) is a pair of permutations (α, ω), α acting on M, ω acting on W, such that

$$j = \pi(i) \quad \text{or} \quad i = \pi(j)$$

is equivalent to

$$\alpha(j) = (\omega \circ \pi)\alpha(i) \quad \text{or} \quad \alpha(i) = (\omega \circ \pi)\alpha(j).$$

It is easily seen that α and ω are conjugate and by relabelling the permutations we may assume $\omega = \alpha$. Then α is an *automorphism* of (M, W) if $(\alpha, \omega) = (\alpha, \alpha)$ is one.

There is almost nothing known without additional assumptions on the automorphism structure. Only one cyclically balanced pair with six blocks is known, having the trivial automorphism only. It is the following one:

$$\begin{aligned}
\pi_0 &= (0)\,(1\ 2\ 3\ 4\ 5),\\
\pi_1 &= (1)\,(0\ 2\ 4\ 3\ 5),\\
\pi_2 &= (2)\,(0\ 1\ 4\ 5\ 3),\\
\pi_3 &= (3)\,(0\ 2\ 5\ 1\ 4),\\
\pi_4 &= (4)\,(0\ 1\ 3\ 2\ 5),\\
\pi_5 &= (5)\,(0\ 3\ 1\ 2\ 4).
\end{aligned}$$

This example is due to L. Danzer (private communication).

An interesting case is the assumption that there exists an automorphism α such that α fixes exactly one block and acts cyclically on the remaining ones. In this case all the cyclically balanced pairs (M, W) of order $m \leqslant 14$ up to isomorphism have been obtained by a computer program. Moreover, the program shows the existence of at least one such balanced pair of order $m \leqslant 36$.

We give an example: $M = \{\infty, 0, 1, 2, 3, 4, 5\}$:

$$\pi_\infty = (\infty)(0\ 1\ 2\ 3\ 4\ 5),$$

$$\pi_i = (i)(\infty\ i+1\ i+4\ i+2\ i+3\ i+5),$$

$$0 \leqslant i \leqslant 5, \quad \text{with addition modulo 5.}$$

Table 1 gives the number $c(m)$ of nonisomorphic cyclically balanced pairs (M, W) of order m, having an automorphism α such that α fixes one block and acts cyclically on the remaining ones.

Table 1

m	4	5	6	7	8	9	10	11	12	13
$c(m)$	1	1	1	1	7	12	31	76	377	1064

Moreover, as already mentioned,

$$c(m) \geqslant 1, \quad \text{for } 14 \leqslant m \leqslant 36.$$

It is somewhat embarrassing to observe that on the one hand the number of nonisomorphic cyclically balanced pairs seems to grow rapidly with m, whereas we are unable to prove that for every m there exists at least *one*. However, as we have shown in [3, Proposition 9, p. 213] we have the following result: When m is a power of a prime, then $c(m) \geqslant 1$.

There sometimes exist cyclically balanced pairs (M, W) having an automorphism α acting cyclically on M and on W. An example is obtained by $M = \{0\ 1 \cdots 6\}$, $\alpha = (0\ 1 \cdots 6)$:

$$\pi_0 = (0) \quad (1\ 3\ 2\ 6\ 4\ 5), \qquad \pi_i = \alpha^i \pi \alpha^{-i}, \qquad i = 0, 1, \ldots, 6,$$

that is

$$\pi_i = (i) \quad (1+i\ \ 3+i\ \ 2+i\ \ 6+i\ \ 4+i\ \ 5+i),$$

where addition is modulo 7.

If $d(m)$ denotes the number of nonisomorphic cyclically balanced pairs of order m having a cyclic automorphism, then a computer program gives the list shown in Table 2 (and also π_0 in each case explicitly).

Table 2

m	4	5	6	7	8	9	10	11	12	13	14
$d(m)$	1	1	0	2	0	0	0	13	110	79	0

We have no conjecture on $d(m)$. The only general result is $d(p) > 0$ for every prime $p > 3$, as we shall show in the Theorem of Section 3.

3. An existence theorem

Theorem. *For every prime $p > 3$ there exists a balanced pair having an automorphism of order p.*

Proof. Suppose $a \in \mathrm{GF}(p)$ is primitive and define permutations π and σ on $\mathrm{GF}(p)$ by

$$\pi(x) = ax + 1; \qquad \sigma(x) = x + 1.$$

We show that with

$$M = \{0, 1, \ldots, p-1\}$$

and

$$W = \{\sigma^{-u}\pi\sigma^{u} : u = 0, 1, \ldots, p-1\},$$

(M, W) is a balanced pair.

To prove property (i) of Definition 1, let $u, v \in M$, $u \neq v$. We first remark that the equations

$$j = \sigma^{-u}\pi\sigma^{u}(i) \quad \text{and} \quad j = \sigma^{-v}\pi\sigma^{v}(i)$$

have no solution $i, j \in M$. Indeed, they are equivalent to

$$j = a(i+u) + 1 - u \quad \text{and} \quad j = a(i+v) + 1 - v.$$

It follows that $(a-1)(v-u) = 0$. That a is primitive implies $a \neq 1$, so that we have reached the contradiction $u = v$. On the other hand, the equations

$$j = \sigma^{-u}\pi\sigma^{u}(i) \quad \text{and} \quad i = \sigma^{-v}\pi\sigma^{v}(j)$$

have the unique solution

$$i = \frac{a(u-v)}{a+1} - u - \frac{1}{a-1} \quad \text{and} \quad j = -\frac{a(u-v)}{a+1} - v - \frac{1}{a-1},$$

and here $u \neq v$ implies $i \neq j$. This proves (i) in Definition 1. To prove property (ii) we only have to remark that the last two equations have, for given $i, j \in M$, exactly one solution u, v and that $i \neq j$ implies $u \neq v$. Obviously σ is an automorphism of (M, W). □

An interesting special case is obtained when $a = (p-1)/2$ in the Theorem, which leads to a geometrical representation.

Lemma. *Let p be a prime and $(p-1)/2$ is a primitive element of $GF(p)$ if and only if one of the following two conditions hold:*

(i) $p \equiv 5 \bmod 8$ *and* 2 *is primitive in* $GF(p)$, *or*

(ii) $p \equiv 7 \bmod 8$ *and* $2^{(p-1)/2} = 1$, $2^i \neq 1$ *for* $i = 1, \ldots, (p-3)/2$.

We call the property of 2 in (ii) *half-primitive.*

Proof. From $(p-1)^i = (-1)^i$ follows the equivalence of $((p-1)/2)^i = 1$ and $(-2)^i = 1$, that is $(p-1)/2$ is primitive if and only if -2 is primitive. Let $Q = \{x^2 : 0 = x \in \mathrm{GF}(p)\}$. Q is the multiplicative subgroup of quadratic residues. Then (compare [2, Theorem 82, p. 64 and Theorem 95, p. 75]):

$$2 \in Q, \quad \text{if and only if } p \equiv 1 \bmod 8 \text{ or } p \equiv 7 \bmod 8,$$

$$-1 \in Q, \quad \text{if and only if } p \equiv 1 \bmod 8 \text{ or } p \equiv 5 \bmod 8.$$

This implies:

$$-2 \in Q, \quad \text{if and only if } p \equiv 1 \bmod 8 \text{ or } p \equiv 3 \bmod 8.$$

Now, if $a \in Q$, then a is not primitive, for $x^2 = a$ implies $a^{(p-1)/2} = x^{p-1} = 1$. Moreover, if $p \equiv 5 \bmod 8$, then -2 is primitive if and only if 2 is primitive, and if $p \equiv 7 \bmod 8$, then -2 is primitive if and only if 2 is half-primitive. □

Corollary. *The permutation* $\pi : \pi(x) = ((p-1)/2)x + 1$ *on* $GF(p)$ *is balanced if and only if the conditions* (i) *or* (ii) *of the Lemma hold.*

The permutation of the corollary has a geometric representation, as exhibited in Figs. 1 and 2.

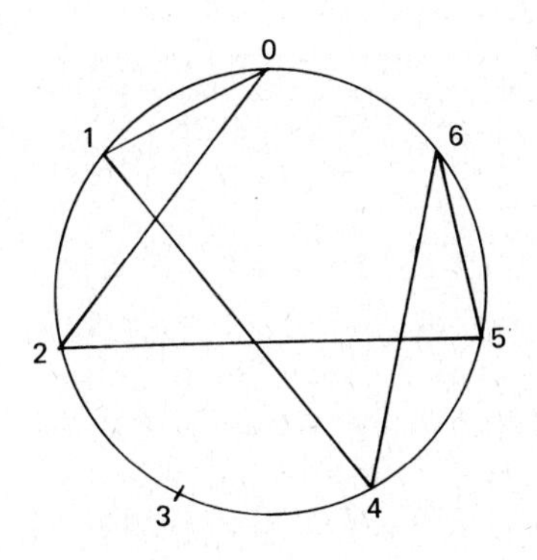

Fig. 1. $p = 7 \equiv 1 \bmod 8$
2 is half-primitive

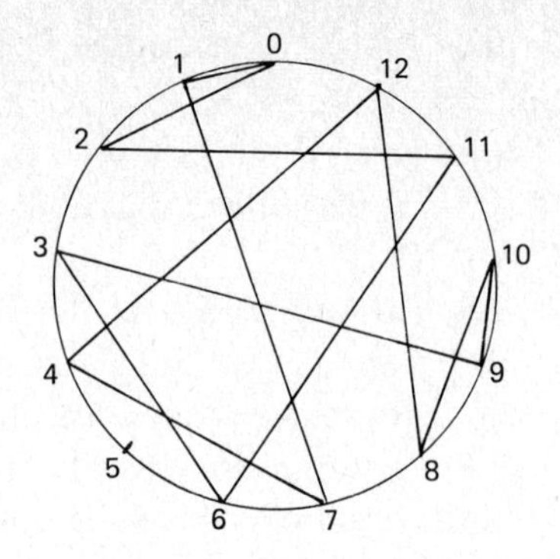

Fig. 2. $p = 13 \equiv 5 \bmod 8$
2 is primitive

4. Concluding remarks

If we omit condition (ii) in Definition 1, then the so obtained structures are closely related to solutions of the so-called Oberwolfach Problem which was posed by G. Ringel and investigated by Köhler [4].

If, instead, condition (∗) in Definition 1(i) is replaced by $j = \pi(i)$ and $i = \sigma(j)$, then the so obtained problem can be reformulated for directed graphs. The so obtained problem has been studied by Alspach, Heinrich and Rosenfeld in [1].

References

[1] B. Alspach, K. Heinrich and M. Rosenfeld, Edge partitions of the complete symmetry directed graph and related designs, Israel J. Math. (to appear).
[2] G.H. Hardy and E.M. Wright, An Introduction to the Theory of Numbers, 4th edn. (Oxford, 1960).
[3] F. Hering, Block designs with cyclic block structure, Ann. Discr. Math. 6 (1980) 201–214.
[4] E. Köhler, Das Oberwolfacher Problem, Mitt. Math. Ges. Hamburg 10 (1973) 124–129.

Received March 1981

Annals of Discrete Mathematics 20 (1984) 183–188
North-Holland

ON AUTOMORPHISMS OF GRAPHS WITH FORBIDDEN SUBGRAPHS

Wilfried IMRICH*

Institute of Mathematics and Applied Geometry, Montanuniversität Leoben, A-8700 Leoben, Austria

This paper investigates the cycle structure, the number of fixed points, and the number of fixed ends in m-connected graphs without subgraphs contractible to $K_{m,n}$.

1

The structure of automorphisms of finite, three-connected planar graphs is well known (see, for example, Fleischner [3]). In particular, every orientation-preserving automorphism has at most two fixed points and all other points are in cycles of the same length. All orientation-reversing automorphisms, on the other hand, are of even order $2k$. They have no fixed points and at most one two-cycle for $k > 1$. Moreover, all other points are in cycles of lengths $2k$ for even $k \geqslant 2$ and in cycles of lengths $2k$ or k for odd $k \geqslant 3$.

The proof of these results in [3] is also valid for automorphisms of infinite, three-connected planar graphs that contain a cycle of finite length. Thus, automorphisms of infinite order in infinite, three-connected planar graphs have cycles of infinite lengths and the problem arises as to how many cycles of finite length such automorphisms admit. This was investigated by Seifter, who derived the following more general result:

Theorem 1 (Seifter [7]). *Let X be an m-connected graph and n a fixed integer. Suppose X contains no subdivision of the complete bipartite graph $K_{m,n}$ and let α be an automorphism of X with a cycle of infinite length. Then at most $m-1$ vertices of X are in cycles of α with finite lengths.*

By Kuratowski's Theorem this implies that an automorphism with a cycle of infinite length in a three-connected planar graph has at most two vertices in cycles of finite lengths.

* The author wishes to thank L. Babai, C. Godsil and N. Seifter for interesting discussions.

Interestingly, the proof of Theorem 1 is of a purely combinatorial nature, whereas the aforementioned result makes use of topological properties of graphs embedded into the plane.

For finite graphs a similar result is due to Babai:

Theorem 2 (Babai [1]). *Let X be a three-connected graph which has no topological $K_{3,p}$ subgraphs, where p is an odd prime divisor of the order of an automorphism α of X. Then α fixes at most two vertices of X.*

We note that the statements 'X has no topological $K_{m,n}$ subgraphs' and 'X has no subgraphs contractible to $K_{m,n}$' are equivalent for $m = 3$. For $m > 3$ this is not the case.

Theorem 3. *Let X be an m-connected graph and n a fixed integer. If no subgraph of X is contractible to $K_{m,n}$, then any automorphism α of X of prime order $p \geqslant n$ has at most $m-1$ fixed points.*

We remark that Theorem 3 is the best possible since $K_{m,n}$ admits an automorphism of order n with m fixed points.

First proof. Suppose α satisfies the assumptions of Theorem 3 but has at least m fixed points $a_1, a_2, \ldots, a_m$. Let b be any other vertex of X not fixed by α. Since X is m-connected there exists a subtree of X containing b whose pendant vertices are $a_1, a_2, \ldots, a_m$ and possibly b.

Let T be such a tree with the minimum number of edges. We claim that no two vertices of T are equivalent under the action of α^i, $1 \leqslant i < p$. For, let x and y be two vertices of T with $\alpha^i x = y$ and let P be the unique path in T from x to y. Removal of any edge of P separates T into two subtrees, R and S. Without loss of generality we can assume $b, y \in S$. But then $R \cup \alpha^i S$ has a subtree that contradicts the minimality of T.

Thus, $\bigcup_{k=1}^{p} \alpha^k T$ is a subgraph of X contractible to $K_{m,p}$.

Second proof (Due to L. Babai). Suppose α satisfies the assumptions of Theorem 3 and fixes the points $a_1, \ldots, a_k$, where $k \geqslant m$. Let Y be a component of $X \setminus \{a_1, \ldots, a_k\}$. Since X is m-connected at least m fixed points, say $a_1, \ldots, a_m$, must be adjacent to vertices of Y, for otherwise X has a separating set of less than m points.

Since Y is connected there exists a subtree T^* of Y any two points of which are inequivalent under the action of α^i, $1 \leqslant i < p$. We claim that T^* is adjacent to $a_1, \ldots, a_m$. For, let $b \in Y$ be adjacent to a_i. If $b \notin T^*$, then some $\alpha^i b$ is in T^* and T^* is adjacent to a_i.

Let T be the tree consisting of T^* and m edges connecting T^* with the vertices $a_1, \ldots, a_m$. Then $\bigcup_{k=1}^{p} \alpha^k T$ is contractible to $K_{m,p}$.

Corollary 1. *Let X be a p-connected graph no subgraph of which is contractible to $K_{p,p+1}$, where p is prime. Then any automorphism of order p has at most p fixed points.*

Proof. Suppose α is an automorphism of order p with at least $p+1$ fixed points $a_1, \ldots, a_{p+1}$. Let T be the tree constructed in either proof of Theorem 3. If there exists a path P from a_{p+1} to T which does not meet $a_1, \ldots, a_p$, then there is such a path which contains no interior point x such that $\alpha^i x \in T$ for some i, $1 \leqslant i \leqslant p$, and $\bigcup_{k=1}^{p} \alpha^k (T \cup P)$ is contractible to $K_{p,p+1}$.

On the other hand, suppose every path from a_{p+1} to T meets some a_i, $1 \leqslant i \leqslant p$. Then $\{a_1, \ldots, a_p\}$ separates $T \backslash \{a_1, \ldots, a_p\}$ from a_{p+1}. Since X is p-connected this implies the existence of a tree S containing $a_1, \ldots, a_{p+1}$ but no inner vertices of any $\alpha^k T$. Then, $S \cup \bigcup_{k=1}^{p} \alpha^k T$ is contractible to $K_{p,p+1}$.

It has been shown by Wagner [9] that a graph X is planar if and only if the following conditions are satisfied:

(a) X contains no subdivision of the Kuratowski graphs $K_{3,3}$ and K_5.

(b) X has at most countably many vertices of degree $\geqslant 3$.

(c) The cardinality of the set of all vertices of X is at most that of the continuum.

Even if we restrict attention to graphs satisfying conditions (b) and (c) there is a large class of non-planar graphs not containing subdivisions of $K_{3,3}$. For three-connected graphs, however, the situation is different. It is not hard to see that any three-connected graph which is different from K_5 but contains a subdivision of K_5 also contains a subdivision of $K_{3,3}$. Since subdivisions of $K_{3,3}$ are contractible to $K_{3,3}$, this means that K_5 is the only three-connected non-planar graph satisfying (b) and (c) which does not contain a subgraph contractible to $K_{3,3}$.

Corollary 2. *Let X be a three-connected planar graph and α an automorphism of order $o(\alpha)$. Then the following assertions hold*:

(i) *If $o(\alpha)$ is odd all cycles in the cycle decomposition of α have the same length with the possible exception of at most two cycles of length one.*

(ii) *If $o(\alpha)$ is not a power of 2 there are at most two vertices in cycles of lengths $\leqslant 2$.*

Proof. Let X be a three-connected graph without subgraphs contractible to $K_{3,3}$ and let α be an automorphism of X.

Suppose α is of odd order and has at least three fixed points. Then the automorphism $\alpha^{o(\alpha)/p}$, where p is a prime $\geqslant 3$ dividing $o(\alpha)$, has order p and also at least three fixed points, in contradiction to Theorem 2.

Furthermore, suppose an automorphism α of odd order contains cycles of lengths r and s with $2 < r < s$. Then α^r is a non-trivial automorphism of odd order with at least three fixed points (the elements of the cycle of length r). As we have just shown this is not possible.

Finally, let $o(\alpha)$ be of the form $2^k m$, where m is odd, and suppose there are at least three vertices in cycles of lengths $\leqslant 2$. Then these vertices are fixed by $\beta = \alpha^{o(\alpha)/m}$ and β has order m, in contradiction to (i).

Corollary 3. *Let X be a three-connected graph no subgraph of which is contractible to $K_{3,4}$ and let α be an automorphism of X whose order is not a power of 2. Then there exist at most three vertices which are in cycles of lengths $\leqslant 2$ in the cycle decomposition of α.*

If $o(\alpha)$ is divisible by a prime $\geqslant 5$ there are at most two such vertices and if $o(\alpha)$ is coprime to 6 all vertices not fixed by α are in cycles of the same length.

Proof. Let α be an automorphism with at least s vertices in cycles of lengths $\leqslant 2$. If α is of odd order these cycles have to be of length 1. If α is of order $2^k m$, where m is odd, then these vertices are fixed points of $\beta = \alpha^{o(\alpha)/m}$ and β has odd order.

Without loss of generality we can therefore assume that α is of odd order and has at least s fixed points. Let p be an odd prime dividing $o(\alpha)$. Then $\alpha^{o(\alpha)/p}$ has order p and at least s fixed points. For $p = 3$ and $s > 3$ this contradicts Corollary 1 and for $p \geqslant 5$ and $s > 2$ Theorem 3.

The assertion about the cycle lengths is proved as in Corollary 2.

On the same lines one can prove the following Corollary:

Corollary 4. *Let X be a three-connected toroidal graph and let α be an automorphism whose order is divisible by some prime $\geqslant 5$. Then there exist at most 2 vertices which are in cycles of lengths $\leqslant 2$ in the cycle decomposition of α.*

If the order of α is coprime to 6 all vertices not fixed by α are in cycles of the same length.

2

Ends of graphs can be considered as points at infinity and are defined as equivalence classes of one-sided infinite paths. (Two one-sided infinite paths are called equivalent if there is a third one which meets both of them infinitely often.) They have been investigated from a graph-theoretic point of view by Halin [4] and Polat [6] and have independently been used extensively in combinatorial group theory (see, for example, Cohen [2]). For trees Tits [8] has shown that every automorphism fixes a vertex, an edge or two ends. (For injective endomorphisms of trees, Halin [5, Theorem 5] proved a similar theorem: Every endomorphism of a tree fixes a vertex, an edge or one end. In fact, it is not hard to show that this is also true for edge-preserving mappings of trees.)

For graphs containing no subdivisions of $K_{m,n}$ or no subgraphs contractible to $K_{m,n}$, the following theorems hold:

Theorem 4 (Seifter [7]). *Let X be an m-connected graph and n a fixed integer. Suppose X contains no subdivision of $K_{m,n}$ and let α be an automorphism of X with a cycle of infinite length. Then α fixes at most two ends.*

If X is locally finite the conclusion holds for connected graphs without further contractibility assumptions, as has first been shown by Halin [4, Theorem 8].

Theorem 5. *Let X be an m-connected graph and n a fixed integer. Suppose X has no subgraphs contractible to $K_{m,n}$ and let α be an automorphism of prime order p, $n \leqslant p$. Then the sum of the number of fixed points and fixed ends is at most $m-1$.*

Proof. Suppose α satisfies the conditions of Theorem 5 and fixes the points $a_1, \ldots, a_k$ as well as the ends $E_{k+1}, \ldots, E_m$.

As in the proof of Theorem 3 one can show the existence of a tree T with the following properties:

(i) $a_1, \ldots, a_k$ are endpoints of T;

(ii) for every end E_j, $k < j \leqslant m$, T contains a one-sided infinite path P_j belonging to E_j;

(iii) $\alpha T \cap T = \{a_1, \ldots, a_k\}$;

(iv) there exist disjoint α-invariant, connected subgraphs A_j, $k < j \leqslant m$, such that $T \cap A_j \subset P_j$.

Let T^* be the finite subtree of T containing $a_1, \ldots, a_k$ and exactly one vertex of every A_j, $k < j \leqslant m$. Then the subgraph $\bigcup_{i=1}^{k} \alpha^i T^* \cup \bigcup_{j=k+1}^{m} A_j$ of X can be contracted to $K_{m,n}$, contrary to assumption.

References

[1] L. Babai, Groups of graphs on given surfaces, Acta Math. Acad. Sci. Hung. 24 (1973) 215–221.
[2] D.E. Cohen, Groups of Cohomological Dimension One, Lecture Notes in Mathematics Vol. 245 (Springer, 1972).
[3] H. Fleischner, Die Struktur der Automorphismen speczieller, endlicher, ebener, dreifach-knotenzusammenhaengender Graphen, Compos. Math. 23 (1971) 435–444.
[4] R. Halin, Ueber unendliche Wege in Graphen, Math. Ann. 157 (1964) 125–137.
[5] R. Halin, Automorphisms and Endomorphisms of Infinite Locally Finite Graphs, Abh. Math. Sem. Univ. Hamburg, Vol. 39 (1973) pp. 251–283.
[6] N. Polat, Aspects topologiques de la séparation dans les graphes infinis, I et II, Math. Z. 165 (1979) 73–100, 171–191.
[7] N. Seifter, Automorphisms of infinite order in graphs with forbidden subgraphs, Combinatorica (in print).
[8] J. Tits, Sur le groupe des automorphismes d'un arbre, Essays topol. related topics, Mem. dédiés à Georges de Rham (1970) pp. 188–211.
[9] K. Wagner, Fastplaettbare Graphen, J. Combinat. Theory 3 (1967) 326–365.

Received March 1981

Annals of Discrete Mathematics 20 (1984) 189–190
North-Holland

WEAKLY SATURATED GRAPHS ARE RIGID

Gil KALAI

Institute of Mathematics, Hebrew University, Jerusalem, Israel

We prove that every embedding of a weakly d-saturated graph G into $\mathbb{R}^{d-2}$, such that the vertices are in the general position, is rigid. This result gives another proof to a conjecture of Bollobás.

A graph G of order n is weakly r-saturated if there is a sequence of graphs $G = G_0 \subset G_1 \subset \cdots \subset G_t = K_n$, such that $|e(G_{i+1})| = |e(G_i)| + 1$ and $k_r(G_{i+1}) > k_r(G_i)$, for $0 \leqslant i < t$. (Here $k_r(G)$ is the number of complete subgraphs of G of order r.)

Given a graph G on the vertex set $\{1, 2, \ldots, n\}$, a d-embedding $G(p)$ of G is a sequence of n points in $\mathbb{R}^d$, $p = (p_1, p_2, \ldots, p_n)$, together with the line segments $[p_i, p_j]$, for $\{i, j\} \in E(G)$. We say that $G(p)$ is rigid if any continuous deformation $(p_1(t), \ldots, p_n(t))$ of $(p_1, \ldots, p_t)$ which preserves the distance between every two adjacent vertices, preserves the distance between any two vertices (and thus extends to an isometry of $\mathbb{R}^d$). We call $G(p)$ flexible if it is not rigid. For precise definitions and further information on rigidity see [1, 5]. We denote by K_d^- the graph obtained by deleting an edge from the complete graph of order d.

The purpose of this note is to prove the following theorem:

Theorem 1. *Suppose G is a weakly $(d+2)$-saturated graph, then every embedding of G into $\mathbb{R}^d$, s.t. the vertices are in the general position, is rigid.*

Proof. Suppose $G = G_0 \subset G_1 \subset \cdots \subset G_t = K_n$, $|e(G_{i+1})| = |e(G_i)| + 1$ and $k_{d+2}(G_{i+1}) > k_{d+2}(G_i)$, for $0 \leqslant i < t$. Suppose $p_1, \ldots, p_n$ are n points, in general position in $\mathbb{R}^d$, and regard $G(p)$, $p = (p_1, \ldots, p_n)$. $G_t(p) = K_n(p)$ is rigid. Assume $G_{i+1}(p)$ is rigid, and suppose that $G_i = G_{i+1} - e$, where $e = \{\mu, \nu\}$ belongs to a $(d+2)$-complete subgraph K_{d+2} of G_{i+1}. It is easy to see that every embedding of K_{d+2}^- in $\mathbb{R}^d$, with vertices in the general position, is rigid. It follows that any continuous deformation of $G_i(p)$ preserves the distance between p_μ and p_ν, and thus it is a continuous deformation of $G_{i+1}(p)$, and by the assumption it preserves the distances between any two vertices of $G_i(p)$. Repeated application of this argument shows that $G(p) = G_0(p)$ is rigid. $\square$

It is well known that if G is a graph of order n and fewer than $dn - \binom{d+1}{2}$ edges, then $G(p)$ is flexible for almost all $p = (p_1, \ldots, p_n) \in \mathbb{R}^{n \times d}$ (see [1] or [5]). Roughly speaking, this follows from the fact that there are nd degrees of freedom to move the vertices, and the dimension of the group of affine rigid motions of $\mathbb{R}^d$ is $\binom{d+1}{2}$. This implies:

Theorem 2. *If G is a weakly $(d+2)$-saturated graph of order n, then* $|e(G)| \geq dn - \binom{d+1}{2}$.

This theorem was conjectured by Bollobás [2, 3]. We prove it by a different method in [4].

References

[1] L. Asimow and B. Roth, The rigidity of graphs, Trans. Amer. Math. Soc. 245 (1978) 279–289.
[2] B. Bollobás, Weakly k-saturated graphs, in: Beiträge zur Graphentheorie (Leipzig, 1968) pp. 25–31.
[3] B. Bollobás, Extremal Graph Theory (Academic Press, 1978) p. 362.
[4] G. Kalai, Hyperconnectivity of graphs (in preparation).
[5] B. Roth, Rigidity of frameworks, Amer. Math. Monthly 88 (1981) 6–21.

Received 9 June 1981

Annals of Discrete Mathematics 20 (1984) 191–196
North-Holland

DECOMPOSABILITY OF POLYTOPES IS A PROJECTIVE INVARIANT

Michael KALLAY*
Hebrew University, Jerusalem, Israel

Let P be a polytope in E^d, and T a projective transformation of E^d admissible for P. The sets of summands of P and TP are shown to be of the same dimension, hence TP is nondecomposable iff P is. An example is given of a polytope P and a projective transformation T, such that the sets of summands of P and TP are not combinatorially equivalent.

1.

Let P be a nonempty convex polytope in E^d. A polytope Q is a *summand* of P if $P = Q + R$ (i.e. $P = \{x + y: x \in Q,\ y \in \mathbb{R}\}$) for some polytope $\mathbb{R}$. Following [5], denote by $\mathcal{S}(P)$ the convex cone $\mathcal{S}(P) = \{Q: \lambda Q$ is a summand of P for some $\lambda > 0\}$.

It has been shown in [2, 3] that $Q \in \operatorname{relint} \mathcal{S}(P)$ iff Q and P are *locally similar* ($Q \sim P$), i.e. iff there exists a combinatorial equivalence φ between P and Q such that for every proper face F of P, $\varphi(F)$ and F have the same set of outward normals. (The original definition of local similarity in [6] is slightly different, but the equivalence of the two definitions has been established in [5] and [3].) Note that if there exists a local similarity between two polytopes, then it is unique. P is *decomposable* if $\mathcal{S}(P)$ contains polytopes which are not homothetic to P. It is easy to see that this happens iff $\dim \mathcal{S}(P) > d + 1$.

Let P and Q be two polytopes in $\mathbb{R}^d$, and suppose they are equivalent in some sense. Does this equivalence imply $\dim \mathcal{S}(P) = \dim \mathcal{S}(Q)$? A negative answer is known for combinatorial equivalence [3, 4], and this paper deals with the case of projective equivalence.

A *projective transformation* T of E^d is one of the form $Tx = (Ax + b)/(\langle x, c\rangle + \delta)$, where A is a $d \times d$ matrix, b and c are d-vectors, δ is a real number, and the matrix $\begin{pmatrix} A & b \\ c & \delta \end{pmatrix}$ is nonsingular. The hyperplane $\{x: \langle x, c\rangle + \delta = 0\}$ has no image in E^d. (It is mapped onto the hyperplane at infinity.) This is why

* This paper is a revised version of a part of a Ph.D. thesis [2] written by the author under the supervision of Professor M.A. Perles at the Hebrew University of Jerusalem, and submitted in 1979. The author is indebted to M.A. Perles for reading the early version of this paper and suggesting many fruitful ideas.

T is said to be *admissible* for the polytope P if $\langle c, x\rangle + \delta \neq 0$ for all $x \in P$, [1, p. 4]. In this case TP is a convex polytope, and the mapping $F \to TF$ (for $F \in \mathcal{F}(P)$) is a combinatorial equivalence between P and TP.

The set of convex bodies (i.e. nonempty compact convex sets) in E^d is linearly embedded as a convex cone in the Banach space of continuous real functions on S^{d-1}, with the norm $\|f\| = \max_{x \in S^{d-1}} |f(x)|$, by identifying each body with its support function, restricted to S^{d-1}. $\mathcal{S}(P)$ is a convex cone in that space [5], and our main theorem asserts that $\dim \mathcal{S}(P)$ (and therefore the decomposability or nondecomposability of P) is invariant under every projective transformation T of E^d admissible for P. An example will show that $\mathcal{S}(P)$ and $\mathcal{S}(TP)$ are in general not linearly (and not even combinatorially) equivalent, and thus the equality of their dimensions is not at all obvious.

In the following we write $[S]$ for the convex hull of S.

2.

Theorem. *Let $P \subset E^d$ be a convex polytope, and let $Tx = (Ax + b)/(\langle c, x\rangle + \delta)$ be a projective transformation of E^d admissible for P. Then* $\dim \mathcal{S}(TP) = \dim \mathcal{S}(P)$.

Proof. T is admissible for P, hence $\langle c, x\rangle + \delta \neq 0$ for all $x \in P$. Assume, without loss of generality, that $\langle c, x\rangle + \delta > 0$ for all $x \in P$, and let α be a number such that $\langle c, x\rangle + \delta < \alpha$ for all $x \in P$. Denote by e the vector $(0, \ldots, 0, 1)$ in E^{d+1}, and identify E^d with the hyperplane $\langle x, e\rangle = 0$ in E^{d+1} (by regarding each point $x \in E^d$ as the point $(x, 0)$ in E^{d+1}). $\mathcal{S}(P)$ still denotes the cone spanned by the set of summands of P in E^d. Define a projective transformation T of E^{d+1} by:

$$Tx = \frac{\begin{pmatrix} & & & 0 \\ & A & & \vdots \\ & & & 0 \\ 0 & \cdots & 0 & \alpha \end{pmatrix} x + \begin{pmatrix} b \\ 0 \end{pmatrix}}{\langle (c, 0), x\rangle + \delta} \qquad (\text{for } x \in E^{d+1}),$$

and note that the restriction of the new T to E^d is our original T. $\langle (c, 0), e\rangle = 0$, hence $\langle (c, 0), x + \lambda e\rangle > 0$ for all λ and all $x \in P$. It follows that the extended T is admissible for the prism $[P, P + e]$. Every point in $P + e$ is of the form $(x, 1)$ with $x \in P$, and the $(d+1)$st coordinate of its image under T equals $[\alpha/(\langle c, x\rangle + \delta)] > 1$. Thus, $T(P + e)$ lies entirely above the hyperplane $H = \{x \in E^{d+1}: \langle x, e\rangle = 1\}$.

Let N be a neighbourhood of P in $\mathcal{S}(P)$ such that for all $Q \in N$, the inequalities $0 < \langle c, x\rangle + \delta < \alpha$ still hold for all $x \in Q$. This means that the extended transformation T is still admissible for the prism-like polytope $[P, Q + e]$, and $T(Q + e)$ still lies entirely above H. For any $Q \in N$ let $\bar{Q}$ denote the polytope $[TP, T(Q+e)]$ $(=T[P, Q+e])$ in E^{d+1}, and let φQ be the polytope $(\bar{Q} \cap H) - e$ in E^d. We shall now prove that φ is a 1–1 affine mapping of N into $\mathcal{S}(TP)$, hence $\dim \mathcal{S}(P) \leq \dim \mathcal{S}(TP)$.

Let Q be a polytope in N, and let F be any face of P. Q is locally similar to P, so let F' denote the face of Q that corresponds to F under the local similarity. Let $\bar{F}$ denote the set $T[F, F'+e]$. If K is a supporting hyperplane of P in E^d such that $K \cap P = F$, and x is any point in $F'+e$, then $\operatorname{aff}(K \cup \{x\})$ is a supporting hyperplane of $[P, Q+e]$, with $[P, Q+e] \cap \operatorname{aff}(K \cup \{x\}) = [F, F'+e]$. It follows that $[F, F'+e]$ is a face of $[P, Q+e]$, and $\bar{F} = T[F, F'+e]$ is a face of $\bar{Q}$. Conversely, let G be any face of $\bar{Q}$, such that $G \cap TP \neq \emptyset$ and $G \cap T(Q+e) \neq \emptyset$. The face $T^{-1}G$ of $[P, Q+e]$ is not contained in one of the bases P and $Q+e$. Let L be a supporting hyperplane of $[P, Q+e]$ such that $L \cap [P, Q+e] = T^{-1}G$. $L \cap E^d$ and $L \cap H$ are parallel $d-1$-planes, supporting P and $Q+e$, respectively, with the same outward normal. Let $F = L \cap P$; then $L \cap (Q+e) = F'+e$, and $T^{-1}G = [F, F'+e]$. Thus, we have $G = T[F, F'+e] = \bar{F}$, proving that every face of $\bar{Q}$ which is not contained in one of the bases TP and $T(Q+e)$ equals $\bar{F}$ for some face F of P.

Now let ψ: $\mathcal{F}(TP) \to \mathcal{F}(\varphi Q)$ be the mapping defined by $\psi(TF) = (\bar{F} \cap H) - e$ $(F \in \mathcal{F}(P))$. We shall prove that ψ is a local similarity, hence $\varphi Q \in \mathcal{S}(TP)$.

$\psi(TF)$ is clearly a face of φQ, and ψ maps $\mathcal{F}(TP)$ onto $\mathcal{F}(\varphi Q)$, since, by the preceding argument, every nonempty face of φQ is of the form $(\bar{G} \cap H) - e$ for some face G of P, and $\psi(\emptyset) = \emptyset$.

Let TF and TG be two faces of TP. If $TF \subseteq TG$, then $\bar{F} \subseteq \bar{G}$, and $\bar{F} \cap H \subseteq \bar{G} \cap H$, hence $\psi(TF) \subseteq \psi(TG)$. Conversely, if $\bar{F} \cap H \subseteq \bar{G} \cap H$ and $F \neq \emptyset$, then $\bar{G}$ contains some points of relint $\bar{F}$, hence $\bar{F} \subseteq \bar{G}$. Intersecting both $\bar{F}$ and $\bar{G}$ with E^d, we obtain $TF \subseteq TG$, thus proving that ψ is a combinatorial equivalence.

Now choose a nonzero vector $u \in E^d$, and let K_u be a supporting hyperplane of TP with outward normal u. Let TF be the face $K_u \cap TP$ of TP. In order to prove that ψ is a local similarity we shall show that $L_u \cap \varphi Q = \psi(TF)$, where L_u is the supporting hyperplane of φQ with outward normal u. For that we look back at P and $Q+e$. Let p be any vertex of $F'+e$, and let $L = \operatorname{aff}(T^{-1}K_u \cup \{p\})$. The local similarity between P and $Q+e$ implies that $F'+e = L \cap (Q+e)$ and L is a supporting hyperplane of $[P, Q+e]$ with $L \cap [P, Q+e] = [F, F'+e]$. The projective transformation T maps L into a supporting hyperplane TL of $\bar{Q}$, with $TL \cap \bar{Q} = \bar{F}$. Now consider the intersection $TL \cap H$. Since H is parallel to E^d, and the polytopes $TP(=\bar{Q} \cap E^d)$ and $\bar{Q} \cap H$ lie on the same side of TL, $TL \cap H$ is a supporting hyperplane (in H) of $\bar{Q} \cap H$, with outward normal u. It follows that $TL \cap H - e = L_u$ is the supporting hyperplane of φQ with outward normal u, and $L_u \cap \varphi Q = (TL \cap H) \cap (\bar{Q} \cap H) - e = (\bar{F} \cap H) - e = \psi(TF)$.

Finally, we show that φ is 1–1 and affine. Let Q be a polytope in N, and let

$(p_1, \ldots, p_n)$ be the vertices of P in some preset order. We shall list the corresponding vertices of Q and φQ in the following manner:

For Q: $(s_1(Q), \ldots, s_n(Q))$, where $s_i(Q)$ is the image of p_i under the local similarity between P and Q.

For φQ: $(t_1(Q), \ldots, t_n(Q))$, where $t_i(Q) = \psi(Tp_i) = (T[p_i, s_i(Q)+e] \cap H) - e$ (i.e. $t_i(Q)$ is the image of Tp_i under the local similarity ψ between TP and φQ). We shall write s_i and t_i in short for $s_i(Q)$ and $t_i(Q)$, in case there is only one Q in consideration.

The mappings $Q \leftrightarrow (s_1(Q), \ldots, s_n(Q))$ and $\varphi Q \leftrightarrow (t_1(Q), \ldots, t_n(Q))$ are 1–1 and linear. Let Q and R be two distinct polytopes in N. $Q \neq R$ implies $s_i(Q) \neq s_i(R)$ for at least one $1 \leqslant i \leqslant n$, hence the semi-closed intervals $]p_i, s_i(Q)+e]$ and $]p_i, s_i(R)+e]$ are disjoint. It follows that $T[p_i, s_i(Q)+e] \cap T[p_i, s_i(R)+e] \cap H = \emptyset$, hence $t_i(Q) \neq t_i(R)$ and $\varphi Q \neq \varphi R$.

To establish the affinity of φ we calculate t_i for each $1 \leqslant i \leqslant n$ and some $Q \in N$:

$$T(s_i + e) = \frac{As_i + b}{\langle c, s_i \rangle + \delta} + \frac{\alpha e}{\langle c, s_i \rangle + \delta} = Ts_i + \frac{\alpha e}{\langle c, s_i \rangle + \delta},$$

and the $(d+1)$st coordinate of $T(s_i + e)$ equals $\alpha / (\langle c, s_i \rangle + \delta)$. The point $\bar{p}_i \cap H = T[p_i, s_i + e] \cap H$ is a convex combination of Tp_i and $T(s_i + e)$. Since its $(d+1)$st coordinate equals 1, this convex combination must be:

$$\bar{p}_i \cap H = Tp_i + \frac{\langle c, s_i \rangle + \delta}{\alpha} (T(s_i + e) - Tp_i)$$

(see Fig. 1). Thus:

$$t_i = \bar{p}_i \cap H - e = Tp_i + \frac{\langle c, s_i \rangle + \delta}{\alpha} (T(s_i + e) - Tp_i) - e$$

$$= Tp_i + \frac{\langle c, s_i \rangle + \delta}{\alpha} Ts_i - \frac{\langle c, s_i \rangle + \delta}{\alpha} Tp_i,$$

and

$$t_i - Tp_i = \frac{\langle c, s_i \rangle + \delta}{\alpha} Ts_i - \frac{\langle c, p_i \rangle + \delta}{\alpha} Tp_i$$

$$- \frac{\langle c, s_i - p_i \rangle}{\alpha} Tp_i = \frac{1}{\alpha} (A(s_i - p_i) - \langle c, s_i - p_i \rangle Tp_i).$$

Therefore $t_i - Tp_i$ is linear in the variable $s_i - p_i$ and the mapping $s_i \to t_i$ is affine for every $1 \leqslant i \leqslant n$.

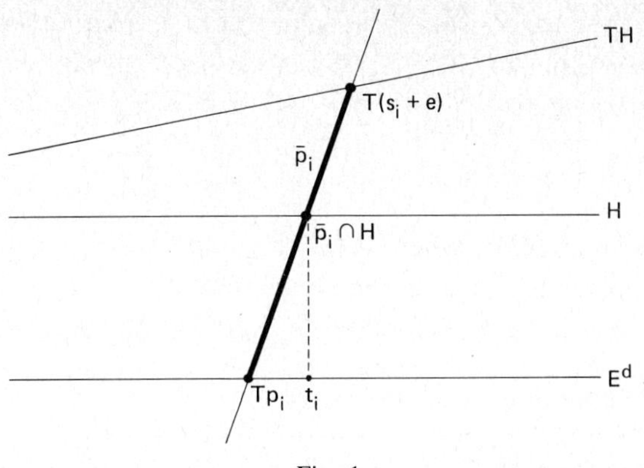

Fig. 1.

We have found a 1-1 affine mapping $\varphi: N \to \mathcal{S}(TP)$, hence $\dim \mathcal{S}(P) \leqslant \dim \mathcal{S}(TP)$. But T is invertible; hence, by the same argument, $\dim \mathcal{S}(TP) \leqslant \dim \mathcal{S}(P)$, and thus $\dim \mathcal{S}(TP) = \dim \mathcal{S}(P)$. □

3.

The following is an example of a polytope P and a projective transformation T such that $\mathcal{S}(P)$ and $\mathcal{S}(TP)$ are not combinatorially equivalent.

First we note that for every polytope P in E^d, $\mathcal{S}(P)$ is the direct sum of E^d and $\mathcal{S}_0(P)$, the set of all elements of $\mathcal{S}(P)$ with Steiner point at the origin. A polytope Q in $\mathcal{S}_0(P)$ is indecomposable iff either $Q = \{0\}$, or the set $\{\lambda Q: \lambda \geqslant 0\}$ is an extremal ray of $\mathcal{S}_0(P)$. (See [5].)

Now let P be a regular pentagon. No summand of P is a line segment, because no two sides of P are parallel. Thus, every indecomposable summand of P is a triangle whose sides have the same outward normals as three of the sides of P. It is easy to see that there are, up to positive homothety, exactly five such triangles, hence $\mathcal{S}_0(P)$ has exactly five extremal rays. ($\mathcal{S}_0(P)$ is in fact a three-dimensional cone with pentagonal cross section.)

P has a projective image TP which resembles an I.B.M. card (Fig. 2). TP has,

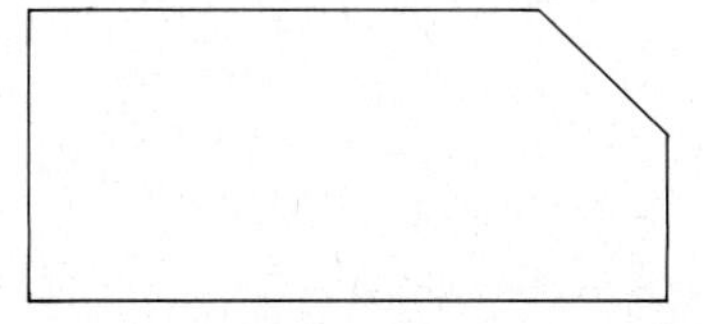

Fig. 2.

up to positive homothety, just three nondecomposable summands: two line segments and one triangle. $\mathcal{S}_0(TP)$ therefore has three extremal rays, and its cross section is triangular. Thus, $\mathcal{S}(P)$ and $\mathcal{S}(TP)$ are not combinatorially equivalent.

References

[1] B. Grünbaum, Convex Polytopes (John Wiley & Sons, London, 1967).
[2] M. Kallay, Indecomposability of convex polytopes, Ph.D. thesis, Hebrew University of Jerusalem (1979) [Hebrew, with English summary].
[3] M. Kallay, Indecomposable polytopes, Israel J. Math. (to appear).
[4] W. Meyer, Minkowski addition of convex sets, Ph.D. dissertation, University of Wisconsin, Madison (1969).
[5] W. Meyer, Indecomposable polytopes, Trans. Amer. Math. Soc. 190 (1974) 77–86.
[6] G.C. Shephard, Decomposable convex polyhedra, Mathematika 10 (1963) 89–95.

Received 11 May 1981

Annals of Discrete Mathematics 20 (1984) 197–201
North-Holland

ADDITIVE SEQUENCES OF PERMUTATIONS

Anton KOTZIG

Centre de Recherche de Mathématiques Appliquées, Université de Montréal, C.P. 6128, Succursale 'A', Montréal, Québec H3C 3J7, Canada

Jean M. TURGEON

Département de Mathématiques et de Statistique, Université de Montréal, C.P. 6128, Succursale 'A', Montréal, Québec H3C 3J7, Canada

1. Definitions, notations, purpose of the paper

Let r and n be positive integers. Let $X = [x_1, \dots, x_r]$ be an ordered set of distinct integers. For $j = 1, \dots, n$, let

$$X^{(j)} = [x_1^{(j)}, \dots, x_r^{(j)}]$$

be a permutation of X. An ordered set

$$X^{(1)}, X^{(2)}, \dots, X^{(n)}$$

of such permutations is called an *additive sequence of permutations of degree r*, or simply an *additive sequence* if, for every subsequence of consecutive permutations of the set, their vector-sum is again a permutation of X. The number n is then called the *length* of the sequence. If the length is 2, we simply speak of *additive permutations*. The cardinality $|X|$ of X is of course the degree of the permutations.

The set X is called the *basis* of the additive sequence. An arbitrary ordered set X is called an $A(n)$-*basis* if there exists an additive sequence of length n with X as its basis. *If X is an $A(n)$-basis, then X is an $A(h)$-basis for every $h < n$*, since every subsequence of consecutive permutations of an additive sequence is itself additive.

Except in [3], the $A(n)$-bases considered so far for $n > 2$ were integer intervals. The purpose of the present paper is to discuss some general properties of additive sequences whose bases are not necessarily integer intervals, and to list some open problems.

In the sequel, we shall call 'bijection of X', rather than 'permutation of X', a 1-1 mapping of X onto itself, reserving 'permutation' to denote an ordered set or vector. A bijection can be decomposed into cycles or cyclic factors (usually called 'cyclic permutations').

2. General properties of additive sequences

It is noted in [2] and easily verified that a necessary condition for a set $X = [x_1, \ldots, x_r]$ to be an $A(n)$-basis, $n \geq 2$, is that $x_1 + \cdots + x_r = 0$. Our first two statements are consequences of this condition.

Theorem 1. *If X is an $A(n)$-basis, then X contains an even number of odd elements; cf.* [1]. *In particular, $\sum_{i=1}^{r} x_i^2$ is even; cf. Theorem 3, below.*

Theorem 2. *If the $A(n)$-basis X is an integer interval, then $|X|$ is odd, say $|X| = 2s + 1$, and the interval is $[-s, \ldots, s]$.*

Let $X^{(j)}$ and $X^{(k)}$ be two, not necessarily distinct, permutations of an additive sequence. We define

$$X^{(j)} \cdot X^{(k)} = \sum_{i=1}^{r} x_i^{(j)} x_i^{(k)}.$$

Theorem 3. *Let $X = [x_1, \ldots, x_r]$ and suppose $\sum_{i=1}^{r} x_i^2 = 2S$; cf. Theorem 1. If $X^{(1)}, \ldots, X^{(r)}$ is an additive sequence, then, for $1 \leq j \leq j + k \leq n$, we have*

$$X^{(j)} \cdot X^{(j+k)} = \begin{cases} 2S, & \text{if } k = 0, \\ -S, & \text{if } k = 1, \\ 0, & \text{if } k > 1. \end{cases}$$

Proof. Theorem 3 is proved in [4] for the case where X is an integer interval, say $[-s, \ldots, s]$. Clearly, in this case we have

$$S = \frac{s}{6}(s + 1)(2s + 1).$$

But this particular value of S plays no role in the proof, which is thus valid also for the general case, where X is an arbitrary $A(n)$-basis.

Theorem 4. *The length of an additive sequence of permutations of degree r never exceeds $r - 1$.*

Proof. Theorem 4 follows from Theorem 3, as indicated in [4].

Theorem 5. *If an additive sequence of length n is generated by repeated applications of a unique bijection, then every cyclic factor of this bijection contains more than n elements, or exactly one.*

Proof. Suppose $(x_1, \ldots, x_h)$ is a cyclic factor of the bijection and $2 \leq h \leq n$. Then the element $x_1+x_2+\cdots+x_h$ of X appears h times in the vector-sum of the first h permutations of the sequence — a contradiction.

The following example shows that the case $h=1$ can occur: the additive sequence

$$X^{(1)}=[-3,-2,-1,0,1,2,3],$$

$$X^{(2)}=[3,1,-1,-3,2,0,-2],$$

$$X^{(3)}=[-2,2,-1,3,0,-3,1],$$

is generated by two applications of the following product of two cycles:

$$(-1)(-3,3,-2,1,2,0).$$

Theorem 6. *Let p be a prime number, $p>2$. Let $A=[a_1,\ldots,a_{p-1}]$ be a finite sequence of positive integers such that the partial sums of consecutive elements of A are all different*:

$$\left|\left\{\sum_{i=j}^{k} a_i \,\middle|\, 1\leqslant j\leqslant k\leqslant p-1\right\}\right|=\tfrac{1}{2}p(p-1). \tag{1}$$

Let

$$a_p=-\sum_{i=1}^{p-1} a_i. \tag{2}$$

For $j=1,\ldots,p$ and $k=1,\ldots,p-1$, let

$$c_j(k)=\sum_{i=j}^{j+k-1} a_i,$$

where the index i is always taken modulo p. Let

$$X=\{c_j(k)\,|\,j=1,\ldots,p;\ k=1,\ldots,p-1\}$$

with its elements taken, say, in their natural order. Let f: $X\to X$ be defined by

$$f(c_j(k))=c_{j+k}(k). \tag{3}$$

Then the sequence $X^{(1)},\ldots,X^{(p-1)}$ defined by

$$X^{(1)}=X;\ X^{(h)}=[f^{h-1}(x_1^{(1)}),\ldots,f^{h-1}(x_{p(p-1)}^{(1)})],\quad h=2,\ldots,p-1, \tag{4}$$

is an additive sequence of permutations of degree $p(p-1)$ and length $p-1$.

The proof of Theorem 6 is given in [3].

Theorem 7. *One more application of the function f defined in* (3) *would not make a longer additive sequence out of* (4).

Proof. We define

$$D_k = \{c_j(k) \mid j = 1, \ldots, p\}; \qquad k = 1, \ldots, p-1.$$

The sets D_k form a partition of X. Since p is prime, the restriction of f to D_k is a cycle of length p:

$$(c_1(k), c_{k+1}(k), \ldots, c_{(p-1)k+1}(k)).$$

The mapping f is thus the product of $p-1$ such cycles. By Theorem 5, the length of the additive sequence generated by repeated applications of f cannot exceed $p-1$.

Examples of sequences satisfying (1) and (2) are given in [3].

3. Some open problems

(1) Is there an additive sequence whose basis is an integer interval and whose length is greater than 4? An example of length 4 is given below, in paragraph (3).

(2) The construction of Theorem 6 provides sequences of length $p-1$, with p prime. If another length, say n, is desired, one can take the first n permutations of an additive sequence of length $p-1$, where p is some prime number greater than $n+1$. But then the degree of the permutations can be relatively very large. *Problem*: To somehow extend the construction of Theorem 6 to lengths n such that $n+1$ is not necessarily prime.

(3) The upper bound given in Theorem 4, for the length of an additive sequence of permutations of given degree, is reached in the following two examples:

an additive sequence with $r = 3$ and $n = 2$:

$$X^{(1)} = [-1, 0, 1],$$

$$X^{(2)} = [0, 1, -1];$$

and one with $r = 5$ and $n = 4$:

$$X^{(1)} = [-2, -1, 0, 1, 2],$$

$$X^{(2)} = [0, 1, 2, -2, -1],$$

$$X^{(3)} = [2, -2, -1, 0, 1],$$

$$X^{(4)} = [-1, 0, 1, 2, -2].$$

In this last case, the sequence is generated by three applications of the cycle $(-2, 0, 2, -1, 1)$. For higher values of r, this upper bound seems very difficult to reach. J. Abrham conjectured that, for a given length n, the minimal possible degree of the permutations is a quadratic, rather than linear, function of n. Our Theorem 6 seems to confirm this conjecture for prime values of $n+1$. So we ask the question: Can the upper bound given in Theorem 4 be improved?

References

[1] J. Abrham and A. Kotzig, Bases of additive permutations with a given number of odd elements, Util. Math. 18 (1980) 283–288.

[2] A. Kotzig, Existence theorems for bases of additive permutations, in: Proc. Eleventh Southeastern Conference on Combinatorics, Graph Theory and Computing, held at Florida Atlantic University, Boca Raton, 3–7 March 1980; Congressus Numerantium, Vol. 28 (1980) pp. 175–185.

[3] J.M. Turgeon, Construction of additive sequences of permutations of arbitrary lengths, in: Theory and Practice of Combinatorics, Annals of Discrete Mathematics, Vol. 12 (North-Holland, Amsterdam, 1982) 239–242.

[4] J.M. Turgeon, An upper bound for the length of additive sequences of permutations, Util. Math. 17 (1980) 189–196.

Received 11 May 1981

Annals of Discrete Mathematics 20 (1984) 203–208
North-Holland

ON PAIRS OF DISJOINT SEGMENTS IN CONVEX POSITION IN THE PLANE

Yakov Shimeon KUPITZ*
Hebrew University, Jerusalem, Israel

1

Consider graphs $\langle V, E\rangle$, where V is a finite point set in $\mathbb{R}^2$, in general position, and E is a collection of non-degenerate closed straight line segments with endpoints in V. We call such graphs g-graphs (geometric graphs). Denote by D a configuration consisting of two disjoint segments, and by CD a D-configuration whose segments are sides of a convex quadrilateral. A g-graph is D-free (CD-free) if it has no subgraph of type $D(CD)$.

The Turán number $T(n, D)(T(n, CD))$ is defined to be the maximum possible number of edges in a D-free (CD-free) g-graph on n vertices.

A g-graph $G = (V, E)$ is *maximal D-free* (*maximal CD-free*) if it is D-free (CD-free) and has the maximum possible number of edges $T(n, D)\,(T(n, CD))$.

It is known that $T(n, D) = n$ for $n \geq 3$, and the structure of maximal D-free g-graphs has been fully described. (See [1, esp. Theorem 2]; also [2, pp. 11–23].)

Fig. 1 is an example of a CD-free g-graph of order 7 with 12 edges. (The points c, d_1, d_2, d_3, d_4, b are in convex position.) Similar examples show that

$$T(n, CD) \geq 2n - 2, \quad \text{for all } n \geq 4.$$

Conjecture 1. $T(n, CD) = 2n - 2$ for $n \geq 4$. (The cases $n = 4,5$ can easily be verified.)

An *n-wheel* W_n $(n \geq 4)$ is an (abstract) graph that is obtained by connecting all vertices of an n-circuit C_n by edges to a common central vertex (hub) v.

Fig. 2. shows that a $(2m + 1)$-wheel can be realized as a CD-free g-graph in more than one way.

We now describe an operation that transforms a g-graph with n vertices and e edges into one with $n + 1$ vertices and $e + 2$ edges.

* My thanks are due to Professor Micha A. Perles for help with revisions to the text.

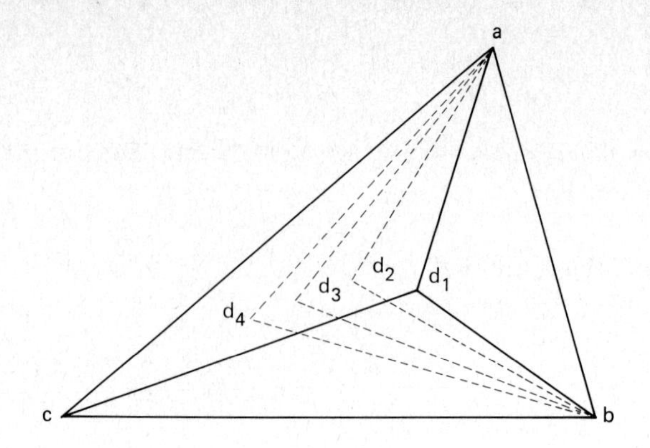

Fig. 1.

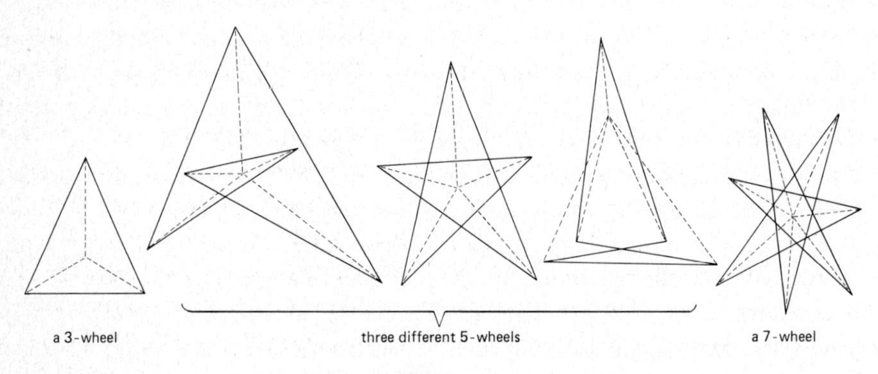

Fig. 2.

Operation A. Suppose $G=\langle V,E\rangle$ is a g-graph, $v\in V$, vx_1, $vx_2\in E$ $(x_1\neq x_2)$ and the edges of G emanating from v that lie within the angle $\measuredangle x_1vx_2$ are $vy_1,\ldots,vy_j$ (j may be zero). Let v' be a point near v such that:

(a) v lies within the angle $\measuredangle x_1v'x_2$, and

(b) $V\cup\{v'\}$ is in general position.

Put $G'=\langle V',E'\rangle$, where $V'=V\cup\{v'\}$, and

$$E'=E\setminus\{vy_1,\ldots,vy_j\}\cup\{v'x_1,v'y_1,\ldots,v'y_j,v'x_2\}.$$

Proposition 1. *If G is CD-free, G′ is obtained from G by Operation A, and the new vertex chosen v′ is sufficiently near v, then G′ is CD-free as well.*

The verification of Proposition 1 is left to the reader.

The following conjecture, if true, clearly implies Conjecture 1.

Conjecture 2. Every maximal CD-free g-graph is obtained from a CD-free $(2m+1)$-wheel by a sequence of operations of type A.

The following theorem was conjectured by the author and proved by Micha A. Perles.

Theorem 1. *Suppose* $\Gamma_i = \langle V_i, E_i \rangle$ $(i = 1,2)$ *are two polygonal cricuits of odd lengths* m *and* n, *respectively,* $V_1 \cap V_2 = \emptyset$ *and* $V_1 \cup V_2$ *is in general position. (Self intersections and mutual intersections of* Γ_1 *and* Γ_2 *are allowed.) Then there is a CD-configuration consisting of one edge of* Γ_1 *and one edge of* Γ_2. *(This may fail if even one of the circuits* Γ_1, Γ_2 *is even.)*

Remark. The condition that $V_1 \cup V_2$ be in general position can be weakened as follows. No line in $\mathbb{R}^2$ contains a vertex of Γ_i and an edge of $\Gamma_{3-i} (i = 1,2)$.

Proof (M.A. Perles). Let $e_1, \ldots, e_m$ and $f_1, \ldots, f_n$ be the edges of Γ_1 and Γ_2, respectively. For line segments a, b, whose four endpoints are in general position, define the crossing number $a \mid b$ as follows. If the line though a crosses the segment b, then $a \mid b = 1$, otherwise $a \mid b = 0$. Note that the segments a and b form a CD-configuration iff $a \mid b = b \mid a = 0$. They cross each other iff $a \mid b = b \mid a = 1$.

Now consider the $m \times n$ matrix A whose (i,j)th entry is the ordered pair $(e_i \mid f_j, f_j \mid e_i)$. Since e_i and f_j form a CD-configuration iff $(e_i \mid f_j, f_j \mid e_i) = (0,0)$, we only have to show that at least one entry of A is $(0,0)$. Define for $1 \leq i \leq m$ and $1 \leq j \leq n$:

$$S_i = \{1 \leq j \leq n : e_i \mid f_j = 1\},$$

$$T_j = \{1 \leq i \leq m : f_j \mid e_i = 1\},$$

$$S = \{(i,j) : 1 \leq i \leq m, j \in S_i\},$$

$$T = \{(i,j) : 1 \leq j \leq n, i \in T_j\}.$$

The closed circuit Γ_2 intersects the line through e_i an even number of times. It follows that $|S_i|$ is even for $1 \leq i \leq n$, and therefore $|S|$ is even as well. A similar argument shows that $|T|$ is even.

Now $|S \cap T|$ is precisely the number of times that the circuits Γ_1 and Γ_2 intersect. This number is again even. This follows from the standard argument used to prove Jordan's Curve Theorem for poylgons (see [3, pp. 267–269]). (Another argument: If z_0 is a variable point on Γ_1, then each time z_0 crosses Γ_2, the index of Γ_2 with respect to z_0 changes by ± 1.)

The number of $(0,0)$ entries of A, which is $m \cdot n - (|S| + |T| - |S \cap T|)$, is therefore odd, and in particular non-zero. □

Corollary. *If n is odd, and G is a g-graph of order $n+1$ with $2n+1$ edges that contains an n-wheel, then G includes a CD-configuration.*

Proof. By the preceding theorem it is enough to show that G contains two odd vertex-disjoint circuits, Γ_1 and Γ_2. But G is obtained from an n-wheel W_n (consisting of a circuit C_n and a hub v), by adding one diagonal pq of C_n. Since n is odd there exists an odd circuit Γ_1 in $G \setminus v$ that uses the edge pq. The other odd circuit Γ_2 may be chosen to be any triangle in $G \setminus \Gamma_1$. □

2. Asterisks

This section is motivated by the problem of finding a g-circuit of maximum Euclidean length with a specified finite set V of vertices. We shall solve this problem when $|V|$ is odd and V is in convex position (i.e. V is the set of extreme points of the convex hull of V).

A *g-multigraph* is a g-graph in which every edge has an assigned (positive integral) multiplicity. A *g-multicircuit* is a regular g-multigraph of degree 2, or, equivalently, a g-multigraph whose graphical components are circuits (of length ≥ 3) and/or double edges ($=2$-circuits). An *asterisk* is a D-free g-multicircuit. (Of course, a double edge is not a D-configuration.)

It is quite easy to show that an asterisk of odd order consists of a single circuit (Fig. 3.), and an asterisk of even order $2k$ consists of k vertex-disjoint double edges (Fig. 4).

It is not hard to show (see, for example, [2, Theorem 3.1.1] for the odd case) that there is at most one asterisk on any given finite vertex set V. Of course, not every finite set $V \subset \mathbb{R}^2$ in general position is the vertex set of an asterisk. We say that V is *in asterisk position* if V is the vertex set of an asterisk. It is easily seen that if V is in convex position, then it is in asterisk position. (The main diagonals form an asterisk.)

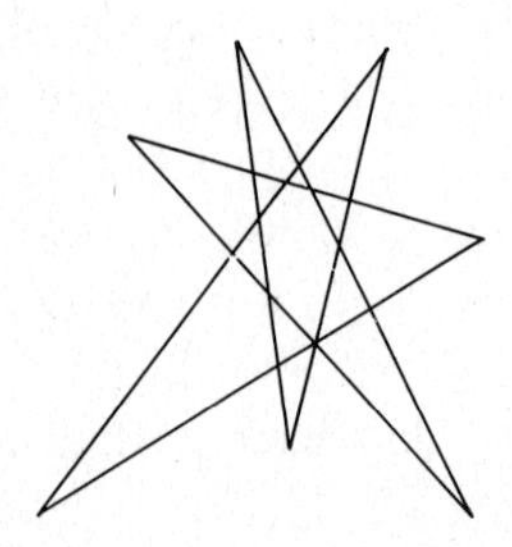

Fig. 3.

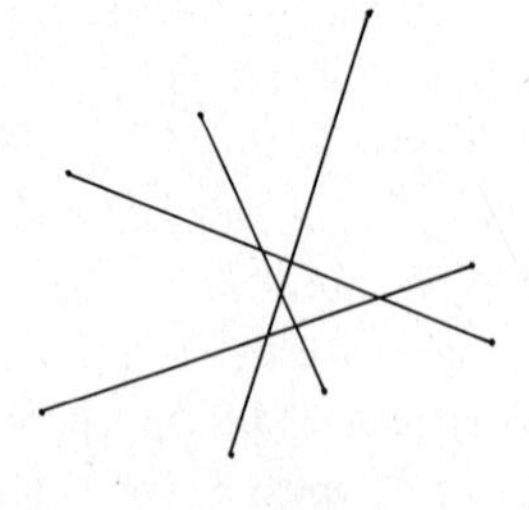

Fig. 4.

Conjecture 3. If V is in asterisk position, then every g-multicircuit on V that is not an asterisk contains a CD-configuration.

Remark. This is obviously true if V is in convex position. Note that even the following simple special case of Conjecture 3 does not seem to be easy. If $|V|=2k$ and V is in asterisk position, and E is a set of k vertex-disjoint segments with endpoints in V, then either 'double E' is the asterisk on V, or there is a CD-configuration in E.

The *Euclidean length* of a g-multigraph G is defined as the sum of the Euclidean lengths of the edges of G, where each edge is counted according to its multiplicity in G.

Conjecture 4. If V is in asterisk position, then the asterisk on V is the (unique) g-multicircuit on V of maximum Euclidean length.

We shall see that the truth of Conjecture 3 for a set V implies the truth of Conjecture 4 for the same set V.

Proof. Let G be a g-multicircuit on V that is not an asterisk. By Conjecture 3, G contains a CD-configuration consisting of two opposite edges $[p,q]$, $[r,s]$ of a convex quadrilateral $[p,q,r,s]$ (Fig. 5). Replace the edges $[p,q]$, $[r,s]$ by the diagonals $[p,r]$, $[q,s]$. This replacement does not alter the valences of the vertices of G, and therefore the resulting g-multigraph G' is again a g-multicircuit on V.

The Euclidean length of G' is clearly larger than that of G. This shows that if G is not an asterisk, then its Euclidean length is not maximal. □

Since Conjecture 3 holds if V is in convex position, we have actually proved the following corollary.

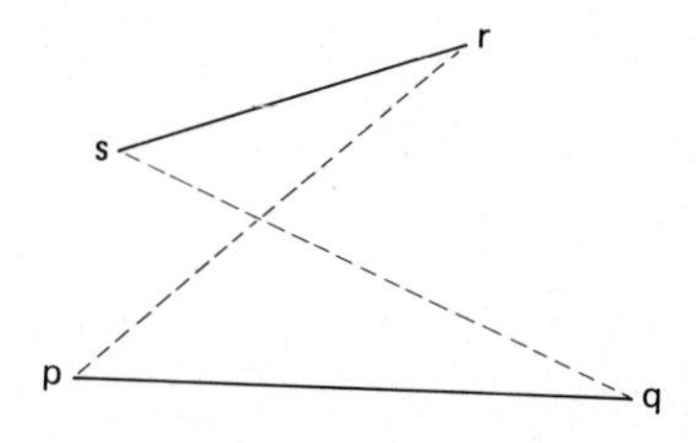

Fig. 5.

Corollary. If V is in convex position, then the asterisk on V is strictly longer than any other g-multicircuit on V.

This solves the problem stated at the beginning of this section if $|V|$ is odd and V is in convex position.

References

[1] D.R. Woodall, Thrackles and deadlock, in: Combinatorial Mathematics and its Applications, Proc. Conf. Oxford, 1969.
[2] Y. Kupitz, Extremal problems in combinatorial geometry, Aarhus Univ. Lecture Notes Series, No. 53 (1979).
[3] R. Courant and H. Robbins, What is Mathematics? (Oxford University Press, 1948).

Received June 1981

Annals of Discrete Mathematics 20 (1984) 209–214
North-Holland

THE DECOMPOSITION OF THE *n*-SPHERE AND THE BOUNDARIES OF PLANE CONVEX DOMAINS

D.G. LARMAN and N.K. TAMVAKIS
Department of Mathematics, University College London, London WC1E 6BT, U.K.

One of the most famous problems in the study of convex sets is to decide whether or not it is possible to cover any set of diameter 1 in E^n by $n+1$ closed sets, each of diameter less than 1. For an excellent survey of this problem, see Grünbaum [2]. Again it has often been conjectured that the sphere S^{n-1} of centre O and unit diameter is the extremal case, i.e. if S^{n-1} can be covered by $n+1$ closed sets of diameter at most d, then so can any other set of diameter 1. However, it is not known what values d can take, even for S^{n-1}. More precisely, let $d(S^{n-1}, n+1)$ be the greatest lower bound of reals d such that S^{n-1} may be partitioned into $n+1$ subsets, each of diameter less than or equal to d. Grunbaum [2] notes that the obvious simplicial decomposition of S^{n-1} shows that $d(S^{n-1}, n+1)$ does not exceed:

$$\left(\frac{n+1}{n+2}\right)^{1/2}, \quad \text{for even } n,$$

$$\left(\frac{1}{2}+\frac{1}{2}\left(\frac{n-1}{n+3}\right)^{1/2}\right)^{1/2}, \quad \text{for odd } n.$$

Hadwiger [3] proved that equality holds in these estimates for $n \leqslant 3$ but established only

$$d(S^{n-1}, n+1) \geqslant \left(\frac{1}{2}+\frac{1}{2}\left(\frac{n-1}{2n}\right)^{1/2}\right)^{1/2}, \qquad n \geqslant 4.$$

Here, we find a better estimate when n is sufficiently large, which shows that $d(S^{n-1}, n+1)$ tends to 1 as $n \to \infty$. More precisely we prove:

Theorem 1. *The n-dimensional sphere of diameter* 1 *cannot be decomposed into* $n+1$ *subsets, each of diameter less than*

$$1-\frac{3}{2n}\log n + \mathrm{O}\left(\frac{1}{n}\right).$$

In order to prove the above theorem, we shall use the following result obtained from the extensive work published by Schmidt [6].

Lemma 1. *Let V_r denote the r-dimensional volume function. If K is a subset of the n-sphere S^{n-1}, having diameter d, with $0<d<1$, then*

$$V_{n-1}(K) \leq V_{n-1}(C_d),$$

where C_d is a spherical cap cut from S^{n-1}, having diameter d.

One generalisation of the above problem is to find the number $p_n(t)$ defined as the least positive number such that any convex body K in E^n of diameter 1, can have its boundary ∂K divided into t sets, each of diameter at most $p_n(t)$, and a closely related problem (Theorem 3) to find the largest possible perimeter of a convex t-polygon of fixed diameter. Somewhat surprisingly, the extreme case in Theorem 3 are not the regular t-polygons, t even, in contrast to maximising the areas of t-polygons of fixed perimeter, as in Fejes-Toth [1]. Studies on the number $d_2(t)$, defined as the least positive number such that any convex domain K, in E^2 of diameter 1, can be divided into t sets, each of diameter at most $d_2(t)$, has been carried out by Lenz [4, 5].

Theorem 2.

$$\sin\frac{\pi}{t} \leq p_2(t) \leq 2\sin\left(\frac{\pi}{2t}\right), \qquad t = 5, 6, \ldots,$$

$$p_2(3) = \tfrac{1}{2}\sqrt{3}, \qquad p_2(4) = (\tfrac{1}{2})^{1/2}.$$

Remark. For a convex set K of constant width 1, let $p_2(K, t)$ denote the least number a such that a t-polygon of side a can be inscribed in ∂K. Then, for t large, $tp_2(K, t)$ is a good approximation to the length π of ∂K, i.e. $p_2(K, t) \triangleq \pi/t$. As $p_2(t) = \sup p_2(K, t)$, there will be a widely differing class of convex sets on which $p_2(K, t)$ is approximately $p_2(t)$. Notice also that $\sin(\pi/t)$ and $2\sin(\pi/2t)$ only differ by an amount of order $\pi^3/12t^3$ when t is large.

We give the following definitions:

(i) A convex t-polygon P of diameter d is said to be a (d, t)-polygon if, for each edge of P, there is a vertex of P which is at a distance d from each of the two end-points of the given edge.

(ii) A *regular* (d, t)-polygon is a (d, t)-polygon with equal edges.

In the case that $t = m2^n$, where n is a non-negative integer and m is an odd integer greater than 1, we may construct regular (d, t)-polygons as follows.

Let k be a divisor of m ($k > 1$ and possibly $k = m$), and let P_k denote the regular k-polygon of diameter d. On each edge of P_k we construct the circular arc, with radius d, briefly called a d-arc, exterior to P_k. The resulting union of these arcs forms a set of constant width d. Next we divide each of the d-arcs into

$(m/k)2^n$ equal arcs. The vertices, corresponding to the end-points of these d-arcs, form the vertices of a regular (d, t)-polygon. Certainly, regular $(d, 4)$-polygons do not exist, although we do not know if regular $(d, 2^n)$ exist for $n \geq 3$. If they do, then Theorem 3 holds for all values of $t \geq 3$.

Theorem 3. *If $t = m2^n$, where n is a non-negative integer and m is an odd integer greater than 1, among all convex t-polygons of a given diameter, the regular (d, t)-polygons are exactly those convex t-polygons with the largest perimeter. If $t = 4$, the square has the largest perimeter amongst the 4-polygons of given diameter.*

Proof of Theorem 1. Let $S^{k-1}(r)$ denote the k-sphere with diameter $2r$. If $n+1$ sets cover $S^{n-1}(\frac{1}{2})$, then at least one of them will have $(n-1)$-dimensional volume greater than or equal to

$$\frac{1}{n+1} V_{n-1}(S^{n-1}(\tfrac{1}{2})).$$

Let d be the diameter of that set, then by Lemma 1, its $(n-1)$-dimensional volume will be at most $V_{n-1}(C_d)$, where C_d is a cap of $S^{n-1}(\frac{1}{2})$, with diameter d. Thus, no $n+1$ sets cover $S^{n-1}(\frac{1}{2})$, if they have diameter at most d and

$$V_{n-1}(C_d) < \frac{1}{n+1} V_{n-1}(S^{n-1}(\tfrac{1}{2})). \tag{1}$$

Also,

$$\begin{aligned} V_{n-1}(C_d) &= V_{n-2}(S^{n-2}(1)) \int_{(\frac{1}{4}-\frac{1}{4}d^2)^{1/2}}^{1/2} (\tfrac{1}{4} - x^2)^{(n-3)/2} \, dx \\ &< V_{n-2}(S^{n-2}(1))(\tfrac{1}{2}d)^{n-2} \\ &= V_{n-2}(S^{n-2}(\tfrac{1}{2}))d^{n-2}. \end{aligned} \tag{2}$$

From (1) and (2) we have:

$$d^{n-2} > \frac{1}{n+1} \frac{V_{n-1}(S^{n-1}(\frac{1}{2}))}{V_{n-2}(S^{n-2}(\frac{1}{2}))}. \tag{3}$$

Now

$$V_k(S^k(\tfrac{1}{2})) = \frac{(k+1)\pi^{(k+1)/2}(\frac{1}{2})^k}{\Gamma(1+\frac{1}{2}(k+1))},$$

and so (3) becomes:

$$d > \frac{n\pi^{1/2}}{2(n-1)(n+1)} \frac{\Gamma(\frac{1}{2}+\frac{1}{2}n)^{1/(n-2)}}{\Gamma(1+\frac{1}{2}n)}. \tag{4}$$

We also have the estimate

$$\left(\frac{2}{n}\right)^{1/2} > \frac{\Gamma(\frac{1}{2}+\frac{1}{2}n)}{\Gamma(1+\frac{1}{2}n)} > \left(\frac{2}{n+1}\right)^{1/2}.$$

So

$$\frac{\Gamma(\frac{1}{2}+\frac{1}{2}n)}{\Gamma(1+\frac{1}{2}n)} = \left(\frac{2}{n}\right)^{1/2}\left(1+\mathrm{O}\left(\frac{1}{n}\right)\right)$$

and

$$\frac{n}{(n-1)(n+1)} = \frac{1}{n}\left(1+\mathrm{O}\left(\frac{1}{n}\right)\right).$$

Hence, using these estimates in (4):

$$d > \left(\frac{\pi}{2}\right)^{1/2} n^{-3/2}\left(1+\mathrm{O}\left(\frac{1}{n}\right)\right)^{(1/n)(1+\mathrm{O}(1/n))}$$

$$= 1 - \frac{3}{2}\frac{\log n}{n} + \mathrm{O}\left(\frac{1}{n}\right),$$

as required.

In order to prove Theorems 2 and 3, we give the following definition.

Definition. If P is a t-polygon of diameter d and AB is an edge of P, let C be the point at distance d from both A and B, which lies on the same side of AB as does P. Then the *d-arc* on AB is the minor arc of the circle with centre C and radius d, which joins A to B. We define P^* as the union of all the d-arcs of P, although P^* may not form the boundary of a convex domain.

We shall prove two lemmas.

Lemma 2. *Let K be a plane convex domain of constant width* 1. *Let A, B be points of ∂K at distance less than* 1 *apart. Then the diameter of the minor arc of ∂K from A to B is attained exactly at the chord AB.*

Note. By the term 'minor arc' we mean that arc of ∂K, determined by A and B, of diameter less than 1.

Proof of Lemma 2. Let $\overline{AB}$ denote the minor arc and suppose that the diameter of AB is attained on CD, where we may assume that C is the point A, but D is not B. Then there is a tangent line to K at D which is perpendicular to CD. Consequently, CD has length 1, which is impossible, since C, D lie on the minor arc $\overline{AB}$.

Lemma 3. *Let Q be a t-polygon of diameter d with sides all equal to* 1 *and let P be a t-polygon with perimeter at least t and each edge of length less that $2d$. Let P^* and Q^* denote the union of the d-arcs erected on the sides of P and Q, respectively. Then the length of Q^* is at most that of P^* with equality if, and only if, P has sides all equal to* 1.

Proof. Let $a_1, \ldots, a_t$ denote the lengths of the sides of P. We may assume that

$$a_1 + \cdots + a_t = t.$$

The length of the minor arc of the circle of radius d and chord a_i is

$$2d \sin^{-1}(a_i/2d), \qquad i = 1, \ldots, t.$$

Thus, the length of P^* is

$$2d \sum_{i=1}^{t} \sin^{-1}(a_i/2d).$$

The function $\sin^{-1} x$, $0 \leqslant x < 1$ has derivative $(1 - x^2)^{-1/2} > 0$, and so $\sin^{-1} x$ is a strictly convex function for $0 < x < 1$. Hence, if $0 \leqslant x < y < 1$, we have

$$\sin^{-1} x + \sin^{-1} y > 2 \sin^{-1} \frac{x+y}{2}.$$

Therefore, the length of P^* has its minimum value precisely when

$$a_1 = a_2 = \cdots = a_t = 1,$$

as required.

Proof of Theorem 3. Let P be a t-polygon of diameter d with maximal perimeter. Then P is contained in a convex domain R of constant width d. Let $A_1, \ldots, A_t$ be the vertices of P and $\boldsymbol{O}$ be the centre of gravity of P. The positive half-line from $\boldsymbol{O}$ through A_i will intersect ∂R at A'_i, say, for $i = 1, \ldots, t$. Let M denote the t-polygon determined by the vertices $A'_i, \ldots, A'_t$. Then, by Cauchy's formula, the length of M is at least that of P with equality if, and only if, P is M. So $P = M$ and $A'_i = A_i$, for $i = 1, \ldots, t$.

Let P^* denote the union of the d-arcs erected on the sides of P. Although P^* may not be the boundary of a convex domain, we may apply Cauchy's formula to each convex set formed individually by the edge A_iA_{i+1}, for $i = 1, \ldots, t$ (with $A_1 = A_{t+1}$) and the corresponding d-arc, which lies in ∂R, to conclude that the length $L(P^*)$ of P^* is at most that of ∂R, which is equal to πd, with equality if and only if P^* is ∂R, i.e.

$$L(P^*) \leqslant \pi d, \tag{5}$$

with equality if and only if $P^* = \partial R$.

Let Q be a regular (d, t)-polygon, then Q^* has constant width d and, as P has maximal perimeter, we have, by Lemma 3:

$$\pi d = L(Q^*) \leq L(P^*), \tag{6}$$

with equality if and only if P has equal sides of length $2d \sin(\pi/2t)$. So, combining (5) and (6), $L(P^*) = \pi d$ and $P^* = \partial R$. Also, P has t equal sides of length $2d \sin(\pi/2t)$, i.e. P is a regular (d, t)-polygon.

Proof of Theorem 2. Let K be a plane convex domain of diameter 1. Then K is contained in a plane convex domain R, say, of constant width 1. Furthermore, if ∂R can be divided into t sets of diameter less than or equal to a, say, then using the nearest point map, so can ∂R. Starting from P_0 we choose successive points $P_1(a), P_2(a), \ldots, P_t(a)$ on ∂R in an anti-clockwise direction, with $\|P_i(a) - P_{i+1}(a)\| = a$, $i = 0, 1, \ldots, t-1$. In view of Lemma 2, $P_t(a)$ is a continuous function of a, and so we may choose a such that $P_t(a) = P_0(a)$. Hence, the points $P_0, P_1, \ldots, P_{t-1}$ corresponding to this value of a, form a t-polygon Q, say, inscribed to ∂R, with

$$\|P_i - P_{i+1}\| = a, \qquad i = 0, 1, \ldots, t-1.$$

By Theorem 3, a is at most $2\sin(\pi/2t)$. Furthermore, by Lemma 2, the minor arc $\partial R_0, \partial R_1, \ldots, \partial R_{t-1}$ forms the required division of ∂R into t sets of diameter at most $2\sin(\pi/2t)$. Hence, using the nearest point map, ∂K can be divided into t sets of diameter at most $2\sin(\pi/2t)$, $1 \cdot cp_2(t) \leq 2\sin(\pi/2t)$. On the other hand, the circumference of the unit circle cannot be divided into t sets of diameter less than $\sin(\pi/2)$. Hence, $\sin(\pi/2) \leq p_2(t)$.

It is well known that any plane set of diameter 1 can be divided into three sets of diameter at most $\sqrt{3}/2$, and hence $p_2(3) = \frac{1}{2}\sqrt{3}$. Also, when $t = 4$, the square is the only 4-polygon of given diameter with greatest perimeter and equal sides and hence $p_2(4) = (\frac{1}{2})^{1/2}$.

References

[1] L. Fejes-Toth, Regular Figures (Macmillan, 1963).

[2] B. Grünbaum, Borsuk's problem and related questions, Proc. Amer. Math. Soc. Symp. Pure Math. (Convexity) 7 (1963).

[3] H. Hadwiger, Von der Zerlegung der Kugel in kleinere Teile, Gaz. Mat. Lisboa 15 (1954) 1–3.

[4] H. Lenz, Uber die Bedeckung Ebener Punktmengen in Solche kleineren Durchmessers, Arch. Math. 7 (1956) 34–40.

[5] H. Lenz, Zerlegung ebener Bereiche in konvexe Zellen von moglichst kleinem Durchmesser, Jber. Deutsch Math. Verein 58 (1956) 87–97.

[6] E. Schmidt, Die Brunn-Minkowskiche Ungleichung und ihr Spiegelbild so wie die isoperimetrische Eigenschaft der Kugel in der euklidischen und nicht euklidischen Geometrie, Math. Nachr. 1 (1948) 81–157.

Received July 1980

Annals of Discrete Mathematics 20 (1984) 215–232
North-Holland

BOUNDING THE NUMBERS OF FACES OF POLYTOPE PAIRS AND SIMPLE POLYHEDRA

Carl W. LEE*

IBM Thomas J. Watson Research Center, Yorktown Heights, New York 10598, U.S.A.
Department of Mathematics, University of Kentucky, Lexington, Kentucky 40506, U.S.A.

Let P be a simplicial d-polytope with ν vertices and $\Sigma(P)$ be the simplicial $(d-1)$-complex associated with the boundary of P. Suppose, for a given face F of P, that we know the numbers of faces of various dimensions of $\mathrm{lk}_{\Sigma(P)}F$. Then we are able to determine upper and lower bounds for the possible numbers of faces of all dimensions of P and of $\Sigma(P)\setminus F$. As a consequence, we can bound the numbers of faces of a simple d-polyhedron P if the numbers of bounded and unbounded facets of P and the dimension of the recession cone of P are specified.

1. Introduction

Klee [7] in 1966 proved that every simple d-polyhedron P with ν facets has at least $\nu - d + 1$ vertices. Grünbaum [6, Section 10.2] speculated whether this result might be improved upon if one specified both the number of bounded and of unbounded facets of P. In 1974 Klee [8] approached problems of this form from the point of view of pairs of simple polytopes while investigating the efficiency of a proposed algorithm to enumerate the vertices of a simple polytope defined by linear inequalities. By taking advantage of the recently established McMullen's conditions for the numbers of faces of simplicial polytopes, Billera and Lee [5, 9] were able to strengthen Klee's results. In this paper we offer some further extensions to the above results, and as a consequence will be able to provide (often tight) bounds on the numbers of faces of a simple unbounded polyhedron with recession cone of a specified dimension.

In the next three sections we review some preliminary material that is largely (though not entirely) discussed also in [5]. The main results of this paper are presented in the remaining two sections.

2. Polyhedra and f-vectors

Let P be a d-polyhedron (convex d-dimensional polyhedron). Faces of P of dimension 0, 1, $d-2$ and $d-1$ will be called *vertices, edges, ridges* (or *subfacets*)

* This research was supported, in part, by the National Science Foundation under grant MCS77-28392 and by a National Science Foundation Graduate Fellowship.

and *facets* of P, respectively. The set of vertices of P will be denoted $V(P)$. We will assume that any polyhedron under consideration has nonempty vertex set. Recall that every unbounded d-polyhedron P can be decomposed as $P = Q + K$, where Q is a d-polytope and K is a polyhedral cone, called the *recession cone* of P. A d-polyhedron P is *simple* if every vertex of P is contained in exactly d facets of P. A d-polytope P is *simplicial* if every face of P is a geometric simplex. Let $\mathcal{P}_s^d$ be the set of all simplicial d-polytopes. For every simplicial (respectively, simple) d-polytope P one can find a simple (respectively, simplicial) d-polytope P^* that is *dual* to P in the sense that there is an inclusion-reversing bijection between the set of faces of P and the set of faces of P^*. For integer $0 \leq j \leq d-1$, let $f_j(P)$ denote the number of j-faces (j-dimensional faces) of P. The d-vector $f(P) = (f_0(P), f_1(P), \ldots, f_{d-1}(P))$ is called the *f-vector* of P. Define $f(\mathcal{P}_s^d) = \{f(P)\colon P \in \mathcal{P}_s^d\}$. When $P \in \mathcal{P}_s^d$ we will also set $f_{-1}(P) = 1$, and $f_j(P) = 0$ if $j < -1$ or $j > d-1$.

3. Simplicial complexes

A *simplicial complex* Δ on the finite set $V = V(\Delta)$ is a nonempty collection of subsets of V with the property that $\{v\} \in \Delta$ for all $v \in V$ and that $F \in \Delta$ whenever $F \subseteq G$ for some $G \in \Delta$. For $F \in \Delta$ we say F is a *face* of Δ and the *dimension* of F, $\dim F$, equals j if $\operatorname{card} F = j+1$. In this case we call F a *j-face* of Δ. The *dimension* of Δ, $\dim \Delta$, is defined to be $\max\{\dim F\colon F \in \Delta\}$. If $\dim \Delta = d$, we will refer to Δ as a *simplicial d-complex*. Analogously to polyhedra, for simplicial $(d-1)$-complex Δ we define the *vertices*, *edges*, *ridges*, *facets* and *f-vector* of Δ.

Two simplicial complexes, Δ_1 and Δ_2, are *isomorphic*, denoted $\Delta_1 \cong \Delta_2$, if there is a bijection between $V(\Delta_1)$ and $V(\Delta_2)$ which induces a bijection between Δ_1 and Δ_2.

We will often write $v_1 v_2 \cdots v_k$ for the set $F = \{v_1, v_2, \ldots, v_k\}$ and $\bar{F}$ or $\overline{v_1 v_2 \cdots v_k}$ for the power set of F.

By $|\Delta|$ is meant the underlying topological space of the simplicial complex Δ. If $|\Delta|$ is a topological d-ball (respectively, d-sphere), we say Δ is a *simplicial d-ball* (respectively, *simplicial d-sphere*). For simplicial d-ball Δ, write $\partial\Delta$ for the simplicial $(d-1)$-sphere associated with $\partial|\Delta|$. This complex is called the *boundary* of Δ and it is known that $\partial\Delta = \bigcup\{\bar{F}\colon F$ is a ridge of Δ contained in exactly one facet of $\Delta\}$.

Let Δ be a simplicial complex. For any $F \in \Delta$, the *link* of F in Δ is the simplicial complex $\operatorname{lk}_\Delta F = \{G \in \Delta\colon G \cap F = \emptyset,\ G \cup F \in \Delta\}$. Furthermore, if $F \neq \emptyset$, the *deletion* of F from Δ is the simplicial complex $\Delta \backslash F = \{G \in \Delta\colon F \not\subseteq G\}$.

For simplicial $(d-1)$-complex Δ, define the polynomials

$$f(\Delta, t) = \sum_{j=-1}^{d-1} f_j(\Delta) t^{j+1}$$

and

$$h(\Delta, t) = (1-t)^d f\left(\Delta, \frac{t}{1-t}\right).$$

The *h-vector* of Δ is the $(d+1)$-vector $h(\Delta) = (h_0(\Delta), h_1(\Delta), \ldots, h_d(\Delta))$ determined by the polynomial relation

$$h(\Delta, t) = \sum_{i=0}^{d} h_i(\Delta) t^i.$$

We also set $h_i(\Delta) = 0$ if $i < 0$ or $i > d$. (McMullen and Shephard [11] write $g_{i-1}^{(d)}(\Delta)$ instead of $h_i(\Delta)$, and use

$$f(\Delta, t) = \sum_{j=-1}^{d-1} (-1)^{j+1} f_j(\Delta) t^{j+1}$$

and

$$g^{(d)}(\Delta, t) = (1-t)^d f\left(\Delta, \frac{t}{t-1}\right) = \sum_{i=-1}^{d-1} g_i^{(d)}(\Delta) t^{i+1}$$

to define the $g_i^{(d)}(\Delta)$.) The $h_i(\Delta)$ can be written explicitly as linear combinations of the $f_j(\Delta)$ by

$$h_i(\Delta) = \sum_{j=0}^{i} (-1)^{i-j} \binom{d-j}{d-i} f_{j-1}(\Delta), \qquad 0 \leq i \leq d,$$

and $f(\Delta)$ can be recovered from $h(\Delta)$ by

$$f(\Delta, t) = (1+t)^d h\left(\Delta, \frac{t}{1+t}\right),$$

or by

$$f_j(\Delta) = \sum_{i=0}^{j+1} \binom{d-i}{d-j-1} h_i(\Delta), \qquad -1 \leq j \leq d-1.$$

We also define $g_i(\Delta) = h_i(\Delta) - h_{i-1}(\Delta)$ for all integer i.

4. Polyhedral complexes

Let P be a simplicial d-polytope. The simplicial $(d-1)$-complex $\Sigma(P)$ associated with P is defined to be $\Sigma(P) = \{F \subset V(P) : \operatorname{conv} F \text{ is a face of } \partial P\}$, where conv F denotes the convex hull of F and ∂P means the boundary of P. It will be natural sometimes to abuse notation and refer to an $F \in \Sigma(P)$ itself as a face of P, but it should always be clear from the context whether by F we mean a face of P or a face of $\Sigma(P)$. Because such a simplicial complex is a simplicial sphere, we will call it a *polyhedral* $(d-1)$*-sphere*. We will write $f(P,t)$, $h(P,t)$, etc. for $f(\Sigma(P),t)$, $h(\Sigma(P),t)$, etc. and call $h(P)$ the h*-vector* of P. Define also $h(\mathcal{P}_s^d) = \{h(P): P \in \mathcal{P}_s^d\}$.

Now suppose $1 \leq k \leq d$ are integers, P is a simplicial d-polytope and F is a $(k-1)$-face of P. Then the simplicial complex $\Delta = \Sigma(P) \backslash F$ is a simplicial $(d-1)$-ball, and will be called a *polyhedral* $(d-1)$*-ball*. Let P^* be a simple d-polytope that is dual to P and F^* be the $(d-k)$-face of P^* corresponding to F. Define $Q^* = P^* \sim F^*$ to be the unbounded d-polyhedron obtained from P^* by applying a projective transformation [6, Section 1.1; 11, Section 1.2] that sends a supporting hyperplane defining F^* onto the hyperplane at infinity. Then Q^* will have a recession cone of dimension $d-k+1$, and Δ is *dual* to Q^* in the sense that there is an inclusion-reversing bijection between the faces of Δ and the nonempty faces of Q^*. In fact, every unbounded, simple d-polyhedron is dual to $\Sigma(P) \backslash v$ for some simplicial d-polytope P with vertex v, and every unbounded, simple d-polyhedron with $(d-k+1)$-dimensional recession cone is dual to $\Sigma(P) \backslash F$ for some simplicial d-polytope P with $(k-1)$-face F.

If Δ is a polyhedral sphere or ball, we say that Δ is a *polyhedral complex.* We now summarize some operations that can be performed on these complexes.

Let Δ_1 and Δ_2 be simplicial complexes on disjoint vertex sets. The *join* of Δ_1 and Δ_2 is the simplicial complex $\Delta_1 \cdot \Delta_2 = \{F_1 \cup F_2 : F_1 \in \Delta_1, F_2 \in \Delta_2\}$. In this case $h(\Delta_1 \cdot \Delta_2, t) = h(\Delta_1, t) h(\Delta_2, t)$. If Δ_1 and Δ_2 are polyhedral complexes, then so is $\Delta_1 \cdot \Delta_2$ [12, 5].

Let Δ be a simplicial $(d-1)$-complex and F be a facet of Δ. For $v \notin V(\Delta)$, the *stellar subdivision* of F in Δ is the simplicial complex $\mathrm{st}(v,F)[\Delta] = (\Delta \backslash F) \cup \bar{v} \cdot \partial \bar{F}$. In this case, $h_i(\mathrm{st}(v,F)[\Delta]) = h_i(\Delta)$ if $i=0$ or $i=d$, and $h_i(\mathrm{st}(v,F)[\Delta]) = h_i(\Delta)+1$ if $1 \leq i \leq d-1$. If Δ is a polyhedral $(d-1)$-sphere (respectively, polyhedral $(d-1)$-ball), then $\mathrm{st}(v,F)[\Delta]$ is a polyhedral $(d-1)$-sphere (respectively, polyhedral $(d-1)$-ball) [12, 5].

For Δ a simplicial $(d-1)$-complex, $v \in V(\Delta)$ and $u \notin V(\Delta)$, the *simplicial wedge* of Δ on v is the simplicial complex $w(u,v)[\Delta] = \{\phi, u, v\} \cdot (\Delta \backslash v) \cup \overline{uv} \cdot \mathrm{lk}_\Delta v$. If Δ is a polyhedral $(d-1)$-sphere, then $w(u,v)[\Delta]$ is a polyhedral d-sphere, $\mathrm{lk}_{w(u,v)[\Delta]} u = \Delta$, and $h_i(w(u,v)[\Delta]) = h_{i-1}(\Delta) + h_i(\Delta) - h_{i-1}(\mathrm{lk}_\Delta v)$, $0 \leq i \leq d+1$ [12, 5].

By repeating the wedging operation, we can, for $k \geq 1$, define the *simplicial* k*-wedge* of a polyhedral $(d-1)$-sphere Δ on a vertex v of Δ, given $u_1, u_2, \ldots, u_k \notin V(\Delta)$, by taking $w^k(u_1 u_2 \cdots u_k, v)[\Delta]$ to be $w(u_1, v)[\Delta]$ if $k=1$ and to be $w(u_k, u_{k-1})[w^{k-1}(u_1 \cdots u_{k-1}, v)[\Delta]]$, if $k \geq 2$. In this case,

$w^k(u_1\cdots u_k, v)$ is a polyhedral $(d+k-1)$-sphere, $\mathrm{lk}_{w^k(u_1\cdots u_k,v)[\Delta]}u_1\cdots u_k=\Delta$, and

$$h_i(w^k(u_1\cdots u_k, v)[\Delta]) = \sum_{j=i-k}^{i} h_j(\Delta) - \sum_{j=i-k}^{i-1} h_j(\mathrm{lk}_\Delta v), \qquad 0\leq i\leq d+k.$$

Two kinds of simplicial d-polytopes figure prominently in the history of f-vector problems. The first is $C(\nu, d)$, the *cyclic polytope* of dimension d with ν vertices [6, Section 4.7; 11]. The Upper Bound Theorem [10, 11] states that $f_j(P)\leq f_j(C(\nu,d))$, $0\leq j\leq d-1$, for all d-polytopes P with ν vertices, and its proof uses properties of the h-vector of the cyclic polytope. It is known that

$$h_i(C(\nu,d)) = \begin{cases} \binom{\nu-d+i-1}{i}, & 0\leq i\leq [d/2], \\ \binom{\nu-i-1}{d-i}, & [d/2]+1\leq i\leq d, \end{cases}$$

where $[x]$ denotes the greatest integer not exceeding x. In fact, the Upper Bound Theorem was proved by showing that $h_i(P)\leq h_i(C(\nu,d))$, $1\leq i\leq d$, for all simplicial d-polytopes P with ν vertices. Adopting the convention that $\binom{-1}{0}=1$ and otherwise $\binom{a}{b}=0$ if $a<b$ or $b<0$, we may formally define $h_i(C(d,d))$; $h_i(C(\nu,0))$ for $\nu>1$; and $h_i(C(\nu,1))$ for $\nu>2$ using the above formula, even though such polytopes do not exist. Consistent with this definition we may take

$$f_j(C(d,d)) = \begin{cases} \binom{d}{j+1}, & -1\leq j\leq d-2, \\ 2, & j=d-1; \end{cases}$$

$f_{-1}(C(\nu,0))=1$ for $\nu>1$; and $f_{-1}(C(\nu,1))=1$ and $f_0(C(\nu,1))=2$ for $\nu>2$. Finally, we define $h_i(C(\nu,d))=0=f_j(C(\nu,d))$ for all i and j if $d<0$.

The other type of simplicial d-polytope of interest is $P(\nu,d)$, defined recursively as follows: $P(d+1,d)$ is any geometric d-dimensional simplex, and for $\nu>d+1$, $P(\nu,d)$ is obtained from $P(\nu-1,d)$ by building a 'pyramidal cap' over any one of the facets of $P(\nu-1,d)$. (Thus, the boundary complex of $P(\nu,d)$ is derived from the boundary complex of a geometric d-simplex by a sequence of $\nu-d-1$ stellar subdivisions.) Such polytopes are usually called *stacked* polytopes. The simplicial d-polytopes of type $P(\nu,d)$ are not combinatorially equivalent, but they all have ν vertices (if $d>1$) and the same numbers of j-faces, and the Lower Bound Theorem [1,2] asserts that $f_j(P)\geq f_j(P(\nu,d))$, $0\leq j\leq d-1$, for all simplicial d-polytopes P with ν vertices. The h-vector of $P(\nu,d)$ has a particularly simple form:

$$h_i(P(\nu,d)) = \begin{cases} 1, & i=0 \text{ or } i=d, \\ \nu-d, & 1\leq i\leq d-1, \end{cases}$$

and the f-vector of this polytope is

$$f_j(P(\nu,d)) = \begin{cases} \binom{d}{j+1} + (\nu-d)\binom{d}{j}, & 0 \leq j \leq d-2, \\ (\nu-d)(d-1)+2, & j = d-1. \end{cases}$$

Again, we formally define $h(P(\nu,0)) = (1)$ and $f_{-1}(P(\nu,0)) = 1$ if $\nu > 1$; $h(P(\nu,1)) = (1,1)$, $f_{-1}(P(\nu,1)) = 1$ and $f_0(P(\nu,1)) = 2$ if $\nu > 2$; and $h_i(P(\nu,d)) = 0 = f_j(P(\nu,d))$ for all i and j if $d < 0$. As a consequence of McMullen's conditions [3, 4, 13] which we are about to describe, we know also that $h_i(P) \geq h_i(P(\nu,d))$, $1 \leq i \leq d$, for all simplicial d-polytopes P with ν vertices.

For positive integers k and i, k can be written uniquely in the form

$$k = \binom{n_i}{i} + \binom{n_{i-1}}{i-1} + \cdots + \binom{n_j}{j},$$

where $n_i > n_{i-1} > \cdots > n_j \geq j \geq 1$, from which is defined

$$k^{\langle i \rangle} = \binom{n_i+1}{i+1} + \binom{n_{i-1}+1}{i} + \cdots + \binom{n_j+1}{j+1}.$$

We also define $0^{\langle i \rangle} = 0$ for positive integer i. A $(d+1)$-vector of integers $(h_0, h_1, \ldots, h_d)$ is called an *O-sequence* (or *M-vector*) if $h_0 = 1$, $h_i \geq 0$, $1 \leq i \leq d$, and $h_{i+1} \leq h_i^{\langle i \rangle}$, $1 \leq i \leq d-1$.

Our primary tool in the study of f-vectors and h-vectors of polyhedral complexes is the characterization of $h(\mathcal{P}_s^d)$ given by

Theorem 1 (McMullen's Conditions). *Let $h = (h_0, h_1, \ldots, h_d)$ be a $(d+1)$-vector of integers, $g_0 = h_0$, and $g_i = h_i - h_{i-1}$, $1 \leq i \leq n = [d/2]$. Then $h \in h(\mathcal{P}_s^d)$ if and only if the following two conditions hold*:

(1) $h_i = h_{d-i}$, $0 \leq i \leq n$ *(the Dehn–Sommerville equations)*;

(2) $(g_0, g_1, \ldots, g_n)$ *is an O-sequence.*

In particular, we remark that (2) implies $h_0 \leq h_1 \leq \cdots \leq h_n$, and hence the h-vector of a simplicial d-polytope must be *unimodal.*

5. Polytope pairs

A *polytope pair* (P, F) of type (d, ν, k, h), where $1 \leq k \leq d < \nu$ and $h \in h(\mathcal{P}_s^{d-k})$, is a simplicial d-polytope P with ν vertices and a $(k-1)$-face $F \in \Sigma(P)$ such that $h = h(\mathrm{lk}_{\Sigma(P)}F)$. We remark that there exists a simplicial $(d-k)$-polytope Q such that $\Sigma(Q)$ is isomorphic to $\mathrm{lk}_{\Sigma(P)}F$ (take Q to be the 'quotient polytope' P/F [11, Section 2.2]); hence the necessity of $h \in h(\mathcal{P}_s^{d-k})$.

Naturally, McMullen's conditions also hold for $h(P)$. It is easy to see that $f_0(\mathrm{lk}_{\Sigma(P)}F) \leqslant v - k$ and hence that $h_1 \leqslant v - d$. Let $\Delta = \Sigma(P) \setminus F$. Then a simple calculation yields $f_j(P) = f_j(\Delta) + f_{j-k}(\mathrm{lk}_{\Sigma(P)}F)$ for all j and $h_i(P) = h_i(\Delta) + h_{i-k}(\mathrm{lk}_{\Sigma(P)}F)$ for all i.

Now let P^* be a simple d-polytope dual to P, F^* be the $(d-k)$-face of P^* corresponding to F, and $Q^* = P^* \sim F^*$. Duality then yields

$$
\begin{aligned}
f_j(P) &= f_{d-j-1}(P^*), && 0 \leqslant j \leqslant d-1,\\
f_j(\Delta) &= f_{d-j-1}(Q^*), && 0 \leqslant j \leqslant d-1,\\
f_j(\partial\Delta) &= f^{(u)}_{d-j-1}(Q^*), && 0 \leqslant j \leqslant d-2,\\
f_j(\mathrm{lk}_{\Sigma(P)}F) &= f_{d-k-j-1}(F^*), && 0 \leqslant j \leqslant d-k-1,
\end{aligned}
$$

where $f^{(u)}_j(Q^*)$ denotes the number of unbounded j-faces of Q^*. In particular, the number of facets of Q^* is $v-1$ if $k=1$ and v if $2 \leqslant k \leqslant d$, and the number of unbounded facets of Q^* is $f_0(\mathrm{lk}_{\Sigma(P)}F)$ if $k=1$ and $f_0(\mathrm{lk}_{\Sigma(P)}F)+k$ if $2 \leqslant k \leqslant d$.

First, let us consider the easy cases of $k = d-1$ and $k = d$. If F is any facet of any simplicial d-polytope P, then $\mathrm{lk}_{\Sigma(P)}F$ is a simplicial (-1)-sphere and $h(\mathrm{lk}_{\Sigma(P)}F) = (1)$. Similarly, if F is any ridge of any simplicial d-polytope P, then $\mathrm{lk}_{\Sigma(P)}F$ is a simplicial 0-sphere and $h(\mathrm{lk}_{\Sigma(P)}F) \times (1,1)$. McMullen's conditions therefore allow us to characterize completely $\{h(P)\}$ and $\{h(\Sigma(P)\setminus F)\}$ for all polytope pairs (P,F) of type (d,v,k,h), when $k=d-1$ or $k=d$. As a result of duality, the Upper Bound Theorem and the Lower Bound Theorem, we have

Theorem 2. *Let $3 \leqslant d < v$. Assume that either $k=1$ and $r=d$; or else that $k=2$ and $r=d+1$. As P ranges over all simple d-polyhedra with recession cone of dimension k, and v facets, r of which are unbounded, then*

$$
\text{(i)} \qquad \min f_j(P) = \begin{cases} f_{d-1}(P(v,d)) - k, & \text{if } j=0,\\ f_{d-2}(P(v,d)) - 1, & \text{if } j=1 \text{ and } k=2,\\ f_{d-j-1}(P(v,d)), & \text{if } k \leqslant j \leqslant d-1; \end{cases}
$$

$$
\text{(ii)} \qquad \max f_j(P) = \begin{cases} f_{d-1}(C(v,d)) - k, & \text{if } j=0,\\ f_{d-2}(C(v,d)) - 1, & \text{if } j=1 \text{ and } k=2,\\ f_{d-j-1}(C(v,d)), & \text{if } k \leqslant j \leqslant d-1. \end{cases}
$$

Moreover, for either value of k, there exist simple d-polyhedra P_1 and P_2 satisfying the above conditions such that $f(P_1)$ achieves all of the values in (i) and $f(P_2)$ achieves all of the values in (ii).

Now fix $3 \leqslant d < v$, $1 \leqslant k \leqslant d-2$ and $h \in h(\mathscr{P}^{d-k}_s)$ such that $1 \leqslant h_1 \leqslant v-d$. As (P,F) ranges over all polytope pairs of type (d,v,k,h) define

$$
\lambda^1_i(d,v,k,h) = \min h_i(P), \qquad 0 \leqslant i \leqslant d,
$$

$$\lambda_i^2(d, v, k, h) = \min h_i(\Sigma(P)\backslash F), \qquad 0 \leq i \leq d,$$

$$\mu_i^1(d, v, k, h) = \max h_i(P), \qquad 0 \leq i \leq d,$$

$$\mu_i^2(d, v, k, h) = \max h_i(\Sigma(P)\backslash F), \qquad 0 \leq i \leq d.$$

Of course, the Dehn–Sommerville equations imply that

$$\lambda_i^1(d, v, k, h) = \lambda_{d-i}^1(d, v, k, h), \qquad 0 \leq i \leq [d/2],$$

$$\mu_i^1(d, v, k, h) = \mu_{d-i}^1(d, v, k, h), \qquad 0 \leq i \leq [d/2].$$

Our goal is to provide bounds on these values and to investigate the tightness of these bounds. In [5] we consider the case $k = 1$ and prove the following three results:

Theorem 3. *Let $3 \leq d < v$ and $h \in h(\mathcal{P}_s^{d-1})$ such that $1 \leq h_1 \leq v - d$. Put $n = [d/2]$ and $m = [(d-1)/2]$. Then*

(i) $$\lambda_i^1(d, v, 1, h) = \begin{cases} 1, & i = 0, \\ v - d - h_1 + h_i, & 1 \leq i \leq n; \end{cases}$$

(ii) $$\lambda_i^2(d, v, 1, h) = \begin{cases} 1, & i = 0, \\ v - d - h_1 + h_i - h_{i-1}, & 1 \leq i \leq m, \\ v - d - h_1, & m+1 \leq i \leq d-1, \\ 0, & i = d; \end{cases}$$

(iii) $$\mu_i^1(d, v, 1, h) = \binom{v-d+i-2}{i} + h_{i-1}, \qquad 0 \leq i \leq n;$$

(iv) $$\mu_i^2(d, v, 1, h) = \begin{cases} \binom{v-d+i-2}{i}, & 0 \leq i \leq n, \\ \binom{v-i-2}{d-i} + h_i - h_{i-1}, & n+1 \leq i \leq d. \end{cases}$$

Moreover, there exist polytope pairs (P_1, v_1) and (P_2, v_2) of type $(d, v, 1, h)$ such that

$$h_i(P_1) = \lambda_i^1(d, v, 1, h), \qquad 0 \leq i \leq d,$$

$$h_i(\Sigma(P_1)\backslash v_1) = \lambda_i^2(d, v, 1, h), \qquad 0 \leq i \leq d,$$

$$h_i(P_2) = \mu_i^1(d, v, 1, h), \qquad 0 \leq i \leq d,$$

$$h_i(\Sigma(P_2)\backslash v_2) = \mu_i^2(d, v, 1, h), \qquad 0 \leq i \leq d.$$

Corollary 1. *Let $3 \leq d \leq r < v$ and put $n = [d/2]$ and $m = [(d-1)/2]$. Then as P*

ranges over all simplicial d-polytopes with ν vertices, one of which, v, is on exactly r edges, we have the following minima and maxima:

	function	*minimum*
(i)	$h_i(\mathrm{lk}_{\Sigma(P)} v)$	$\begin{cases} 1, & i=0, \\ r-d+1, & 1 \leqslant i \leqslant m; \end{cases}$
(ii)	$h_i(P)$	$\begin{cases} 1, & i=0, \\ \nu-d, & 1 \leqslant i \leqslant n; \end{cases}$
(iii)	$h_i(\Sigma(P)\setminus v)$	$\begin{cases} 1, & i=0, \\ \nu-d-1, & i=1, \\ \nu-r-1, & 2 \leqslant i \leqslant d-1, \\ 0, & i=d; \end{cases}$
(iv)	$f_j(\mathrm{lk}_{\Sigma(P)} v)$	$\begin{cases} \binom{d-1}{j+1} + (r-d+1)\binom{d-1}{j}, & 0 \leqslant j \leqslant d-3, \\ (r-d+1)(d-2)+2, & j=d-2; \end{cases}$
(v)	$f_j(P)$	$\begin{cases} \binom{d}{j+1} + (\nu-d)\binom{d}{j}, & 0 \leqslant j \leqslant d-2, \\ (\nu-d)(d-1)+2, & j=d-1; \end{cases}$
(vi)	$f_j(\Sigma(P)\setminus v)$	$\begin{cases} \binom{d}{j+1} + (\nu-d-1)\binom{d-1}{j} + (\nu-r-1)\binom{d-1}{j-1}, & 0 \leqslant j \leqslant d-2, \\ (\nu-r-1)(d-2)+\nu-d, & j=d-1; \end{cases}$
	function	*maximum*
(vii)	$h_i(\mathrm{lk}_{\Sigma(P)} v)$	$\binom{r-d+i}{i}, \quad 0 \leqslant i \leqslant m;$
(viii)	$h_i(P)$	$\binom{\nu-d+i-2}{i} + \binom{r-d+i-1}{i-1}, \quad 0 \leqslant i \leqslant n;$
(ix)	$h_i(\Sigma(P)\setminus v)$	$\begin{cases} \binom{\nu-d+i-2}{i}, & 0 \leqslant i \leqslant n, \\ \binom{\nu-i-2}{d-i}, & n+1 \leqslant i \leqslant d-2, \\ \nu-r-1, & i=d-1, \\ 0, & i=d; \end{cases}$

(x) $f_j(\mathrm{lk}_{\Sigma(P)}v)$ $f_j(C(r,d-1)),\qquad 0\leqslant j\leqslant d-2;$

(xi) $f_j(P)$ $f_j(C(\nu-1,d))+f_j(C(r+1,d))-f_j(C(r,d)),\ 0\leqslant j\leqslant d-1;$

$$\text{(xii)}\quad f_j(\Sigma(P)\backslash v)\begin{cases} f_j(C(\nu-1,d)), & 0\leqslant j\leqslant d-3,\\ f_{d-2}(C(\nu-1,d))+d-r, & j=d-2,\\ f_{d-1}(C(\nu-1,d))+d-r-1, & j=d-1.\end{cases}$$

Moreover, there exist polytopes P_1, P_2 and P_3 of the above form with vertices v_1, v_2 and v_3, respectively, such that $h_i(\mathrm{lk}_{\Sigma(P_1)}v_1)$, $h_i(P_1)$, $h_i(\Sigma(P_1)\backslash v_1)$, $f_j(\mathrm{lk}_{\Sigma(P_1)}v_1)$, $f_j(P_1)$ and $f_j(\Sigma(P_1)\backslash v_1)$ are the values given in (i) through (vi), respectively; $h_i(\mathrm{lk}_{\Sigma(P_2)}v_2)$, $h_i(P_2)$, $f_j(\mathrm{lk}_{\Sigma(P_2)}v_2)$ and $f_j(P_2)$ are the values given in (vii), (viii), (x) and (xi), respectively; and $h_i(\Sigma(P_3)\backslash v_3)$ and $f_j(\Sigma(P_3)\backslash v_3)$ are the values given in (ix) and (xii), respectively.

Corollary 2. *Let $3\leqslant d\leqslant r\leqslant \nu$. As P ranges over all simple d-polyhedra with ν facets, exactly r of which are unbounded, then*

$$\text{(i)}\qquad \min f_j(P)=\begin{cases}(\nu-r)(d-2)+\nu-d+1, & j=0,\\[2mm] \binom{d}{j}+(\nu-d)\binom{d-1}{j}+(\nu-r)\binom{d-1}{j+1}, & 1\leqslant j\leqslant d-1;\end{cases}$$

$$\text{(ii)}\qquad \max f_j(P)=\begin{cases} f_{d-1}(C(\nu,d))+d-r-1, & j=0,\\ f_{d-2}(C(\nu,d))+d-r, & j=1,\\ f_{d-j-1}(C(\nu,d)), & 2\leqslant j\leqslant d-1.\end{cases}$$

Moreover, there exist simple d-polyhedra P_1 and P_2 satisfying the above conditions such that $f(P_1)$ achieves all of the values in (i) and $f(P_2)$ achieves all of the values in (ii).

We now turn to the polytope pairs with $2\leqslant k\leqslant d-2$. We can determine the values of $\lambda_i^1(d,\nu,k,h)$ and $\lambda_i^2(d,\nu,k,h)$, but at the present time must content ourselves with placing upper bounds on $\mu_i^1(d,\nu,k,h)$ and $\mu_i^2(d,\nu,k,h)$ in most cases.

Theorem 4. *Let $4\leqslant d<\nu$, $2\leqslant k\leqslant d-2$ and $h\in h(\mathscr{P}_s^{d-k})$ such that $h_1\leqslant \nu-d$. Let $n=[d/2]$, $m=[(d-k)/2]$ and $p=[(d-k+1)/2]$. Then (with the convention that $h_i=0$ if $i<0$):*

$$\text{(i)}\qquad \lambda_i^1(d,\nu,k,h)=\begin{cases}1, & i=0,\\ \nu-d-h_1+h_i, & 1\leqslant i\leqslant m,\\ \nu-d-h_1+h_m, & m+1\leqslant i\leqslant n;\end{cases}$$

$$\text{(ii)}\quad \lambda_i^2(d,\nu,k,h)=\begin{cases}1, & i=0,\\ \nu-d-h_1+h_i-h_{i-k}, & 1\leqslant i\leqslant m,\\ \nu-d-h_1+h_m-h_{i-k}, & m+1\leqslant i\leqslant d-m-1,\\ \nu-d-h_1, & d-m\leqslant i\leqslant d-1,\\ 0, & i=d;\end{cases}$$

$$\text{(iii)}\quad \mu_i^1(d,\nu,k,h)\leqslant\binom{\nu-d+i-1}{i}-\binom{\nu-d+i-k-1}{i-k}+h_{i-k},\ 0\leqslant i\leqslant n;$$

$$\text{(iv)}\quad \mu_i^2(d,\nu,k,h)\leqslant\begin{cases}\binom{\nu-d+i-1}{i}-\binom{\nu-d+i-k-1}{i-k}, & 0\leqslant i\leqslant n,\\ \binom{\nu-i-1}{d-i}-\binom{\nu-i-k-1}{d-i-k}+h_{d-i-k}-h_{i-k}, & \\ & n+1\leqslant i\leqslant d.\end{cases}$$

Moreover, there exists a polytope pair (P^*,F^*) of type (d,ν,k,h) such that

$$h_i(P^*)=\lambda_i^1(d,\nu,k,h),\qquad 0\leqslant i\leqslant n,$$

$$h_i(\Sigma(P^*)\backslash F^*)=\lambda_i^2(d,\nu,k,h),\qquad 0\leqslant i\leqslant d.$$

Finally, the upper bounds in (iii) and (iv) are achievable
(1) if $0\leqslant i\leqslant p$ or $d-p\leqslant i\leqslant d$,
(2) for all i if $h_j=\binom{\nu-d+j-1}{j}$, $0\leqslant j\leqslant m$, or
(3) for all i if $k\leqslant[(d+1)/2]$ and $h_j=h_{[(d+1)/2]-k}$, $[(d+1)/2]-k+1\leqslant j\leqslant m$,
in each case there being a polytope pair (P^*,F^*) of type (d,ν,k,h) achieving these upper bounds simultaneously.

Remark. If $k\geqslant n$, then the inequalities in (iii) and (iv) say simply $\mu_i^1(d,\nu,k,h)\leqslant\binom{\nu-d+i-1}{i}$, $0\leqslant i\leqslant n$; $\mu_i^2(d,\nu,k,h)\leqslant\binom{\nu-d+i-1}{i}$, $0\leqslant i\leqslant n$; and $\mu_i^2(d,\nu,k,h)\leqslant\binom{\nu-i-1}{d-i}-h_{i-k}$, $n+1\leqslant i\leqslant d$.

Proof. Note that the Dehn–Sommerville equations imply that $h_i(P)=h_{d-i}(P)$, $0\leqslant i\leqslant d$, and $h_i=h_{d-k-i}$, $0\leqslant i\leqslant d-k$, for every polytope pair of type (d,ν,k,h).

Establishing the bounds. We will determine lower bounds for $\lambda_i^1(d,\nu,k,h)$ and upper bounds for $\mu_i^1(d,\nu,k,h)$ by induction on k, from which we obtain bounds for $\lambda_i^2(d,\nu,k,h)$ and $\mu_i^2(d,\nu,k,h)$. First observe that we have equality in (i) and (ii) and we have equality with the upper bounds in (iii) and (iv) when we set $k=1$ by Theorem 3 (using for (i) and (ii) that $m=n$ if d is odd and $h_m=h_n$ by the Dehn–Sommerville equations if d is even). So assume $k\geqslant 2$ and (P,F) is a polytope pair of type (d,ν,k,h). Let v be any vertex of P in F, take Q to be a vertex figure of P at v, i.e. $Q = H \cap P$ for some hyperplane H that strictly separates v from the remaining vertices of P, and put $V' = V(F)\backslash v$. Then $G = \operatorname{conv} V'$ corresponds to a $(k-2)$-face of Q, which we will call G also. Note that $\operatorname{lk}_{\Sigma(P)}F = \operatorname{lk}_{\operatorname{lk}_{\Sigma(P)}v}G \cong \operatorname{lk}_{\Sigma(Q)}G$. Thus (Q, G) is a polytope pair of type $(d-1, \nu', k-1, h)$ for some $\nu' \leqslant \nu - 1$, and (P, v) is a polytope pair of type $(d, \nu, 1, h(Q))$.

Now $h_i(P) \geqslant \lambda_i^1(d, \nu, 1, h(Q))$, $h_1(Q) = \nu' - d + 1$ and $h_i(Q) \geqslant \lambda_i^1(d-1, \nu', k-1, h)$. So by Theorem 3 and induction:

$$h_i(P) \geqslant \nu - d - h_1(Q) + h_i(Q), \qquad 1 \leqslant i \leqslant n$$

$$\geqslant \begin{cases} \nu - d - (\nu' - d + 1) + (\nu' - d + 1) - h_1 + h_i, & 1 \leqslant i \leqslant m \\ \nu - d - (\nu' - d + 1) + (\nu' - d + 1) - h_1 + h_m, & m + 1 \leqslant i \leqslant [(d-1)/2] \end{cases}$$

$$= \begin{cases} \nu - d - h_1 + h_i, & 1 \leqslant i \leqslant m \\ \nu - d - h_1 + h_m, & m + 1 \leqslant i \leqslant [(d-1)/2]. \end{cases}$$

Also, as a consequence of McMullen's conditions, $h_n(P) \geqslant h_m(P) \geqslant \nu - d - h_1 + h_m$. Therefore, the values in (i) are lower bounds for the $\lambda_i^1(d, \nu, k, h)$.

Using the remarks in the beginning of this section, we have

$$h_i(\Sigma(P)\backslash F) = h_i(P) - h_{i-k}$$

$$\geqslant \begin{cases} \nu - d - h_1 + h_i - h_{i-k}, & 1 \leqslant i \leqslant m, \\ \nu - d - h_1 + h_m - h_{i-k}, & m + 1 \leqslant i \leqslant n; \end{cases}$$

$$\begin{aligned} h_i(\Sigma(P)\backslash F) &= h_i(P) - h_{i-k} \\ &= h_{d-i}(P) - h_{i-k} \\ &\geqslant \nu - d - h_1 + h_m - h_{i-k}, \qquad n + 1 \leqslant i \leqslant d - m - 1; \end{aligned}$$

and

$$\begin{aligned} h_i(\Sigma(P)\backslash F) &= h_i(P) - h_{i-k} \\ &= h_{d-i}(P) - h_{i-k} \\ &\geqslant \nu - d - h_1 + h_{d-i} - h_{i-k} \\ &= \nu - d - h_1 + h_{i-k} - h_{i-k} \\ &= \nu - d - h_1, \qquad d - m \leqslant i \leqslant d - 1. \end{aligned}$$

Therefore the values in (ii) are lower bounds for the $\lambda_i^2(d, \nu, k, h)$.

For the upper bounds in (iii), $h_i(P) \leq \mu_i^1(d, \nu, 1, h(Q))$ and $h_i(Q) \leq \mu_i^1(d-1, \nu', k-1, h) \leq \mu_i^1(d-1, \nu-1, k-1, h)$. (To justify this last inequality, we observe that if (P, F) is any polytope pair of type (d, ν, k, h), and P' is obtained from P by performing a stellar subdivision of some facet of P not containing F, then (P', F) is a polytope pair of type $(d, \nu+1, k, h)$, and $h_i(P') \geq h_i(P)$ for all i.) So by Theorem 3 and induction, for $0 \leq i \leq n$ we have

$$\begin{aligned} h_i(P) &\leq \binom{\nu-d+i-2}{i} + h_{i-1}(Q) \\ &\leq \binom{\nu-d+i-2}{i} + \binom{\nu-d+i-2}{i-1} - \binom{\nu-d+i-k-1}{i-k} + h_{i-k} \\ &= \binom{\nu-d+i-1}{i} - \binom{\nu-d+i-k-1}{i-k} + h_{i-k}. \end{aligned}$$

Therefore we have the upper bounds in (iii).

The upper bounds in (iv) come immediately from those in (iii) using the relation $h_i(\Sigma(P)\backslash F) = h_i(P) - h_{i-k}$.

Achieving the bounds. Because the bounds were derived using the relation $h_i(\Sigma(P)\backslash F) = h_i(P) - h_{i-k}$, it is sufficient to find a polytope pair (P^*, F^*) of type (d, ν, k, h) such that $h(P^*)$ achieves all of the bounds for $\lambda_i^1(d, \nu, k, h)$ simultaneously, for then we may conclude that $h(\Sigma(P^*)\backslash F^*)$ achieves all of the bounds for $\lambda_i^2(d, \nu, k, h)$ simultaneously.

By [5, Theorem 3.13] there exists a simplicial $(d-k)$-polytope P such that

(i) $h_i(P) = h_i$, $0 \leq i \leq d-k$, and

(ii) there is a vertex v of P such that $h_i(\mathrm{lk}_{\Sigma(P)} v) = h_i$, $0 \leq i \leq [(d-k-1)/2]$.

Let $\Sigma = w^k(u_1 \cdots u_k, v)[\Sigma(P)]$, where $u_1, u_2, \ldots, u_k \notin V(P)$. Now put $\Sigma_0 = \Sigma$ and for integer $j \geq 1$ let $\Sigma_j = \mathrm{st}(v_j, G_{j-1})[\Sigma_{j-1}]$ for some $v_j \notin V(\Sigma_{j-1})$ and some facet G_{j-1} of Σ_{j-1} not containing $u_1 u_2 \cdots u_k$. Let P^* be a simplicial d-polytope such that $\Sigma(P^*) = \Sigma_{\nu-d-h_1}$, and put $F^* = \mathrm{conv}\{u_1, u_2, \ldots, u_k\}$. Using the facts about stellar subdivisions and wedges it can be deduced that (P^*, F^*) is a polytope pair of type (d, ν, k, h) achieving all of the bounds of (i), and hence also of (ii), simultaneously.

To achieve the upper bounds of (iii) and (iv) for case (1), by [5, Theorem 3.13] there is a simplicial $(d-k)$-polytope P and a simplicial $(d-k+1)$-polytope Q such that

(i) $h(P) = h$;

(ii) Q has $\nu - k + 1$ vertices;

(iii) $$h_i(Q) = \begin{cases} \binom{\nu-d+i-2}{i} + h_{i-1}, & 0 \leq i \leq p, \\ \binom{\nu-k-i-1}{d-k-i+1} + h_{d-k-i}, & p+1 \leq i \leq d-k+1; \end{cases}$$

(iv) P is a vertex figure of Q at a vertex $z \in V(Q)$.
Let $\Sigma = w^{k-1}(u_1 \cdots u_{k-1}, z)[\Sigma(Q)]$, where $u_1, \ldots, u_{k-1} \notin V(Q)$, and let P^* be a simplicial d-polytope such that $\Sigma = \Sigma(P^*)$. Then

$$\begin{aligned} \mathrm{lk}_{\Sigma(P^*)} u_1 \cdots u_{k-1} z &= \mathrm{lk}_{\mathrm{lk}_{\Sigma(P^*)} u_1 \cdots u_{k-1}} z \\ &= \mathrm{lk}_{\Sigma(Q)} z \\ &= \Sigma(P). \end{aligned}$$

Put $F^* = u_1 \cdots u_{k-1} z$. Then $h(\mathrm{lk}_{\Sigma(P^*)} F^*) = h(P) = h$. Also, P^* has ν vertices, so (P^*, F^*) is a polytope pair of type (d, ν, k, h). For $0 \leqslant i \leqslant p$,

$$\begin{aligned} h_i(P^*) &= \sum_{j=i-k+1}^{i} h_j(Q) - \sum_{j=i-k+1}^{i-1} h_j(P) \\ &= \sum_{j=i-k+1}^{i} \left[\binom{\nu - d + j - 2}{j} + h_{j-1} \right] - \sum_{j=i-k+1}^{i-1} h_j \\ &= \binom{\nu - d + i - 1}{i} - \binom{\nu - d + i - k - 1}{i - k} + h_{i-k}. \end{aligned}$$

Hence, $h_i(P^*)$ achieves the upper bounds in (iii) if $0 \leqslant i \leqslant p$, and $h_i(\Sigma(P^*) \backslash F^*)$ achieves the upper bounds in (iv) if $0 \leqslant i \leqslant p$ or $d - p \leqslant i \leqslant d$.

Achieving the upper bounds of (iii) and (iv) in the case (2) is straightforward, taking P^* to be $C(\nu, d)$ with ordered vertex set $\{v_1, v_2, \ldots, v_\nu\}$, and $F^* = v_1 v_2 \cdots v_k$. The construction of a polytope pair to achieve the upper bounds in the case (3) is more tedious than illuminating, so we will simply refer the reader to [9, Section 5.3] for the details. □

Corollary 3. *Let* $4 \leqslant d < r \leqslant \nu$ *and* $2 \leqslant k \leqslant d - 2$. *Put* $n = [d/2]$ *and* $m = [(d-k)/2]$. *As* P *ranges over all simplicial* d*-polytopes with* ν *vertices and a* $(k-1)$*-face* F *such that* $f_0(\mathrm{lk}_{\Sigma(P)} F) = r - k$, *we have*

(i) $$\min h_i(\mathrm{lk}_{\Sigma(P)} F) = \begin{cases} 1, & i = 0, \\ r - d, & 1 \leqslant i \leqslant m; \end{cases}$$

(ii) $$\min h_i(P) = \begin{cases} 1, & i = 0, \\ \nu - d, & 1 \leqslant i \leqslant n; \end{cases}$$

(iii) $$\min h_i(\Sigma(P) \backslash F) = \begin{cases} 1, & i = 0, \\ \nu - d, & 1 \leqslant i \leqslant k - 1, \\ \nu - d - 1, & i = k, \\ \nu - r, & k + 1 \leqslant i \leqslant d - 1, \\ 0, & i = d; \end{cases}$$

(iv) $\min f_j(\mathrm{lk}_{\Sigma(P)}F) = f_j(P(r-k, d-k)), \qquad 0 \leq j \leq d-k-1;$

(v) $\min f_j(P) = f_j(P(v, d)), \quad 0 \leq j \leq d-1;$

(vi) $\min f_j(\Sigma(P)\backslash F) = f_j(P(v, d)) - f_{j-k}(P(r-k, d-k)), \qquad 0 \leq j \leq d-1;$

(vii) $\max h_i(\mathrm{lk}_{\Sigma(P)}F) = \binom{r-d+i-1}{i}, \qquad 0 \leq i \leq m;$

(viii) $$\max h_i(P) \leq \binom{v-d+i-1}{i} - \binom{v-d+i-k-1}{i-k} + \binom{r-d+i-k-1}{i-k}, \qquad 0 \leq i \leq n;$$

(ix) $$\max h_i(\Sigma(P)\backslash F) \leq h_i(C(v, d)) - h_{i-k}(C(v-2k, d-2k)) + h_{i-k}(P(r-2k, d-2k)) - h_{i-k}(P(r-k, d-k)), \qquad 0 \leq i \leq d;$$

(x) $\max f_j(\mathrm{lk}_{\Sigma(P)}F) = f_j(C(r-k, d-k)), \qquad 0 \leq j \leq d-k-1;$

(xi) $$\max f_j(P) \leq f_j(C(v, d)) - \sum_{i=0}^{k} \binom{k}{i} [f_{j-k-i+1}(C(v-2k, d-2k)) - f_{j-k-i+1}(C(r-2k, d-2k))], \qquad 0 \leq j \leq d-1;$$

(xii) $$\max f_j(\Sigma(P)\backslash F) \leq f_j(C(v, d)) - f_{j-k}(P(r-k, d-k)) - \sum_{i=0}^{k} \binom{k}{i} [f_{j-k-i+1}(C(v-2k, d-2k)) - f_{j-k-i+1}(P(r-2k, d-2k))], \qquad 0 \leq j \leq d-1,$$

with equality in (viii) *and* (ix) *if* $0 \leq i \leq [(d-k+1)/2]$ *or* $[(d+k)/2] \leq i \leq d$ *and equality in* (xi) *and* (xii) *if* $0 \leq j \leq [(d-k+1)/2]-1$. *Moreover, equality holds in* (viii), (ix), (xi) *and* (xii) *if* $k \leq [(d-1)/2]$, *and equality holds in* (viii) *and* (xi) *if* $r = v$.

Remark. As with Theorem 4, some of the inequalities above simplify if $k \geq n$:

$$\max h_i(P) \leq \binom{v-d+i-1}{i}, \qquad 0 \leq i \leq n;$$

$$\max h_i(\Sigma(P)\backslash F) \leq h_i(C(v, d)) - h_{i-k}(P(r-k, d-k)), \qquad 0 \leq i \leq d;$$

$$\max f_i(P) \leq f_j(C(v, d)), \qquad 0 \leq j \leq d-1;$$

and

$$\max f_j(\Sigma(P)\backslash F) \leq f_j(C(v, d)) - f_{j-k}(P(r-k, d-k)), \qquad 0 \leq j \leq d-1.$$

Proof. Because $f_0(\mathrm{lk}_{\Sigma(P)}F)=r-k$ we know that $h_1(\mathrm{lk}_{\Sigma(P)}F)=r-d$. Define $h_i^{(1)}=h_i(P(r-k,d-k))$, $g_i^{(1)}=h_i^{(1)}-h_{i-1}^{(1)}$ and $h_i^{(2)}=h_i(C(r-k,d-k))$ for all i. Let $h\in h(\mathscr{P}_s^d)$ with $h_1=r-d$, and put $h_i=0$ if $i<0$ or $i>d$.

By Theorem 1, $h_i\geqslant h_i^{(1)}$, $0\leqslant i\leqslant m$, which gives us (i) and (ii) using Theorem 4. For $1\leqslant i\leqslant m$.

$$\begin{aligned}h_i-h_{i-k}&=(h_i-h_{i-1})+(h_{i-1}-h_{i-2})+\cdots+(h_{i-k+1}-h_{i-k})\\&\geqslant g_i^{(1)}+g_{i-1}^{(1)}+\cdots+g_{i-k+1}^{(1)}\\&=h_i^{(1)}-h_{i-k}^{(1)},\end{aligned}$$

and for $m+1\leqslant i\leqslant d-m-1$, the unimodality of h forces $h_m-h_{i-k}\geqslant h_m^{(1)}-h_{i-k}^{(1)}$. Thus we obtain (iii) using Theorem 4.

By Theorem 1 or the Upper Bound Theorem, $h_i\leqslant h_i^{(2)}$, $0\leqslant i\leqslant m$, giving us (vii) and (viii) using Theorem 4. For (ix) we use Theorem 4 (iv) and attempt to make $h_{d-i-k}-h_{i-k}$ as large as possible, i.e. $h_{i-k}-h_{d-i-k}$ as small as possible, $n+1\leqslant i\leqslant d$. If $n+1\leqslant i\leqslant d-m-1$, then $i-k\leqslant d-k-m-1\leqslant m$ and $i-k>d-i-k$, so

$$\begin{aligned}h_{i-k}-h_{d-i-k}&=\sum_{j=d-i-k+1}^{i-k}(h_j-h_{j-1})\\&\geqslant\sum_{j=d-i-k+1}^{i-k}g_j^{(1)}\\&=h_{i-k}^{(1)}-h_{d-i-k}^{(1)}.\end{aligned}$$

If, on the other hand, $d-m\leqslant i\leqslant d$, then $d-i-k<d-i\leqslant m$, so

$$\begin{aligned}h_{i-k}-h_{d-i-k}&=h_{d-i}-h_{d-i-k}\\&=\sum_{j=d-i-k+1}^{d-i}(h_j-h_{j-1})\\&\geqslant\sum_{j=d-i-k+1}^{d-i}g_j^{(1)}\\&=h_{d-i}^{(1)}-h_{d-i-k}^{(1)}\\&=h_{i-k}^{(1)}-h_{d-i-k}^{(1)}.\end{aligned}$$

Therefore we obtain (ix).

The minima and the maxima for the f_j in (iv), (v), (vi) and (x) are determined from the facts that the f_j are non-negative linear combinations of the h_i, and that in each of these cases the bounds on the h_i are simultaneously achievable. The upper bounds in (xi) and (xii) are similarly obtained from the upper bounds in (viii) and (ix).

Equality in (viii) and (ix) if $0\leqslant i\leqslant[(d-k+1)/2]$ or $[(d+k)/2]\leqslant i\leqslant d$

comes from case (1) of Theorem 4, from which the equalities in (xi) and (xii) for $0 \leqslant j \leqslant [(d-k+1)/2]-1$ follow, recalling that f_j is a non-negative linear combination of $h_0, h_1, \ldots, h_{j+1}$. Equality for (viii) and (xi) if $r = \nu$ follows by case (2) of Theorem 4 and equality for (viii), (ix), (xi) and (xii) if $k \leqslant [(d-1)/2]$ follows from case (3) of Corollary 2. □

Corollary 4. *Let $4 \leqslant d < r \leqslant \nu$ and $2 \leqslant k \leqslant d-2$. As P ranges over all simple d-polyhedra with recession cone of dimension $d-k+1$, and ν facets, r of which are unbounded, then*

(i) $\min f_j(P) = f_{d-j-1}(P(\nu, d)) - f_{d-j-k-1}(P(r-k, d-k)), \qquad 0 \leqslant j \leqslant d-1;$

(ii)
$$\max f_j(P) \leqslant f_{d-j-1}(C(\nu, d)) - f_{d-j-k-1}(P(r-k, d-k)) - \sum_{i=0}^{k} \binom{k}{i} [f_{d-i-j-k}(C(\nu-2k, d-2k)) - f_{d-i-j-k}(P(r-2k, d-2k))], \qquad 0 \leqslant j \leqslant d-1.$$

Moreover, there exists a simple d-polyhedron P_1 satisfying the above conditions such that $f(P_1)$ achieves all of the values in (i); and if $k \leqslant [(d-1)/2]$, there exists a simple d-polyhedron P_2 satisfying the above conditions such that $f(P_2)$ achieves all of the upper bounds in (ii).

Proof. Immediate from Corollary 3 (vi) and (xii) and duality. □

Whether the inequalities of Theorem 4 (iii) and (iv); Corollary 3 (viii), (ix), (xi) and (xii); and Corollary 4 (ii) are actually equalities remains to be seen.

6. h-Vectors of polyhedral balls

McMullen's conditions provide a characterization of the set of h-vectors of simplicial polytopes, but as yet there is no characterization of the h-vectors of polyhedral balls. Two necessary conditions for a vector of integers $h = (h_0, h_1, \ldots, h_d)$ to be the h-vector of $\Sigma(P) \backslash F$ for some simplicial d-polytope P with $(k-1)$-face F are:

(1) $(h_0 - h_{d+j}, h_1 - h_{d+j-1}, \ldots, h_r - h_{d+j-r})$ is an O-sequence for all integer $0 \leqslant j \leqslant d+1$, where $r = [(d+j-1)/2]$.

(2) There exist vectors $a = (a_0, a_1, \ldots, a_{d-k}) \in h(\mathcal{P}_s^{d-k})$ and $b = (b_0, b_1, \ldots, b_d) \in h(\mathcal{P}_s^d)$ such that $h_i + a_{i-k} = b_i$, $0 \leqslant i \leqslant d$.

Condition (2) is immediate from the discussion in the beginning of Section 5 and the proof of condition (1) is the same as that in [5], once it is observed that the boundary of a polyhedral $(d-1)$-ball is always a polyhedral $(d-2)$-sphere.

Conjecture 1. Let $h = (h_0, h_1, \ldots, h_d)$ be a $(d+1)$-vector of integers. Then $h = h(\Sigma(P)\setminus F)$ for some simplicial d-polytope P with $(k-1)$-face F, $1 \leqslant k \leqslant d$, if and only if h satisfies (1) and (2).

We remark that the conjecture is true for $k = d-1$ or $k = d$ using the sufficiency of McMullen's conditions for h-vectors of simplicial d-polytopes, and that condition (2) is implied by condition (1) in the case that $k = 1$, using [5, Corollary 3.9].

Note added in proof. A construction of Barnette has established that all upper bounds in Corollaries 3 and 4 are tight.

References

[1] D.W. Barnette, The minimum number of vertices of a simple polytope, Israel J. Math. 10 (1971) 121–125.
[2] D.W. Barnette, A proof of the lower bound conjecture for convex polytopes, Pacific J. Math. 46 (1973) 349–354.
[3] L.J. Billera and C.W. Lee, Sufficiency of McMullen's conditions for f-vectors of simplicial polytopes, Bull. (New Series) Amer. Math. Soc. 2 (1980) 181–185.
[4] L.J. Billera and C.W. Lee, A proof of the sufficiency of McMullen's conditions for f-vectors of simplicial convex polytopes, J. Combinat. Theory (A) 31 (1981) 237–255.
[5] L.J. Billera and C.W. Lee, The numbers of faces of polytope pairs and unbounded polyhedra, European J. Combinatorics 2 (1981) 307–322.
[6] B. Grünbaum, Convex Polytopes (Wiley, New York, 1967).
[7] V. Klee, A comparison of primal and dual methods for linear programming, Numer. Math. 9 (1966) 227–235.
[8] V. Klee, Polytope pairs and their relationship to linear programming, Acta Math. 133 (1974) 1–25.
[9] C.W. Lee, Counting the faces of simplicial convex polytopes, Ph.D. Thesis, Cornell University, Ithaca, N.Y. (1981).
[10] P. McMullen, The maximum numbers of faces of a convex polytope, Mathematika 17 (1970) 179–184.
[11] P. McMullen and G.C. Shephard, Convex polytopes and the upper bound conjecture, London Math. Soc. Lecture Note Series 3 (Cambridge, 1971).
[12] J.S. Provan and L.J. Billera, Decompositions of simplicial complexes related to diameters of convex polyhedra, Math. Oper. Res. 5 (1980) 576–594.
[13] R.P. Stanley, The number of faces of a simplicial convex polytope, Adv. Math. 35 (1980) 236–238.

Received 13 April 1981; revised 25 October 1981

Annals of Discrete Mathematics 20 (1984) 233–240
North-Holland

PARITY IN THE REALM OF INFINITE PERMUTATIONS

Gadi MORAN
University of Haifa, Haifa 31999, Israel

1. Introduction

Let S_A denote the symmetric group over a set A. If A is finite, S_A admits a parity homomorphism p onto the two-element group $\mathbb{Z}_2 = \{0, 1\}$, assigning parity $p(\theta) \in \mathbb{Z}_2$ to each $\theta \in S_A$. We say that a subset X of S_A has *distinct parity* (d.p.) if it is nonempty, and $p(\theta) = p(\theta')$ whenever θ and θ' both belong to X. If $X \subseteq S_A$ is of distinct parity, we define its parity $p(X)$ by $p(X) = p(\theta)$, where θ is any element of X. The conjugacy classes (COCs) in S_A are of distinct parity. Also, whenever X, $Y \subseteq S_A$ are of distinct parity, such is $X \cdot Y = \{\xi\eta \mid \xi \in X, \eta \in Y\}$ and we have $p(X \cdot Y) = p(X) \dot{+} p(Y)$, where $0 \dot{+} 0 = 1 \dot{+} 1 = 0$, $0 \dot{+} 1 = 1 \dot{+} 0 = 1$. That is, the family of subsets of S_A of distinct parity forms a semigroup including all COCs, and the parity function p is a homomorphism of this semigroup onto $\mathbb{Z}_2$.

This scenario breaks completely when A is infinite — as we assume in the sequel. Indeed, no homomorphism of S_A onto $\mathbb{Z}_2$ exists. Moreover, every nontrivial homomorphic image of S_A is uncountable (such being the index of any proper normal subgroup of S_A [7]).

There is, however, a maximal normal subgroup S_A^0 of S_A admitting a parity homomorphism p; namely, the subgroup S_A^0 of *finitary* permutations — those $\theta \in S_A$ whose support $M(\theta) = \{a \in A \mid \theta(a) \neq a\}$ is finite. The parity $p(\theta)$ of $\theta \in S_A^0$ is defined then by $p(\theta) = p(\theta \mid M(\theta))$ (where $\theta \mid B$ denotes the restriction of θ to the subset B of A). p is again a class-function (i.e. p has a fixed value on each finitary COC), and so extends to the semigroup generated by the finitary COCs. There is, however, a COC X in S_A satisfying $X^2 = S_A$ [3], and so the parity homomorphism of the semigroup generated by the finitary COCs does not extend to the semigroup generated by all COCs. [Recall that in arbitrary symmetric group S_A, θ and θ' are conjugate if and only if for each $n = 1, 2, 3, \ldots, \aleph_0$ (where $\aleph_0$ is the first infinite cardinal) the set of θ-orbits[1] of cardinality n and the set of θ'-orbits of cardinality n are of equal cardinality [7]; hence the COC $[\theta]$ is completely determined by the cyclic structure of θ.]

[1] The set $\{\theta^m(a) \mid m \in \mathbb{Z}\}$ is called the θ-orbit containing a.

Several results in the literature suggest, however, that in studying products of COCs, some notion of parity for nonfinitary classes might prove useful. Let us say that $X \subseteq S_A$ has *distinct parity* (d.p.) if $X \cap S_A^0$ has distinct parity; that is, $X \cap S_A^0 \neq \emptyset$, and whenever $\theta, \theta' \in X \cap S_A^0$ we have $p(\theta) = p(\theta')$. When X has d.p. we let $p(x) = p(\theta)$, where $\theta \in X \cap S_A^0$. Here are some examples ($[\theta]$ denotes the COC of θ):

Example 1. If A is countable and θ is a complete cycle (with no fixed points) on A, then $[\theta]^2 \cap S_A^0$ is precisely the set of even finitary permutations [1]. Thus, $X = [\theta]^2$ has d.p. and $p([\theta]^2) = 0$.

Example 2. If φ is an involution with no fixed points then $[\varphi]^2$ is the set of 'nicely even' permutations, i.e. having an even[2] number of orbits of cardinality n for each n [4]. Thus $[\varphi]^2$ has d.p., and $p([\varphi]^2) = 0$.

Example 3. More generally, if $\varphi(\psi)$ is an involution with $i(j)$ fixed points, where i, j are non-negative integers, and if for some integer k $i - j = 2k$, then $([\varphi] \cdot [\psi]) \cap S_A^0$ has d.p., and $p([\varphi] \cdot [\psi]) = p(k)$, where $p(k) = 0$ if k is an even integer, $p(k) = 1$ if k is an odd integer ([4, Theorem 2.6, Lemma A.3]). If $i - j$ is an odd integer then $([\varphi] \cdot [\psi]) \cap S_A^0 = \emptyset$ and so $[\varphi] \cdot [\psi]$ has no d.p. ([4, Corollary 2.3(3)]; see also [5]).

Denote by P_i the COC of involutions in S_A that fix i points, and consider the family R of all P_i, where i is a finite integer. It follows from (3) that this family splits into two disjoint subfamilies: R_0 — COCs where the number of fixed points is even — and R_1 — COCs where the number of fixed points is odd. On each subfamily a parity-function $\hat{p}$ can be defined in such a way that $p(X \cdot Y) = \hat{p}(X) \dot{+} \hat{p}(Y)$. Namely, let $\hat{p}(R_{2k+1}) = \hat{p}(R_{2k}) = p(k)$.

Our next example points out a similar phenomenon for another family of COCs in S_A, where A is countably infinite.

Example 4. Let $|A| = \aleph_0$. Call $\xi \in S_A$ *monic* if ξ has a cofinite orbit. Let $\xi \in S_A$ be monic, let $A_1 \subseteq A$ be its cofinite orbit, and let $A_0 = A \setminus A_1$. Then A_0 is finite, and $\xi = \xi_0 \xi_1 = \xi_1 \xi_0$, where $\xi_0, \xi_1 \in S_A$ are defined by the conditions $\xi_i | A_i = \xi | A_i$ and $\xi_i | A_{1-i} = \mathbf{1}_{A_{1-i}}$, $i = 0, 1$ and $\mathbf{1}_B$ denotes the identity function on B. Thus, ξ_0 is a finitary permutation, and ξ_1 is a *cofinite cycle*, i.e. a monic permutation fixing every point not in its cofinite orbit. We refer to this factorization as the *canonical factorization* (c.f.) of ξ.

For a cofinite cycle ξ with i fixed points let $\hat{p}(\xi) = p(i)$; that is, $\hat{p}(\xi) = 0$ if i is

[2] Every infinite cardinal is considered even.

even and $\hat{p}(\xi)=1$ if i is odd. For arbitrary monic ξ let $\hat{p}(\xi)=p(\xi_0)\dot{+}\hat{p}(\xi_1)$. Obviously, $\hat{p}$ is invariant under conjugacy, and we let $\hat{p}([\xi])=\hat{p}(\xi)$ whenever ξ is monic. We are ready to state:

Theorem 1. *Let ξ and η be monic. Then $[\xi]\cdot[\eta]$ has distinct parity, and*

$$p([\xi]\cdot[\eta])=\hat{p}(\xi)\dot{+}\hat{p}(\eta).$$

The interplay relation between monic and finitary permutations is further illuminated by the following theorem:

Theorem 2. *Let M denote the set of all monic permutations. Then,*
(a) *for every $\eta\in M$, $M=S^0_A\cdot[\eta]=[\eta]\cdot S^0_A$.*
(b) *If $\theta\in S^0_A$, $\zeta\in M$ then*

$$\hat{p}(\theta\zeta)=\hat{p}(\zeta\theta)=p(\theta)\dot{+}\hat{p}(\zeta).$$

By part (b), given $\xi\in M$, $\hat{p}(\xi)$ can be computed from arbitrary factorization $\xi=\theta\zeta$, with $\theta\in S^0_A$, $\zeta\in M$.

Let L be the set of permutations that are either finitary or monic, i.e. $L=S^0_A\cup M$. Define $q(\zeta)\in\mathbb{Z}_2$ for $\zeta\in L$ by $q(\zeta)=p(\zeta)$ if $\zeta\in S^0_A$, $q(\zeta)=\hat{p}(\zeta)$ if $\zeta\in M$. Then L is a union of COCs, and q is a class-function defined on L. Assume now that $\xi,\eta,\zeta\in L$ satisfy $\xi\eta=\zeta$. By Theorems 1 and 2 we have $q(\xi)\dot{+}q(\zeta)=q(\eta)$ whenever at least one of ξ, η and ζ is finitary. This restriction is essential. Indeed, it was shown by Bertram [1] that if ζ is an infinite cycle with no fixed points, then ζ is a product of two of its conjugates. It follows that whenever $\xi=\xi_0\xi_1$ and $\eta=\eta_0\eta_1$ are the c.f. of the monic permutations ξ and η, and ξ_1 and η_1 fix the same set A_0 of points, then $[\xi]\cdot[\eta]$ contains a monic permutation ζ with c.f. $\zeta_0\zeta_1$, where A_0 is the set of points fixed by ζ_1, and $\zeta_0=\xi_0\eta_0$. Thus,

$$q(\zeta)=\hat{p}(\zeta)=p(\zeta_0)\dot{+}\hat{p}(\zeta_1)=p(\xi_0)\dot{+}p(\eta_0)\dot{+}\hat{p}(\zeta_1).$$

Hence, if A_0 has odd cardinality, i.e. $p(|A_0|)=\hat{p}(\xi_1)=\hat{p}(\eta_1)=\hat{p}(\zeta_1)=1$, then we have $q(\zeta)=q(\xi)\dot{+}q(\eta)\dot{+}1$.

Define $\check{p}(\xi)$ for monic ξ by $\check{p}(\xi)=\hat{p}(\xi)\dot{+}1$. It is easily checked that $\check{p}$ is another class function satisfying Theorems 1 and 2, and that $\hat{p}$ and $\check{p}$ are the only such class functions (see [6]). Thus, the parity displayed by monic permutations, although restricted for algebraic applications, enjoys more symmetry than the familiar parity displayed by finitary permutations.

Example 1 is due to E.A. Bertram and M.A. Perles [1]. Theorem 1 of Example 4 for monic permutations with trivial finitary part is included in an interesting note by Boccara [2]. In Section 3 we present a detailed proof of Theorems 1 and 2.

2. The general landscape: Parity families of COCs

Taking the hint from the above examples, and in particular from Theorems 1 and 2, we define for arbitrary $\xi, \eta \in S_A$:

$$\xi \equiv \eta, \quad \text{iff } ([\xi] \cdot [\eta]) \cap S^0_A \neq \emptyset.$$

Theorem A. $\equiv$ *is an equivalence relation on* S_A, *partitioning* S_A *into the* $\equiv$*-equivalence classes* $[\xi]_\equiv$ *given by*:

$$[\xi]_\equiv = [\xi] \cdot S^0_A = S^0_A \cdot [\xi].$$

Let us say that $\xi \in S_A$ has *parity features* (p.f.) if $[\xi]^2$ contains no finitary odd permutations. [Thus, every finitary or monic permutation has p.f.; while any ξ with $[\xi]^2 = S_A$ does not have p.f.]

Call a family $\mathcal{F}$ of COCs in S_A a *parity family* (PAF) if a function $\hat{p} : \mathcal{F} \to \{0,1\}$ exists, such that for any $X, Y \in \mathcal{F}$, $X \cdot Y$ has distinct parity, and:

$$p(X \cdot Y) = \hat{p}(X) \dot{+} \hat{p}(Y)$$

holds whenever $X, Y \in \mathcal{F}$.

[The finitary COCs form a PAF, and by Theorem 1 so do the monic COCs.]

For arbitrary $\xi \in S_A$, let:

$$\mathcal{F}(\xi) = \{[\zeta] : \xi \equiv \zeta\}.$$

Theorem B. *Let* $\xi \in S_A$. *Then the following are equivalent*:

(a) ξ *has parity features.*

(b) $\mathcal{F}(\xi)$ *is a maximal parity family.*

[Thus, the finitary COCs and the monic COCs form two distinct maximal PAFs, while the PAFs R_0 and R_1 of Example 3 are not maximal.]

The parity families of S_A are determined by Theorems A and B and the following theorem.

Theorem C. *Let* $\xi \in S_A$. *Then* ξ *has parity features if and only if one of the following holds*:

(a) ξ *has one infinite orbit, and for each finite n only finitely many orbits of length n, or*

(b) ξ *has no infinite orbit, and whenever* ξ *has infinitely many orbits of length m and infinitely many orbits of length n, m and n are divisible by the same powers of two.*

Proofs of Theorems A, B and C will appear in [6].

3. Proof of Theorems 1 and 2

We denote by $\mathbb{Z}$ the set of integers, and by $|B|$ the cardinality of the set B. $\xi\eta$ denotes the permutation obtained by acting *first* with η, *then* with ξ (as follows from our convention that $\eta(a)$ denotes the value of η at a). The notation $[a_1,\ldots,a_n]$ and $[\ldots,a_{-2},a_{-1},a_0,a_1,a_2,\ldots]$, where the symbols displayed are distinct, will be used to denote *finite* and *infinite* cycles, respectively. Thus, for instance, $[\ldots,a_{-2},a_{-1},a_0,a_1,a_2\cdots]$ is the permutation $\zeta\in S_A$ satisfying $\zeta(a_m)=a_{m+1}$ for $m\in\mathbb{Z}$ and $\zeta(a)=a$ for all other a. Note that $[a]$ denotes $\mathbf{1}_A$ for any $a\in A$. We call $[a_1,\ldots,a_n]$ a cycle of *length n*, and a cycle of length 2 is called a *transposition.* We recall that by $[1,\ldots,n]=[1,n]\cdots[1,3]\ [1,2]$ every cycle of length n is a product of $n-1$ transpositions. It follows that every finitary permutation θ is a (finite) product of transpositions. Moreover, the number r of transpositions appearing in any such product must satisfy $p(r)=p(\theta)$.

Proof of Theorem 2. By the previous remarks, part (b) is a corollary of:

Proposition 2.1. *Let ζ be monic, σ a transposition. Then $\sigma\zeta$ and $\zeta\sigma$ are monic, and* $\hat{p}(\sigma\zeta)=\hat{p}(\zeta\sigma)=\hat{p}(\zeta)\dot{+}1$.

Proof. Since $\zeta\sigma=\sigma^{-1}(\sigma\zeta)\sigma$ it suffices to prove that $\sigma\zeta$ is monic and $\hat{p}(\sigma\zeta)=p(\zeta)\dot{+}1$. Let $\sigma=[a,b]$ and let $\zeta=\zeta_0\zeta_1$ be the canonical factorization of ζ, where $\zeta_1=[\ldots,a_{-2},a_{-1},a_0,a_1,a_2,\ldots]$. Then $\zeta_0(a_m)=a_m$ for all $m\in\mathbb{Z}$. Let $A_1=\{a_m\mid m\in\mathbb{Z}\}$, $A_0=A\setminus A_1$. Then A_0 is a finite set, and $\hat{p}(\zeta)=p(\zeta_0)+p(|A_0|)$.

Case 1: $a,b\in A_0$. Then $\sigma\zeta_0$ is a finitary permutation satisfying $p(\sigma\zeta_0)=p(\zeta_0)\dot{+}1$. But then $\sigma\zeta=(\sigma\zeta_0)\zeta_1$ is a monic permutation and $(\sigma\zeta)_0(\sigma\zeta)_1=(\sigma\zeta_0)\zeta_1$ is its c.f. Hence,

$$\hat{p}(\sigma\zeta)=p(\sigma\zeta_0)\dot{+}\hat{p}(\zeta_1)=(p(\zeta_0)\dot{+}1)\dot{+}\hat{p}(\zeta_1)=\hat{p}(\zeta)\dot{+}1.$$

Case 2: $a,b\in A_1$. We may assume $a=a_0$ and $b=a_k$ for some $k>0$. Then $\sigma\zeta_0=\zeta_0\sigma$ and so $\sigma\zeta=\zeta_0(\sigma\zeta_1)$. But $\sigma=[a_0,a_k]$, and:

$$\begin{aligned}&[a_0,a_k][\cdots a_{-2},a_{-1},a_0,a_1,a_2,\ldots,a_k,\ldots]\\&\quad=[a_0,a_1,\ldots,a_{k-1}][\cdots a_{-2},a_{-1},a_k,a_{k+1}\cdots].\end{aligned}\tag{1}$$

So $\sigma\zeta=\zeta_0[a_0,\ldots,a_{k-1}]\ [\cdots a_{-2},a_{-1},a_k,a_{k+1}\cdots]$. Since all permutations on the right have disjoint supports, we conclude that $\sigma\zeta$ is monic, $(\sigma\zeta)_0=\zeta_0[a_0,\ldots,a_{k-1}]$ and that $(\sigma\zeta)_1$ is a cofinite cycle whose set of fixed points is $A_0\cup\{a_0,a_1,\ldots,a_{k-1}\}$. Thus,

$$\begin{aligned}&p((\sigma\zeta)_0)=p(\zeta_0)\dot{+}p(k)\dot{+}1,\\&\hat{p}((\sigma\zeta)_1)=p(|A_0|+k)=\hat{p}(\zeta_1)\dot{+}p(k).\end{aligned}$$

Hence, again:

$$\hat{p}(\sigma\zeta) = p((\sigma\zeta)_0) \dot{+} \hat{p}((\sigma\zeta)_1) = p(\zeta_0) \dot{+} \hat{p}(\zeta_1) \dot{+} 1 = \hat{p}(\zeta) \dot{+} 1.$$

Case 3: $a \in A_0$, $b \in A_1$. Let $[c_1, \ldots, c_k]$ be the finite cycle of ζ_0 where a appears, say $a = c_1$ (the case $k = 1$ is possible!), and assume $b = a_0$. Let $\zeta_0 = \zeta_0'[c_1, \ldots, c_k]$. Then $\zeta_0'(c_1) = c_1$ and $\zeta_0'(a_0) = a_0$ and so ζ_0' commutes with σ, and $\sigma\zeta = \sigma(\zeta_0'[c_1, \ldots, c_k]\zeta_1) = \zeta_0'(\sigma[c_1, \ldots, c_k][\cdots a_{-2}, a_{-1}, a_0, a_1, a_2, \ldots])$. But $\sigma = [c_1, a_0]$ and:

$$\begin{aligned}&[c_1, a_0][c_1, \ldots, c_k][\cdots a_{-2}, a_{-1}, a_0, a_1, a_2, \ldots]\\ &\quad = [\cdots a_{-2}, a_{-1}, c_1, \ldots, c_k, a_0, a_1, a_2, \ldots].\end{aligned} \tag{2}$$

Thus, $\sigma\zeta$ is monic with the c.f. $\sigma\zeta = (\sigma\zeta)_0(\sigma\zeta)_1$, where $(\sigma\zeta)_0 = \zeta_0'$, and $(\sigma\zeta)_1$ is a cofinite cycle, whose set of fixed points is $A_0 \setminus \{c_1, \ldots, c_k\}$. Now

$$p(\zeta_0) = p(\zeta_0'[c_1, \ldots, c_k]) = p(\zeta_0') \dot{+} p(k) \dot{+} 1.$$

So adding $p(k) \dot{+} 1$ we have:

$$p((\sigma\zeta)_0) = p(\zeta_0') = p(\zeta_0) \dot{+} p(k) \dot{+} 1.$$

Also,

$$\begin{aligned}p((\sigma\zeta)_1) &= p(|A_0 \setminus \{c_1, \ldots, c_k\}|) = p(|A_0| - k)\\ &= p(|A_0|) \dot{+} p(k) = \hat{p}(\zeta_1) \dot{+} p(k).\end{aligned}$$

Hence, again:

$$p(\sigma\zeta) = p(\zeta_0) \dot{+} \hat{p}(\zeta_1) \dot{+} 1 = \hat{p}(\zeta) \dot{+} 1. \qquad \square$$

Let us prove Theorem 2(a). Since S_A^0 is normal in S_A, $\zeta \cdot S_A^0 = S_A^0 \cdot \zeta$ for each $\zeta \in S_A$; hence, $[\zeta] \cdot S_A^0 = S_A^0 \cdot [\zeta]$. Now let $\zeta \in M$. Since $[\zeta] \subseteq M$, it follows from Proposition 2.1 that $S_A^0 \cdot [\zeta] \subseteq M$. We obtain equality by showing that given ζ, $\xi \in M$ we can find $\theta \in S_A^0$ and a conjugate ξ' of ξ such that $\theta\zeta = \xi'$. (Since M is closed under conjugacy, $\xi \in M$ follows.) Indeed, let $\zeta = \zeta_0\zeta_1$ and $\xi = \xi_0\xi_1$ be the c.f. of ζ and ξ. Using (1) and (2) we can find finitary θ' such that $\xi_1' = \theta'\zeta_1$ is a cofinite cycle whose set A_0 of fixed points has the cardinality of the set of points fixed by ξ_1. Let ξ_0' be a finitary permutation, conjugate to ξ_0 and fixes every point not in A_0. Let $\theta = \xi_0'\theta'\zeta_0^{-1}$. Then θ is finitary, and $\theta\zeta = \xi_0'\xi_1'$ is a conjugate of ξ.

$\square$

Proof of Theorem 1. By Theorem 2(a), $\xi = \tilde{\eta}\theta$ for some $\tilde{\eta} \in [\eta]$, $\theta \in S_A^0$, and so $\xi\tilde{\eta}^{-1} \in S_A^0$. But $\tilde{\eta}^{-1} \in [\eta]$, and so $\theta \in [\xi] \cdot [\eta]$. Thus, $[\xi] \cdot [\eta] \cap S_A^0 \neq \emptyset$.

Now, let $\xi, \eta \in M$, and let $\theta = \xi\eta \in S_A^0$. We show that $p(\theta) = \hat{p}(\xi) \dot{+} \hat{p}(\eta)$. Let $\xi = \xi_0\xi_1$ and $\eta = \eta_0\eta_1$ be the c.f. of ξ, η. Since θ is finitary, we may assume:

$$\eta_1 = [\cdots a_{-2}, a_{-1}, b_1, \ldots, b_k, a_0, a_1, a_2, \ldots],$$

where $\theta(a_m) = a_m$ for $m \in \mathbb{Z}$.

Since $\theta = \xi\eta$, $\eta(a_m) = a_{m+1}$ for $m \neq -1$, and $\eta(a_{-1}) = b_1$ we conclude that $\xi(a_{m+1}) = a_m$ $(m \neq -1)$ and $\xi(b_1) = a_{-1}$. Since ξ has only one infinite orbit, it must include the cofinite set $\{a_m \mid m \in \mathbb{Z}\} \cup \{b_1\}$, and so:

$$\xi_1 = [\ldots, a_2, a_1, a_0, c_1, c_2, \ldots, c_l, a_{-1}, a_{-2}, \ldots],$$

where $c_l = b_1$.

We also note that if ξ_1 and η_1 fix A_0 and A_0', respectively, then $A_0 \cup \{c_1, \ldots, c_l\} = A_0' \cup \{b_1, \ldots, b_k\} = A \setminus \{a_m \mid m \in \mathbb{Z}\}$, and the unions of the left are disjointed. Hence,

$$\begin{aligned}\hat{p}(\eta_1) \dotplus p(k) &= p(|A_0'|) \dotplus p(k) = p(|A \setminus \{a_m \mid m \in \mathbb{Z}\}|) \\ &= p(|A_0|) \dotplus p(l) = \hat{p}(\xi_1) \dotplus p(l),\end{aligned}$$

whence

$$\hat{p}(\eta_1) \dotplus \hat{p}(\xi_1) = p(k) \dotplus p(l).$$

Let $\sigma = [a_1, b_1] = [a_1, c_l]$ and define $\tilde{\xi}$ and $\tilde{\eta}$ by $\tilde{\xi} = \xi\sigma$ and $\tilde{\eta} = \sigma\eta$. Then, letting $\zeta = [\ldots, a_{-2}, a_{-1}, a_1, a_2, \ldots]$ we have:

(i) $\xi_1\sigma = [a_0, c_1, \ldots, c_l]\,\zeta^{-1}$,

(ii) $\sigma\eta_1 = [a_0, b_1, \ldots, b_k]\,\zeta$,

(iii) $\tilde{\xi}\tilde{\eta} = \xi\eta = \theta$.

Now, by (i) and (ii) we see that

$$\tilde{\xi} = \xi_0[a_0, c_1, \ldots, c_l]\,\zeta^{-1},$$

$$\tilde{\eta} = \eta_0[a_0, b_1, \ldots, b_k]\,\zeta,$$

and since the permutations on the right have disjoint supports, we conclude that:

(iv)
$$\tilde{\xi}_0 = \xi_0[a_0, c_1, \ldots, c_l], \qquad \tilde{\xi}_1 = \zeta^{-1},$$
$$\tilde{\eta}_0 = \eta_0[a_0, b_1, \ldots, b_k], \qquad \tilde{\eta}_1 = \zeta.$$

Also, since $\tilde{\xi}_1 = \tilde{\eta}_1^{-1} = \zeta^{-1}$, we have $\tilde{\xi}_1\tilde{\eta}_0 = \tilde{\eta}_0\tilde{\xi}_1$ and so, by (iii):

(v) $\theta = \tilde{\xi}_0\tilde{\eta}_0$.

Thus, taking parities in (iv) and (v) we obtain:

$$p(\theta) = p(\tilde{\xi}_0) \dotplus p(\tilde{\eta}_0) = p(\xi_0) \dotplus p(k) + p(\eta_0) \dotplus p(l),$$

and by $p(k) \dotplus p(l) = \hat{p}(\eta_1) \dotplus \hat{p}(\xi_1)$ we come to the desired conclusion:

$$p(\theta) = \hat{p}(\xi) + \hat{p}(\eta). \qquad \square$$

References

[1] E.A. Bertram, Permutations as products of conjugate infinite cycles, Pacific J. Math. 39 (1971) 275–284.

[2] G. Boccara, Sur les permutations d'un ensemble infini denombrable, dont toute orbite essentielle est infinie, C.R. Acad. Sci., Paris Série A, 287 (1978) 281–283.

[3] A.B. Gray, Infinite symmetric groups and monomial groups, Dissertation, New Mexico State University, Las Cruces, N.M. (1960).

[4] G. Moran, The product of two reflection classes of the symmetric group, Discr. Math. 15 (1967) 63–77.

[5] G. Moran, The algebra of reflections of an infinite set, Notices A.M.S. 73T (August 1973) A 193.

[6] G. Moran, Parity features for classes of the infinite symmetric group (to appear).

[7] W. R. Scott, Group Theory (Prentice-Hall, Englewood Cliffs, N.J., 1964).

Received 1 April 1981

Annals of Discrete Mathematics 20 (1984) 241–251
North-Holland

NON-HAMILTONIAN SIMPLE 3-POLYTOPES WITH ONLY ONE TYPE OF FACE BESIDES TRIANGLES

P.J. OWENS
Mathematics Department, University of Surrey, Guildford, Surrey, U.K.

It is shown, for $q = 8$, 9 and 10, that the class of simple 3-polytopes whose faces are of only two types, triangles and q-gons, contains non-Hamiltonian members and even has shortness coefficient less than unity.

1. Introduction

For any graph G let $v(G)$ denote the number of vertices, $h(G)$ the length of a maximum cycle and $h^*(G)$ the length (measured by the number of vertices) of a maximum path. The *shortness coefficient* $\rho(\mathcal{G})$ of an infinite class $\mathcal{G}$ of graphs [6] is defined by

$$\rho(\mathcal{G}) = \liminf_{G \in \mathcal{G}} \frac{h(G)}{v(G)},$$

and ρ^* is defined similarly in terms of h^*. Since every cycle can be converted into a path of the same length by deleting one edge, $\rho \leq \rho^*$ for any class of graphs.

Let $G_3(p,q)$ denote the class of 3-connected regular trivalent planar graphs (or, equivalently, simple 3-polytopes) whose faces are of only two types, namely p-gons and q-gons. This class is not infinite unless $3 \leq p \leq 5$ and $q \geq 6$. It has been shown by Goodey that every member of $G_3(p,6)$ is Hamiltonian, for $p = 3$ [5] and $p = 4$ [4]. Non-Hamiltonian members of $G_3(5,q)$ have been given by Zaks for $q = 8$ [11], $q = 9$ [13], all $q \geq 11$ [12] and by the present author for $q = 7$ [8] and $q = 10$ [7]. Non-Hamiltonian members of $G_3(4,q)$ have been given by Walther for $q = 9$ [10] and all odd $q \geq 11$ (to appear).

In the present paper we consider the class $G_3(3,q)$, where $q \geq 8$, and we start with a simple lemma.

Lemma 1. *The class $G_3(3,q)$ is non-empty only if $q \leq 10$.*

Proof. Let $G \in G_3(3,q)$. No two triangles in G have a common edge, otherwise G would not be 3-connected. Hence, a given q-gon is adjacent to at most $[q/2]$

triangles. Let H be the graph obtained from G by shrinking all triangles to single vertices. No face of H has less than $q-[q/2]$ edges. Since every planar graph has some faces with less than 6 edges, $q-[q/2]<6$, and this implies that $q \leq 10$.

We shall prove the following three theorems.

Theorem 1.
(1) *There is a non-Hamiltonian member of* $G_3(3,8)$ *with only* 174 *vertices and a member of* $G_3(3,8)$ *with no Hamiltonian paths and only* 384 *vertices.*
(2) $\rho^*(G_3(3,8)) \leq 139/140 < 1$.

Theorem 2.
(1) *There is a non-Hamiltonian member of* $G_3(3,9)$ *with only* 176 *vertices and a member of* $G_3(3,9)$ *with no Hamiltonian paths and only* 380 *vertices.*
(2) $\rho^*(G_3(3,9)) \leq 135/136 < 1$.

Theorem 3.
(1) *There is a non-Hamiltonian member of* $G_3(3,10)$ *with only* 158 *vertices and a member of* $G_3(3,10)$ *with no Hamiltonian paths and only* 466 *vertices.*
(2) $\rho^*(G_3(3,10)) \leq 174/175 < 1$.

In order to describe our constructions, certain graphs which occur repeatedly as subgraphs will be denoted by capital letters and represented in diagrams by labelled circles. As the first example, Fig. 1 shows a subgraph B which is the graph of the dodecahedron minus two adjacent vertices. The 'dangling' edges are not part of the subgraph but show how it is to be joined into a graph. The numbers around the circumference of the circle are the numbers of vertices which the subgraph contributes to the adjoining faces of any graph in which it occurs. By a path *through* a subgraph we mean a path whose ends are not in the subgraph. Given any two vertices u, v of a graph or subgraph, $P(u,v)$ will denote any path connecting u to v. The property of B which makes it useful to us is stated in the following lemma, equivalent to [3, Lemmas 2.1 and 2.2]. We omit the straightforward, but tedious, proof.

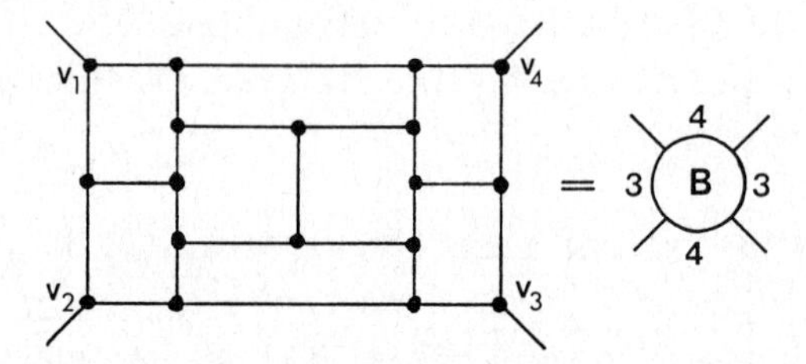

Fig. 1. Subgraph of type B.

Lemma 2. *Let P be a path through B which spans B. Then $P \cap B$ takes one of the following three forms (see Fig. 1):*

(1) $P(v_1, v_2) \cup P(v_3, v_4)$,

(2) $P(v_1, v_2)$ *(or, similarly,* $P(v_3, v_4)$*), or*

(3) $P(v_1, v_3)$ *(or, similarly,* $P(v_2, v_4)$*).*

The next lemma states a well-known result and no proof is offered.

Lemma 3. *Let G be a graph which contains a cycle of length 3 and let H be the graph obtained from G by shrinking this cycle to a vertex. Then G and H are either both Hamiltonian or both non-Hamiltonian.*

The graphs S and T, shown in Fig. 2, form the basis of some of our constructions. The numbers show the numbers of edges of the faces.

Lemma 4. *No spanning cycle in S, or in T, contains both of the edges e_1 and e_2.*

Proof. Suppose that S has a spanning cycle C which contains both e_1 and e_2. By Grinberg's theorem (see [1]), if f'_r and f''_r denote the numbers of r-gons inside C and outside C, respectively, then

$$2(f'_4 - f''_4) + 3(f'_5 - f''_5) + 4(f'_6 - f''_6) = 0.$$

By inspection of S, $f'_6 = 0$ and $f''_6 = 1$. Moreover, since e_1 and e_2 are both contained in C and separate the two pairs of 4-gons, $f'_4 = f''_4 = 2$. Hence,

$$3(f'_5 - f''_5) - 4 = 0,$$

which is impossible because the left-hand side is not even congruent to 0 (modulo 3). This contradiction shows that no such cycle C exists and the proof (for S) is complete. We omit the similar proof for T, since this property of T is well known as a starting point in the construction of Tutte's famous non-Hamiltonian graph [9] (or see [1]).

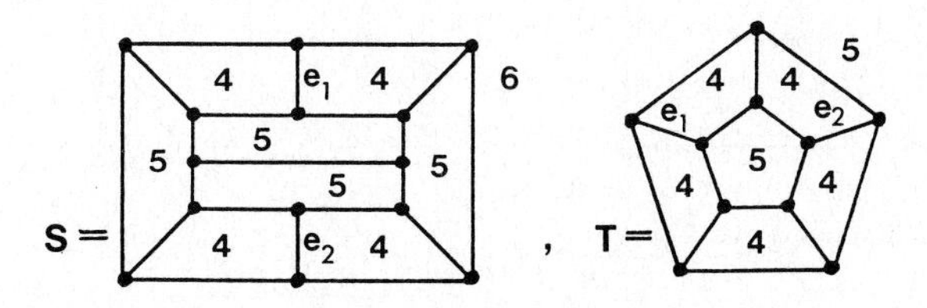

Fig. 2. The graphs S and T.

2. Proof of Theorem 1

Let L denote the subgraph shown in Fig. 3. Here, and in later diagrams, small unlabelled circles represent triangular faces. Note that every face within L is a triangle or 8-gon and that $v(L) = 47$. The following lemma exploits the presence in L of a modified form of B (denoted by B^0 in Fig. 3).

Lemma 5. *Every path through L that spans L contains the edge a.*

Proof. First note that, by Lemma 3, we can apply Lemma 2 to B^0 as well as to B. Let P be a path through L that spans L and suppose that (if possible) the edge a is not in P. Then b and c are both in P. Now consider three cases.

Case 1: Both edges d and e in P. Then $P \cap B^0$ (or one component of $P \cap B^0$) is of the form $P(v_1, v_2)$. Thus, e, b, c, d and $P(v_1, v_2)$ form a cycle in P, which is impossible.

Case 2: Just one of the edges d or e (say d) in P. Then e is not in P, so f is in P. Moreover, $P \cap B^0$ must be of the form $P(v_1, v_3)$, so h is in P. Thus, h, f, b, c, d and $P(v_1, v_3)$ form a cycle in P, which is impossible.

Case 3: Neither d nor e in P. Then f and g are both in P, also $P \cap B^0$ is of the form $P(v_3, v_4)$ and so h and k are both in P. Thus, k, g, c, b, f, h and $P(v_3, v_4)$ form a cycle in P, which is impossible.

Since in each case we get a contradiction, no such path P exists and the lemma follows.

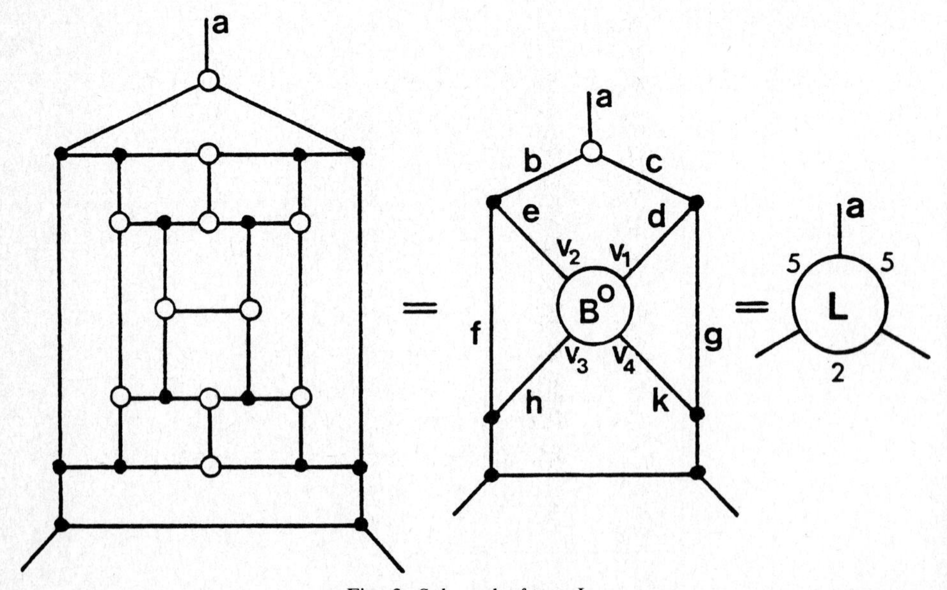

Fig. 3. Subgraph of type L.

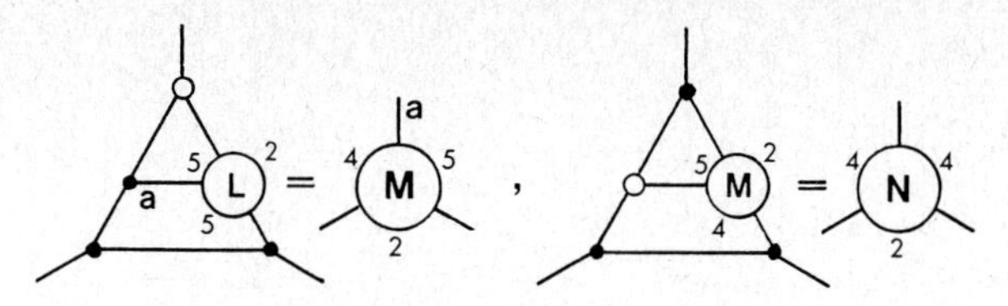

Fig. 4. Subgraphs of types M and N.

After Bosák [2] and others, any edge which has the property of the edge a in Lemma 5 will be called an *a-edge*.

We define subgraphs M and N in terms of L as shown in Fig. 4. The extra faces added to L are triangles and 8-gons only and it is easy to prove that, as indicated, one of the dangling edges of M is an a-edge. We obtain the graph G_0, shown in Fig. 5, from the graph T by replacing four vertices by two copies of M (one in mirror image form), one copy of N and one triangle and we denote by X the subgraph of G_0 that remains when N is deleted. By inspection, $G_0 \in G_3(3,8)$, $v(G_0) = 174$ and $v(X) = 115$.

Lemma 6.

(1) *G_0 is non-Hamiltonian.*

(2) *No path through X spans X.*

Proof.

(1) Let C be (if possible) a spanning cycle of G_0. Thus, C contains the edges e_1 and e_2 since these are the a-edges associated with the two copies of M. By shrinking the copies of M, N and the triangle to single vertices, G_0 is converted into T and C into a spanning cycle of T which contains the edges e_1 and e_2. Since this contradicts Lemma 4 it follows that no such cycle C exists.

(2) Let P be (if possible) a path through X that spans X. Then there is a spanning cycle in the graph obtained from G_0 by shrinking N to a single vertex, which, as in part (1), leads to a contradiction to Lemma 4. Hence no such path P exists.

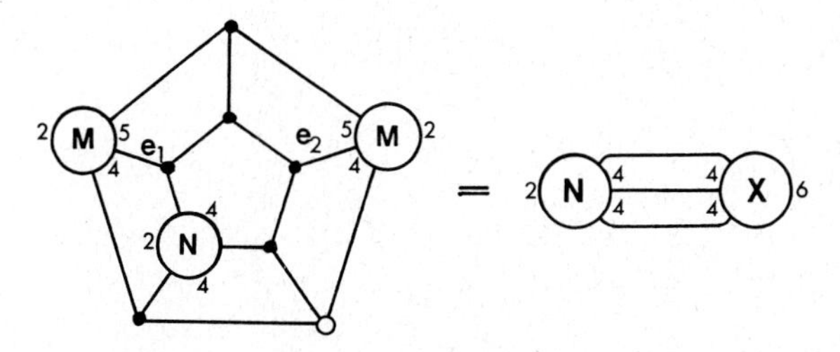

Fig. 5. The graph G_0.

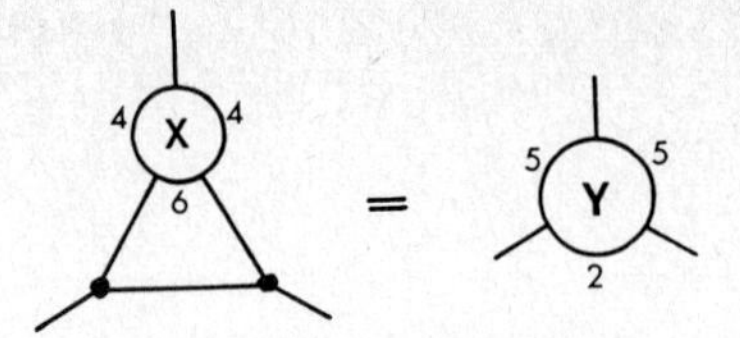

Fig. 6. Subgraph of type Y.

Now let Y be the subgraph, with two more vertices than X, shown in Fig. 6. Since Y and L each contribute 5, 5 and 2 vertices to the three adjoining faces of any graph in which either occurs, $G_3(3,8)$ is closed under replacement of copies of L by similarly oriented copies of Y. Note that $v(Y)=117$.

Let G_1 be the graph obtained from G_0 by replacing all three copies of L by copies of Y.

Lemma 7. *G_1 has no Hamiltonian paths.*

Proof. Since G_1 contains three copies of X and any path P in G_1 has only two ends, at least one copy of X in G_1 does not contain either end of P. By Lemma 6(2) this copy of X is not spanned by P, so the lemma follows.

To conclude the proof of Theorem 1(1) we note that

$$v(G_1)=v(G_0)+3(v(Y)-v(L))=384.$$

Note. Evidently $h(G_1)\leqslant 381$, so $h(G_1)/v(G_1)\leqslant 127/128$. This is rather smaller than the value we shall obtain for ρ.

We now use the fact that Y contains two copies of L to construct an infinite sequence $\langle G_n\rangle$ of non-Hamiltonian members of $G_3(3,8)$, starting with G_1. For $n\geqslant 1$, let G_{n+1} be the graph obtained from G_n when both copies of L in one (any one) of its subgraphs of type Y are replaced by new copies of Y.

For $n\geqslant 1$, G_n contains $n+2$ copies of X, so Lemma 6(2) implies that

$$h(G_n)\leqslant v(G_n)-n-2,$$

and, since at most two copies of X contain ends of a given path in G_n.

$$h^*(G_n)\leqslant v(G_n)-n.$$

Also,

$$v(G_n)=v(G_1)+2(n-1)(v(Y)-v(L))=244+140n,$$

so we obtain:

$$\rho^*(G_3(3,8))\leqslant 139/140<1,$$

and this completes the proof of Theorem 1. Note that our treatment leads to the same upper bound for ρ as for ρ^*.

3. Proof of Theorem 2

As the proof closely resembles that of Theorem 1, a brief treatment will suffice. In place of L we use the subgraph L' shown in Fig. 7. Every face within L' is a triangle or 9-gon and $v(L') = 55$. By Lemma 5 (applied to L') there is an a-edge, as shown. Let G_0' be the graph shown in Fig. 8, which contains three copies of L', one of which is labelled L_0'. Let X' be the subgraph of G_0' that remains when L_0' is deleted. By inspection, $G_0' \in G_3(3,9)$, $v(G_0') = 176$ and $v(X') = 121$. By Lemma 6 (applied to G_0'), G_0' is non-Hamiltonian and no path through X' spans X'. By adding two extra vertices and an edge to X' we obtain a subgraph Y' which, like L', contributes 5, 5 and 2 vertices to the three adjoining faces of any graph in which it occurs. Thus, $G_3(3,9)$ is closed under replacement of copies of L' by similarly oriented copies of Y'. Let G_1' be the graph obtained by replacing all three copies of L' in G_0' by copies of Y'. By Lemma 7 (applied to G_1'), G_1' has no Hamiltonian paths. We complete the proof of Theorem 2(1) by noting that

$$v(G_1') = v(G_0') + 3(v(Y') - v(L')) = 380.$$

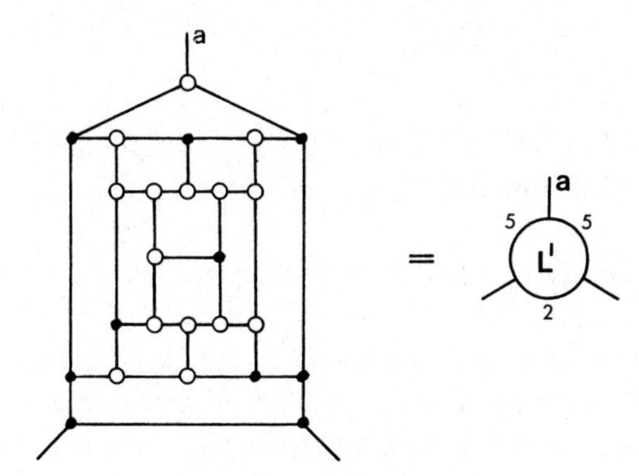

Fig. 7. Subgraph of type L'.

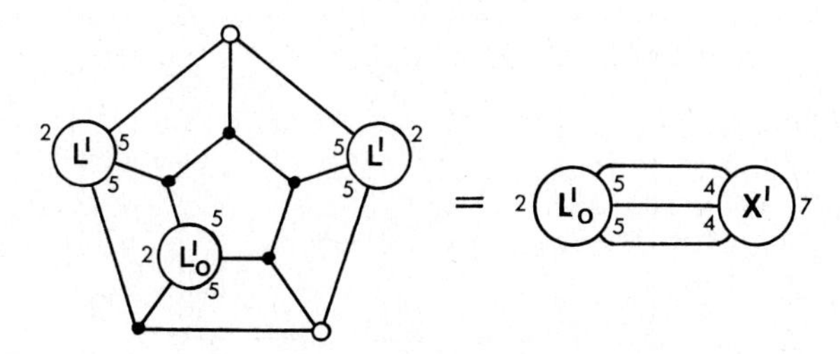

Fig. 8. The graph G_0'.

We now construct an infinite sequence $\langle G'_n \rangle$ of non-Hamiltonian members of $G_3(3,9)$, starting with G'_1. For $n \geqslant 1$, let G_{n+1} be the graph obtained from G'_n when both copies of L' in one of its subgraphs of type Y' are replaced by new copies of Y'. For $n \geqslant 1$, G'_n contains $n+2$ copies of Y' so

$$h^*(G'_n) \leqslant v(G'_n) - n.$$

A simple calculation gives $v(G'_n) = 244 + 136n$ for all $n \geqslant 1$, so we obtain:

$$\rho^*(G_3(3,9)) \leqslant 135/136 < 1,$$

as required.

4. Proof of Theorem 3

We start with a subgraph L'' (shown in Fig. 9) which is analogous to L, L', and has an a-edge, as shown. Every face within L'' is a triangle or 10-gon and $v(L'') = 63$. The rest of our proof is somewhat different from the previous cases.

We obtain the graph G''_0, shown in Fig. 10, from the graph S by replacing 12

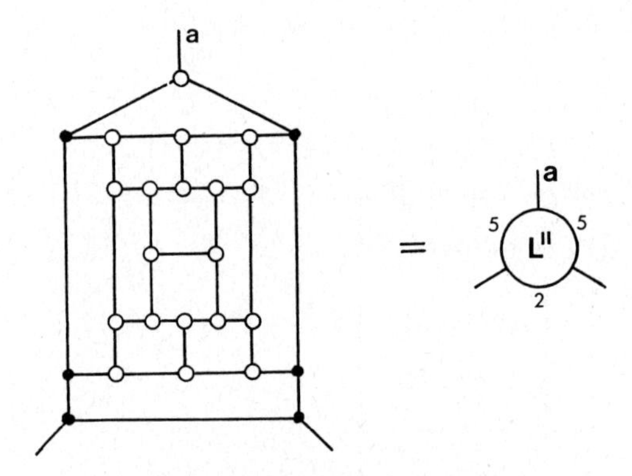

Fig. 9. Subgraph of type L''.

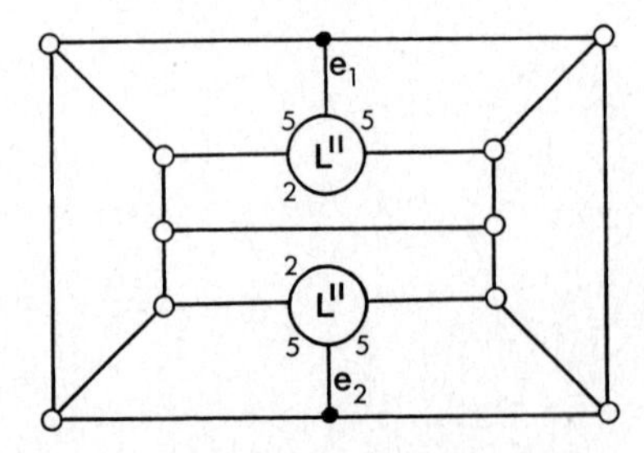

Fig. 10. The graph G''_0.

vertices by 10 triangles and 2 copies of L''. By inspection, $G_0'' \in G_3\ (3,10)$ and $v(G_0'') = 158$. Since e_1 and e_2 are the a-edges associated with the two copies of L'' we can prove (as in Lemma 6) that G_0'' is non-Hamiltonian.

Let H denote the graph shown in Fig. 11 and let Q be the subgraph of H obtained by deleting the vertex v_0. As in Lemma 6 we can prove that H is non-Hamiltonian and that no path through Q spans Q. By inspection, all faces within Q are triangles or 10-gons and $v(Q) = 143$.

Now define a subgraph R in terms of Q as shown in Fig. 12. Every path through R omits at least one vertex and $v(R) = 151$. By inserting three copies of R and 4 triangles in place of 7 vertices of the graph of the cube, we obtain the graph G_1'' shown in Fig. 13. By inspection, $G_1'' \in G_3(3,10)$ and $v(G_1'') = 466$. As in Lemma 7 we can show that G_1'' has no Hamiltonian paths, and this completes the proof of Theorem 3(1).

Let X'' be the subgraph of G_1'' obtained by deleting one copy of L'' (see Fig. 13) and let Y'' be as shown in Fig. 14. Note that $v(Y'') = 413$. The class $G_3(3,10)$

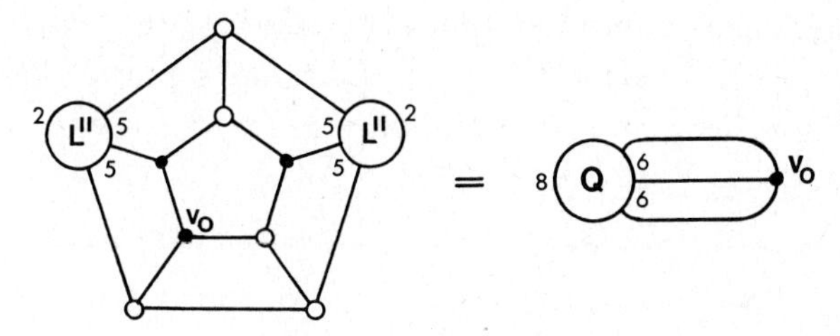

Fig. 11. The graph H.

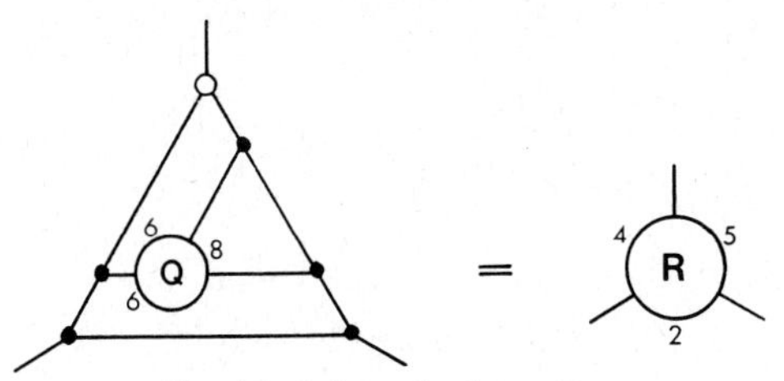

Fig. 12. Subgraph of type R.

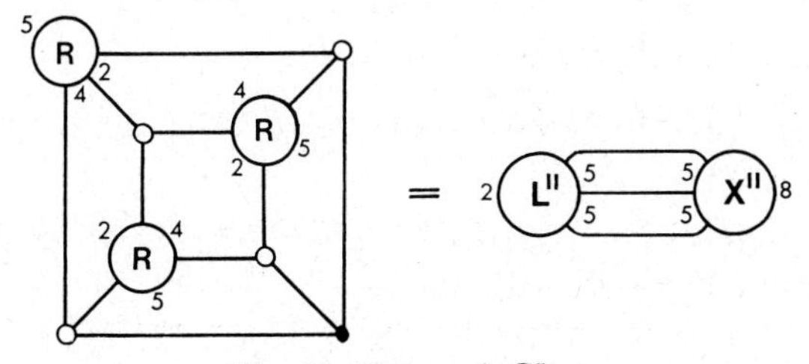

Fig. 13. The graph G_1''.

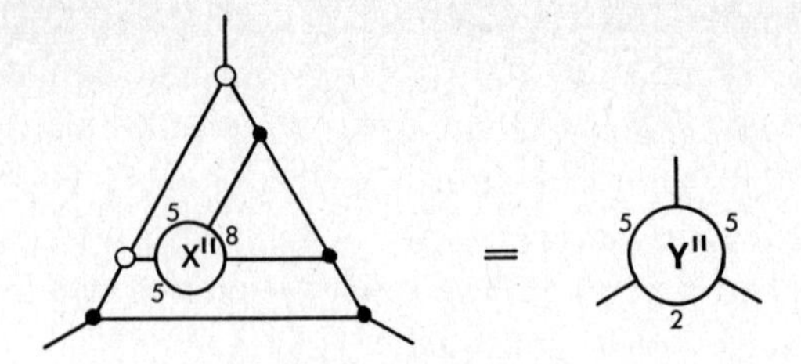

Fig. 14. Subgraph of type Y''.

is closed under replacement of copies of L'' by similarly oriented copies of Y''. We construct an infinite sequence $\langle G''_n \rangle$ of non-Hamiltonian members of $G_3(3,10)$, starting with G''_1. Let G''_2 be the graph obtained from G''_1 by replacing any one of its six copies of L'' by a copy of Y''. Now Y'' contains two subgraphs of type R, each of which contains two copies of L'', and one additional copy of L'' which we denote by L''_0. For $n \geqslant 2$, let G''_{n+1} be the graph obtained from G''_n by replacing L''_0 by a new copy of Y''. The substitution leaves copies of R in G''_n unaltered and introduces two more besides a new L''_0. Thus, for $n \geqslant 1$, G''_n contains $2n$ copies of R, so

$$h^*(G''_n) \leqslant v(G''_n) - 2n + 2.$$

Also,

$$v(G''_n) = v(G''_1) + (n-1)(v(Y'') - v(L'')) = 116 + 350n.$$

Hence,

$$\rho^*(G_3(3,10)) \leqslant 174/175 < 1,$$

and the proof of Theorem 3 is complete.

5. Remarks

It appears to be still unknown whether there are any non-Hamiltonian members of the following classes:

$$G_3(3,7),\ G_3(4,7),\ G_3(5,6) \quad \text{and} \quad G_3(4,2k), \quad \text{for } k \geqslant 4.$$

I should like to thank the organisers for their efforts in making the conference an enjoyable and memorable occasion.

References

[1] J.A. Bondy and U.S.R. Murty, Graph Theory with Applications (American Elsevier, New York; Macmillan, London, 1976).
[2] J. Bosák, Hamiltonian lines in cubic graphs, in: Proceedings of the International Symposium on the Theory of Graphs, Rome, 1966 (Gordon and Breach, New York; Dunod, Paris, 1967).
[3] G.B. Faulkner and D.H. Younger, Non-Hamiltonian cubic planar maps, Discr. Math. 7 (1974) 67–74.
[4] P.R. Goodey, Hamiltonian circuits in polytopes with even sided faces, Israel J. Math. 22 (1975) 52–56.
[5] P.R. Goodey, A class of Hamiltonian polytopes, J. Graph Theory 1 (1977) 181–185.
[6] B. Grünbaum and H. Walther, Shortness exponents of families of graphs, J. Combinat. Theory (A) 14 (1973) 364–385.
[7] P.J. Owens, Shortness parameters of families of regular planar graphs with two or three types of face, Discr. Math. 39 (1982) 199–209.
[8] P.J. Owens, Non-Hamiltonian simple 3-polytopes whose faces are all 5-gons or 7-gons, Discr. Math. 36 (1981) 227–230.
[9] W.T. Tutte, On Hamiltonian circuits, J. London Math. Soc. 21 (1946) 98–101.
[10] H. Walther, Note on two problems of J. Zaks concerning Hamiltonian 3-polytopes, Discr. Math. 33 (1981) 107–109.
[11] J. Zaks, Non-Hamiltonian non-Grinbergian graphs, Discr. Math. 17 (1977) 317–321.
[12] J. Zaks, Non-Hamiltonian simple 3-polytopes having just two types of faces, Discr. Math. 29 (1980) 87–101.
[13] J. Zaks, Non-Hamiltonian simple planar graphs, in: Theory and Practice of Combinatorics, Annals of Discrete Mathematics, Vol. 12 (North-Holland, Amsterdam, 1982) pp. 255–263.

Received 26 April 1981

Annals of Discrete Mathematics 20 (1984) 253–254
North-Holland

AT MOST 2^{d+1} NEIGHBORLY SIMPLICES IN E^d

M.A. PERLES*
Institute of Mathematics, Hebrew University, Jerusalem, Israel

A family of d-simplices in E^d is called *neighborly* ([1–4], for other references see [4]), if every pair of them has a $(d-1)$-dimensional intersection. Let $f(d)$ denote the maximum number of d-simplices in a neighborly family in E^d. $8 \leq f(3) \leq 9$ is due to Baston [2], while it has been repeatedly conjectured that $f(d) = 2^d$ [1–4]; $f(d) \leq (2/3)(d+1)!$ was shown in [3] and $f(d) \geq 2^d$ has been recently established [4].

The purpose of this short note is to significantly improve the upper bound of $f(d)$, as follows.

Theorem. *$f(d) \leq 2^{d+1}$ holds for all d.*

Proof. Let $P_1, P_2, \ldots, P_t$ be a neighborly family of d-simplices in E^d, and let B be the B-matrix (see [2, 3]) of the family, defined as follows. Let $H_1, H_2, \ldots, H_s$ be the collection of all the hyperplanes in E^d which contain a facet of some of the simplices; fix a notation H_i^+ and H_i^- for the two closed half-spaces determined by H_i, $1 \leq i \leq s$; $B = (b_{ij})$ is defined by

$$b_{ij} = \begin{cases} 0, & \text{if } P_i \text{ has no facets on } H_j, \\ 1, & \text{if } P_i \text{ has a facet on } H_j \text{ and } P_i \subset H_j^+, \\ -1, & \text{if } P_i \text{ has a facet on } H_j \text{ and } P_i \subset H_j^-. \end{cases}$$

It follows from the neighborliness that every two rows have a column where the terms are $+1$ and -1, i.e. for every i and j, $1 \leq i < j \leq t$, there is an h such that $b_{ih} = -b_{jh} \neq 0$. Each row has precisely $d+1$ non-zero terms; hence, it has precisely $s-d-1$ zero entries.

Construct a ± 1 matrix C by replacing each and every row of B with the 2^{s-d-1} rows, obtained by putting $+1$ or -1 instead of every zero, and of course doing it in all the 2^{s-d-1} possible different ways. Clearly, C has $t \cdot 2^{s-d-1}$ rows and s columns.

Every two rows of C are different because they have come from either the same row in B or from different rows in B; in the first case they differ in at least

* Written by J. Zaks, and presented to the Conference by him.
All responsibilities for this note lie with the writer, who wishes to thank G. Kalai for drawing his attention to the beautiful proof of M.A. Perles.

one place where there was a zero (in B), and in the second case they differ in a place where the two rows of B have a $+1$ and a -1, which are unchanged in C.

As C has rows of length s, and all of its rows are different, C can have at most 2^s rows, that is to say

$$t \cdot 2^{s-d-1} \leq 2^s,$$

implying that

$$t \leq 2^{s+1},$$

and the proof is complete.

We prove the following corollary in a similar way:

Corollary. *A neighborly family of d-polytopes in E^d, each having at most k facets, can have at most 2^k members.*

The previous corresponding known upper bound was $(3/2)k!$, see [3].

Actually, the proof of the theorem does not make use of the neighborliness, but of a weaker assumption, called *near-neighborliness* in [3]; a family of d-simplices in E^d is called *nearly-neighborly* if for every two of them, say P and Q, there exists a hyperplane H in E^d containing a facet of both P and Q, such that $P \subset H^+$ and $Q \subset H^-$. Let $g(d)$ denote the maximum number of d-simplices in a nearly-neighborly family in E^d; clearly $f(d) \leq g(d)$ because every neighborly family is also nearly-neighborly.

The theorem does, in fact, prove that $g(d) \leq 2^{d+1}$. Combining with the result of [4], we have the following:

Corollary. $2^d \leq f(d) \leq g(d) \leq 2^{d+1}$.

Note added in proof (by J. Zaks). We thank N. Alon for remarking (May 1983) that the essential part of the proof of the Theorem had been done by B. Alspach, O.T. Ollmann and K.B. Reid, Jr., in their paper in Discr. Math. 12 (1975) 205–209.

References

[1] F. Bagemihl, A conjecture concerning neighboring tetrahedra, Amer. Math. Monthly 63 (1956) 328–329.
[2] V.J.D. Baston, Some Properties of Polyhedra in Euclidean Space (Pergamon Press, Oxford, 1965).
[3] J. Zaks, Bounds of neighborly families of convex polytopes, Geometriae Dedicata 8 (1979) 279–296.
[4] J. Zaks, Neighborly families of 2^d d-simplices in E^d, Geometriae Dedicata 11 (1981) 505–507.

Received 1 April 1981

Annals of Discrete Mathematics 20 (1984) 255–262
North-Holland

SOME RECENT RESULTS ON CYCLE TRAVERSABILITY IN GRAPHS

M.D. PLUMMER

Vanderbilt University, Nashville, Tennessee 37235, U.S.A. and University of Bonn, Bonn, Fed. Rep. Germany

1. Introduction

The main concept in this paper is the following. A graph G is $C(m^+, n^-)$ if it has at least $m + n$ points and for every two disjoint point sets $A, B \subseteq V(G)$ with $|A| = m$ and $|B| = n$ there is a cycle in $G - B$ containing all members of A. We shall see that this concept generalizes such well-known and widely studied concepts as the (point-) connectivity of a graph G, $\kappa(G)$, on the one hand, and the existence of hamiltonian cycles on the other. We will elaborate on these (and other) special cases, showing how some classical results fit into our framework and also discuss some very recent results concerning this graphical parameter. A number of conjectures and opcn problems will also be presented.

There are many known and conjectured interrelationships among the ideas treated below. In order to introduce some coherence into our discussion we shall proceed under two topic headings:

(a) $C(m^+, n^-)$ versus connectivity, and

(b) $C(m^+, n^-)$ versus itself.

For any terminology not defined herein we refer the reader to Harary [14], to Holton and Plummer [16] or to Holton, McKay, Plummer and Thomassen [17].

2. $C(m^+, n^-)$ versus connectivity

It is easy to see that (point-) connectivity is a special case of $C(m^+, n^-)$. The proof of the following lemma is a straightforward application of the point-version of Menger's theorem. Independent proofs may be found in Mesner and Watkins [20] and in Halin [13].

Lemma 1. *If $n \geqslant 2$, then G is n-(point-) connected iff G is $C(2^+, n - 2^-)$.*

An important early result in this area is the following due to Dirac [7] which is not as well known in graph theory as it deserves to be.

Theorem 1. *If G is n-connected, then given any 2 lines and $n-2$ points, there is a cycle through all n of these elements.*

The following immediate corollary is much more widely known. Indeed, it seems to have entered the realm of graph theory 'folklore'.

Corollary 1. *If G is n-connected, then G contains a cycle through any set of n points, i.e. G is $C(n^+, 0^-)$.*

This is certainly immediate from a stronger result of Halin [13] which includes Lemma 1 as well. We will return to the result of Halin in Section 3. Corollary 1 also leads naturally to the following question due to Watkins and Mesner [34].

Problem 1. Given integers $n \geq 2$ and $k \geq 0$, characterize those graphs G with $\kappa(G) = n$ which are $C(n+k^+, 0^-)$, but not $C(n+k+1^+, 0^-)$.

They settled the question for $k=0$ and $n \geq 2$, but the general problem remains open.

Let us now consider some important special classes of n-connected graphs. A graph is called *n-polytopal* if it is the skeleton of a polytope of dimension n. Balinski [1] proved the following result about such graphs.

Theorem 2. *If G is n-polytopal, then G is n-connected.*

The complete bipartite graph $K(n, n+1)$ provides a simple example of an n-connected graph which is not $C(n+1^+, 0^-)$, so in a sense Corollary 1 is the best possible. But if G is n-polytopal, we can obtain a bit more.

Theorem 3. *If G is n-polytopal then it is $C(n+1^+, 0^-)$.*

This result was proved by Sallee [23] who, in the same paper, conjectured the following which, so far as we know, remains open.

Problem 2 (Conjecture). If G is n-polytopal, then G is $C(n+2^+, 0^-)$.

If this conjecture is true, $n+2$ is the best possible as Sallee shows by exhibiting for each $n \geq 3$, an n-polytope which is not $C(n+3^+, 0^-)$. For $n=3$, this conjecture was proved true independently by Sallee [23] and by Plummer and Wilson [22]. We remind the reader that by an important result of Steinitz [9, 24, 25] 3-polytopal graphs are precisely the 3-connected planar graphs.

Theorem 4. *If G is 3-connected and planar, then G is $C(5^+, 0)$.*

This result is sandwiched between the following two results, the first of which is trivial and the second of which is a deep result due to Tutte [30–32].

Theorem 5. *If G is 2-connected and planar, G is $C(2^+,0^-)$ and this is the best possible.*

Theorem 6. *If G is 4-connected and planar, then G contains a Hamiltonian cycle.*

Let us interject here for the sake of completeness that attention has been called to the special case $C(m^+,0^-)$ by Chvátal [5]. He defines the *cyclability of G* to be the maximum value of m for which G is $C(m^+,0^-)$. Moreover, we point out that the even more special case of determining which graphs are $C(p^+,0^-)$, where $p=|V(G)|$, is identical to determining which graphs have Hamiltonian cycles. But today the literature in this area is so enormous that we shall shrink from the abyss and content ourselves with mentioning that two particularly good surveys of Hamiltonian graphs are to be found in Bondy [3] and Bermond [2].

Now if we view the preceding three theorems as results about graphs of genus 0, we are immediately tempted to head off in yet another unexplored direction: that of $C(m^+,n^-)$ versus genus. We again restrain ourselves and are content to mention an intriguing unsettled conjecture due to Grünbaum [10].

Problem 3 (Conjecture). If G is 4-connected and toroidal, then G has a Hamiltonian cycle.

But now let us return to general n-connected graphs and proceed on a different tack. Assume for the moment that our graphs are regular. What can be gained through this restriction? Holton [15] has proved the following result.

Theorem 7. *If G is n-connected and n-regular, then G is $C(n+4^+,0^-)$.*

The value $n+4$ is probably not the best possible, as Holton is quick to point out. At the moment we only know that if $N=\max\{k \mid \text{if } G \text{ is } n\text{-connected and } n\text{-regular then } G \text{ is } C(k^+,0^-)\}$ then $N \leqslant 10n-11$. For each value of $n \geqslant 3$ an extremal graph is provided for this bound by an example due to Meredith [19].

For the special case when $n=3$, however, the sharpest possible bound has recently been obtained by Holton, McKay, Plummer and Thomassen [17].

Theorem 8. *If G is 3-connected and cubic then* (a) *G is $C(9^+,0^-)$ and* (b) *there is an infinite family of such graphs* (*based on the Petersen graph*) *which are not $C(10^+,0^-)$.*

Problem 4. What is the best possible function of n which may be substituted for $n+4$ in the conclusion of Theorem 7?

Before ending this section we want to mention another concept similar to (but quite distinct from) $C(m^+, n^-)$ which was formulated by Zamfirescu [36] and investigated by Grünbaum [11] and Walther [33]. Given positive integers j and k, let C_k^j denote the greatest lower bound of the values of p such that there is a k-connected graph G with p points such that given any set of j points in G there is a longest cycle of G missing all j of the prescribed points. An earlier, but related, idea may be found in Mesner and Watkins [20]. Once again we pull back from the brink and refer the interested reader to the references given.

3. $C(m^+, n^-)$ versus itself

First we hasten to say that the title of this section was not chosen (entirely) capriciously. Here we mean to deal with problems of the following type: given non-negative integer variables m, n, r and s, for what values of these variables does the implication $C(m^+, n^-) \to C(r^+, s^-)$ hold?

As we promised in Section 2 of this paper, we return to the result of Halin [13] which says somewhat more than Lemma 1.

Lemma 2. *If G is n-connected, then G is $C(n-k^+, k^-)$ for all k, $0 \leq k \leq n-2$.*

At this point we direct the reader's attention to Table 1 which summarizes in compact form most of the results discussed in this section. First, note that by Lemma 1, the properties in each row of the first two columns are equivalent. Moreover, Lemma 2 in fact says that the property in the left-most position of

Table 1

2-connected	$\leftrightarrow$	$C(2^+,0^-)$								
			$\nwarrow$							
3-connected	$\leftrightarrow$	$C(2^+,1^-) \to$	$C(3^+,0^-)$							
			$\nwarrow$	$\nwarrow$						
4-connected	$\leftrightarrow$	$C(2^+,2^-) \to$	$C(3^+,1^-) \to$	$C(4^+,0^-)$						
			$\nwarrow$	$\nwarrow$	$\nwarrow$					
5-connected		$C(2^+,3^-) \to$	$C(3^+,2^-) \to$	$C(4^+,1^-) \to$	$C(5^+,0^-)$					
			$\nwarrow$	$\nwarrow$	$\nwarrow$	$\nwarrow$				
6-connected		$C(2^+,4^-) \to$	$C(3^+,3^-) \to$	$C(4^+,2^-) \to$	$C(5^+,1^-)$	$C(6^+,0^-)$				
			$\nwarrow$	$\nwarrow$	$\nwarrow$	$\nwarrow$	$\nwarrow$			
7-connected		$C(2^+,5^-) \to$	$C(3^+,4^-) \to$	$C(4^+,3^-) \to$	$C(5^+,2^-)$	$C(6^+,1^-)$	$C(7^+,0^-)$			
			$\nwarrow$	$\nwarrow$	$\nwarrow$	$\nwarrow$	$\nwarrow$	$\nwarrow$		
8-connected		$C(2^+,6^-) \to$	$C(3^+,5^-) \to$	$C(4^+,4^-) \to$	$C(5^+,3^-)$	$C(6^+,2^-)$	$C(7^+,1^-)$	$C(8^+,0^-)$		
			$\nwarrow$	$\nwarrow$	$\nwarrow$	$\nwarrow$	$\nwarrow$	$\nwarrow$	$\nwarrow$	
9-connected		$C(2^+,7^-) \to$	$C(3^+,6^-) \to$	$C(4^+,5^-) \to$	$C(5^+,4^-)$	$C(6^+,3^-)$	$C(7^+,2^-)$	$C(8^+,1^-)$	$C(9^+,0^-)$	
			$\nwarrow$	$\nwarrow$	$\nwarrow$	$\nwarrow$	$\nwarrow$	$\nwarrow$	$\nwarrow$	$\nwarrow$
10-connected	$\leftrightarrow$	$C(2^+,8^-) \to$	$C(3^+,7^-) \to$	$C(4^+,6^-) \to$	$C(5^+,5^-)$	$C(6^+,4^-)$	$C(7^+,3^-)$	$C(8^+,2^-)$	$C(9^+,1^-) \not\to$	$C(10^+,0^-)$

each row implies all others to the right of it in the same row. We omit these implication arrows in Table 1 to reduce clutter. Second, it is obvious that in any given column of the array any property is implied by any property beneath it by definition. Similarly, by definition we have the validity of the diagonal implications shown, i.e.

$$C(m^+, n^-) \to C(m-1^+, n^-), \quad \text{for all } m \geqslant 3,\ n \geqslant 0.$$

But now we come to possible horizontal left-to-right implications and here we soon encounter non-trivial questions. First of all it is easy to see (cf. Holton and Plummer [16]) that if the top-most left-to-right implication holds in any column, then so do all those directly beneath it.

Lemma 3. *Suppose $m \geqslant 2$ and $n \geqslant 1$. If $C(m^+, n^-) \to C(m+1^+, n-1^-)$ is valid, then so is $C(m^+, n+1^-) \to C(m+1^+, n^-)$.*

But now we present a non-trivial result [16].

Theorem 9. *If G is $C(4^+, 1^-)$, then G is $C(5^+, 0^-)$.*

The proof of this result is long and will be published elsewhere. The main tool used in the proof is a very nice (but seldom-used) variation on Menger's Theorem due to Perfect [21].

At first one is tempted to conjecture that the implication

(I) $\quad C(n^+, 1^-) \to C(n+1^+, 0)$

holds for all graphs with at least three points and with $n \geqslant 2$. But any hypohamiltonian graph on $n+1$ points suffices to show that the implication fails for that value of n. The smallest hypohamiltonian graph is the Petersen graph, so we know, for example, that $C(9^+, 1^-) \not\to C(10^+, 0^-)$. In fact, the implication (I) above fails for $n = 9, 12, 14, 15$ and all $n \geqslant 17$ since hypohamiltonian graphs of order $n+1$ are known to exist for all these values of n. Hypohamiltonian graphs have generated a lot of interest; for a survey of such graphs see Collier and Schmeichel [6] and Thomassen [26, 28]. In particular, any counterexample to (I) for $n = 10, 11$, or 13 cannot be hypohamiltonian for it is known that hypohamiltonian graphs on 11, 12, and 14 points do not exist. It is presently unknown if there is a hypohamiltonian graph on 17 points.

4. Concluding remarks

In closing, let us mention a few further results and some open conjectures.

Problem 5 (Conjecture). The implication (I) holds for $5 \leqslant n \leqslant 8$.

Problem 6 (Conjecture). If implication (I) fails to hold, the only counterexamples are hypohamiltonian.

A helpful referee has recently pointed out to us the following conjecture of Grünbaum [11].

Problem 7 (Conjecture). Let $p = |V(G)|$. Then if G is $C(p-2^{+}, 2^{-})$, G must contain a cycle of length $\geqslant p-1$.

Relevant to Theorem 8 above, we state the following:

Problem 8 (Conjecture). If some 10 points of a 3-connected cubic graph G do not lie on a cycle in G, then they are the principal points of a subgraph of G homeomorphic to the Petersen graph.

In relation to Tutte's result (Theorem 6 above) and $C(m^{+}, n^{-})$ we resurrect the following 'old chestnut'.

Problem 9 (Conjecture). If G is planar, 4-connected, and has p points, then G is $C(p-2^{+}, 2^{-})$.

We remark that the fact that G is $C(p-1^{+}, 1^{-})$ follows easily from Tutte's theorem as was shown in 1972 by D. Nelson (unpublished).

The following conjecture and two theorems represent yet another interesting variation on the theme of $C(m^{+}, n^{-})$ and are due to Bondy and Lovász [4].

Theorem 10. *If G is n-connected and not bipartite, then any $n-1$ points of G lie on an odd cycle. The value $n-1$ is sharp.*

This result thus settles a conjecture of B. Toft [29].

Theorem 11. *If $n \geqslant 3$ and G is n-connected, then any $n-1$ points of G lie on an even cycle.*

Problem 10 (Conjecture). If $n \geqslant 3$ and G is n-connected, then any n points of G lie on an even cycle.

Bondy and Lovász have verified this conjecture for $n = 3$.

We bring our remarks to a close by mentioning that there are many analogous questions which can be asked about cycles through specified sets of *lines* and cycles containing specified *paths*, but we hasten to close the lid on this Pandora's box and direct the interested reader to work by Bondy and Lovász [4], Grötschel [8], Häggkvist and Thomassen [12], Lovász [18], Neil Robertson (unpublished), Thomassen [27], Toft [29] and Woodall [35].

Note added in proof. Since this paper was written, Problem 8 has been settled in the affirmative by Ellingham, Holton and Little and independently by Kelmans and Lomonosov.

References

[1] M. Balinski, On the graph structure of convex polyhedra in n-space, Pacific J. Math. 11 (1961) 431–434.
[2] J.-C. Bermond, Hamiltonian graphs, in: L.W. Beineke and R.J. Wilson, Eds., Selected Topics in Graph Theory (Academic Press, New York, 1979) Ch. 6, pp. 127–167.
[3] J.A. Bondy, Hamiltonian cycles in graphs and digraphs, Proc. 9th S.E. Conf. Combinatorics, Graph Theory and Computing, Utilitas Math. (Winnipeg, 1979) pp. 3–28.
[4] J.A. Bondy and L. Lovász, Cycles through specified vertices of a graph, Univ. Waterloo Res. Report CORR 79–14 (1979).
[5] V. Chvátal, New directions in hamiltonian graph theory, in: F. Harary, Ed., New Directions in the Theory of Graphs (Academic Press, New York, 1973) pp. 65–95.
[6] J.B. Collier and E.F. Schmeichel, Systematic searches for hypohamiltonian graphs, Networks 8 (1978) 193–200.
[7] G.A. Dirac, In abstrakten Graphen vorhandene vollständige 4-Graphen und ihre Unterteilungen, Math. Nachr. 22 (1960) 61–85.
[8] M. Gröstschel, Graphs with cycles containing given paths, Ann. Discr. Math. 1 (1977) 233–245.
[9] B. Grünbaum, Convex Polytopes (Wiley, New York, 1967).
[10] B. Grünbaum, Polytopes, graphs and complexes, Bull. Amer. Math. Soc. 76 (1970) 1131–1201.
[11] B. Grünbaum, Vertices missed by longest paths or circuits, J. Combinat. Theory (A) 17 (1974) 31–38.
[12] R. Häggkvist and C. Thomassen, Circuits through specified edges, Aarhus Univ. Math. Inst. Preprint No. 7 (1980/81).
[13] R. Halin, Zur Theorie der n-fach zusammenhängenden Graphen, Abh. Math. Sem. Hamburg 33 (1969) 133–164.
[14] F. Harary, Graph Theory (Addison-Wesley, Reading, 1969).
[15] D.A. Holton, Cycles through specified vertices in k-connected regular graphs (submitted).
[16] D.A. Holton and M.D. Plummer, A nine-point theorem for 3-connected graphs, Combinatorica 2 (1982) 53–62.
[17] D.A. Holton, B.D. McKay, M.D. Plummer and C. Thomassen, Cycles with required and forbidden points, Ann. Discr. Math. 16 (1982) 129–147.
[18] L. Lovász, Problem 5, Period. Math. Hungar. 4 (1974) 82.
[19] G.H.J. Meredith, Regular n-valent n-connected non-Hamiltonian non-n-edge-colorable graphs, J. Combinat. Theory (B) 14 (1973) 55–60.
[20] D.M. Mesner and M.E. Watkins, Some theorems about n-vertex connected graphs, J. Math. Mech. 16 (1966) 321–326.
[21] H. Perfect, Applications of Menger's graph theorem, J. Math. Anal. Appl. 22 (1968) 96–111.

[22] M.D. Plummer and E.L. Wilson, On cycles and connectivity in planar graphs, Canad. Math. Bull. 16 (1973) 283–288.
[23] G.T. Sallee, Circuits and paths through specified nodes, J. Combinat. Theory 15 (1973) 32–39.
[24] E. Steinitz, Polyheder und Raumeinteilungen, Enzykl. Math. Wiss., Vol. 3 (Geometrie), Part 3AB12 (1922) 1–139.
[25] E. Steinitz und H. Rademacher, Vorlesungen über die Theorie der Polyhedra (Springer, Berlin, 1934).
[26] C. Thomassen, On hypohamiltonian graphs, Discr. Math. 10 (1974) 383–390.
[27] C. Thomassen, Note on circuits containing specified edges, J. Combinat. Theory (B) 22 (1977) 279–280.
[28] C. Thomassen, Hypohamiltonian graphs and digraphs, in: Theory and Applications of Graphs, Lecture Notes in Math. No. 642 (Springer-Verlag, Berlin, 1978) pp. 557–571.
[29] B. Toft, Problem 11, in: M. Fiedler and J. Bosák, Eds., Recent Advances in Graph Theory, Proc. Prague Sympos. 1974 (Academia Praha, Prague, 1975) p. 544.
[30] W.T. Tutte, A theorem on planar graphs, Trans. Amer. Math. Soc. 82 (1956) 99–116.
[31] W.T. Tutte, Bridges and hamiltonian circuits in planar graphs, Aeq. Math. 15 (1977) 1–33.
[32] W.T. Tutte, Bridges and hamiltonian circuits in planar graphs, in: Selected Papers of W.T. Tutte, Vol. II (Charles Babbage Research Centre, St. Pierre, Manitoba, 1979) pp. 841–873.
[33] H. Walther, Über die Nichtexistenz zweier Knotenpunkte eines Graphen, die alle längsten Kreise fassen, J. Combinat. Theory 8 (1970) 330–333.
[34] M.E. Watkins and D.M. Mesner, Cycles and connectivity in graphs, Canad. J. Math. 19 (1967) 1319–1328.
[35] D.R. Woodall, Circuits containing specified edges, J. Combinat. Theory (B) 22 (1977) 274–278.
[36] T. Zamfirescu, A two-connected planar graph without concurrent longest paths, J. Combinat. Theory (B) 13 (1972) 116–121.

Received 16 March 1981

Annals of Discrete Mathematics 20 (1984) 263–269
North-Holland

GENERALIZED HAMILTONIAN PATHS IN TOURNAMENTS

Moshe ROSENFELD*
Ben Gurion University of the Negev, Beer Sheva, Israel

1. Introduction

A *tournament* is an orientation of the complete graph. Any two vertices (players) v, w are adjacent by exactly one arc, either $v \to w$ (v beats w) or $w \to v$. A permutation $¶ = v_1, \ldots, v_n$ of all vertices is a *generalized Hamiltonian path.* The pattern of ¶ is the binary sequence $B_{n-1} = e_1, \ldots, e_{n-1}$ in which $e_i = 1$ if $v_i \to v_{i+1}$ and $e_i = 0$ if $v_i \leftarrow v_{i+1}$. We say that ¶ realizes the pattern (binary sequence) B_{n-1} in the tournament T_n. Generalized Hamiltonian circuits are similarly defined. It is well known (folklore) that every tournament T_n has a Hamiltonian path, that is, every tournament T_n realizes the sequences $1, \ldots, 1$ and $0, \ldots, 0$. Various other binary sequences are known to be always realizable [1,4,6,8]. Certain families of tournaments, most notably tournaments with $n = 2^k$ players, realize all possible 2^{n-1} binary sequences B_{n-1} [2]. Obviously, there are $n!$ distinct permutations of the players of T_n and only 2^{n-1} distinct patterns; these facts led the author to conjecture [8]:

Every tournament T_n, $n > 7$, realizes every binary sequence B_{n-1}.

In this paper, we trace the evolution of this conjecture, survey the known results and discuss generalizations to oriented trees, general digraphs, orientations of n-chromatic graphs and complexity of computation. We show that there exists a digraph D_n, having n vertices and only $5n + O(1)$ edges, that realizes every binary sequence B_{n-1}. Bollobás suggested that such a graph can be used to prove that almost all tournaments realize every binary sequence. We denote by T_n a tournament on n players. TT_n denotes the transitive tournament, that is, the tournament in which the relation $v \to w$ is transitive. TR_{2k+1} denotes a regular tournament, i.e. a tournament in which every player beats k players exactly. T_n is nearly regular if the number of losses of any pair of players differs by at most 1. A realization of the binary sequence 1010...(0101...) is an

* I wish to thank Bela Bollobás for stimulating conversations.

anti-directed path (circuit). A tournament T_n is self-complementary if it is isomorphic to the tournament obtained by reversing the orientation of every edge.

2. Generalized Hamiltonian paths

Grünbaum [4] must be held responsible for initiating the search for generalized Hamiltonian paths. He proved that every tournament T_n, with three exceptions, has an anti-directed Hamiltonian path. (The exceptions are TR_3, TR_5 and the regular tournament TR_7 that has no TT_4 as a subtournament.) This theorem was strengthened by the author in [8], where the above conjecture was made. It is easy to show that every binary sequence is realizable in TT_n, actually a trivial inductive proof shows that every orientation of the edges of any tree on n vertices is realizable in TT_n. In [1], it was shown that all binary sequences $B_{n-1} = e_1 \cdots e_{n-1}$ consisting of two blocks, or such that the ith block has at least $i+1$ digits, are realizable. (A block of a binary sequence is a maximal subsequence of consecutive similar digits.) The sequences $1010\cdots01\cdots1$ (Hell and Rosenfeld [6]) and $0\cdots010\cdots0$ (Grünbaum [5]) are also realizable, but the most remarkable result in this direction is Forcade's Theorem [2], that every tournament T_n, with $n = 2^k$ players, realizes every binary sequence B_{n-1}. Actually, Forcade proved considerably more; he proved that the parity of the number of realizations of a given binary number B_{n-1} is independent of T_n. He gave a most elegant formula for computing the parity. This formula is a deep generalization of Redéi's Theorem [7], which proves that every tournament has an odd number of distinct Hamiltonian paths. Without a doubt, Forcade's Theorem will prove most valuable in establishing the realizability of binary sequences. To state this theorem we need the following:

Definition. Let $a, b \in N$, and let their binary representations be $a = \sum_{i \in I_a} 2^i$ and $b = \sum_{i \in I_b} 2^i$. We say that aRb if $I_a \subseteq I_b$. Obviously, R defines a partial order on N, the set of natural numbers.

Forcade's Theorem. *Let* $\#(e_1 \cdots e_{n-1}; T_n)$ *be the number of permutations realizing the binary sequence* $e_1 \cdots e_{n-1}$ *in* T_n, *and let* $A = \{k \mid e_k = 0, kRn\}$. *Then*

$$\#(e_1 \cdots e_{n-1}; T_n) \equiv \operatorname{card}\{B \mid B \subseteq A, B \text{ is } R\text{-linearly ordered}\} \pmod 2$$

(*the empty set is considered* R*-linearly ordered*).

Many more binary sequences can be shown to be realizable using Forcade's Theorem. For example, if $B_{n-1} = e_1 \cdots e_{n-1}$ and if $e_k = 1$ when kRn, then $A = \emptyset$, card$\{B \mid B \subseteq A\} = 1$ and therefore B_{n-1} is realizable in T_n. Also, if $e_k = 0$ whenever kRn, then $e_1 \cdots e_{n-1}$ is realizable in T_n. To see this, observe first that kRn iff $(n-k)Rn$, therefore the sequence $e'_1 \cdots e'_{n-1}$ defined by $e'_i = 1 - e_{n-i}$ is realizable in T_n by the previous argument. But if $v_1 \cdots v_n$ realizes this sequence in T_n, then $v_n \cdots v_1$ realizes $e_1 \cdots e_{n-1}$ in T_n. (A direct inductive proof also shows that card$\{B \mid B \subseteq A, B\ R\text{-linearly ordered}\} \equiv 1 \bmod(2)$ when $A = \{k \mid kRn\}$.) We conclude this section by proving the existence of a digraph with 'few' edges that realizes every binary sequence.

Theorem 1. *There are digraphs D_n, with n vertices and no more than $5n + O(1)$ edges that have generalized Hamiltonian paths realizing every binary sequence B_{n-1}.*

Proof. Let TR_7 be the regular tournament with vertex set $\{0,1,\ldots,6\}$ and $i \to i + \{1,2,3\}$ (mod 7). Observe that the automorphism group of this tournament is transitive on the vertices (actually the automorphism group is C_7 the cyclic group of order 7) and that TR_7 is self-complementary. Therefore, if TR_7 realizes every binary sequence B_6, then B_6 is realized by permutations starting with any preassigned vertex. It is a tedious matter to check that TR_7 realizes all binary sequences; below is shown a computer output verifying this fact. The output is a sequence of 32 numbers, the ith number being the number of permutations starting with 0 realizing the binary sequence B_6 obtained by extending the binary representation of $i-1$ to a binary sequence of length 6 by adding zeros in front of it. Since TR_7 is self-complementary, the number of realizations of $e_1 \cdots e_6$ is equal to the number of realizations of $e'_1 \cdots e'_6$ $(e'_i = 1\, e_i)$. These sequences are obtained from the binary representations of i and $63 - i$, respectively; thus it is enough to give only the first 32 entries in this sequence:

25, 16, 10, 19, 16, 6, 12, 22, 16, 11, 3, 8, 11, 7, 15, 19, 10, 12, 9, 7, 3, 2, 5, 6, 12, 6, 5, 11, 15, 12, 13, 16.

If $n = 7k$, we construct D_n as follows. Let $T^1, \ldots, T^k$ be k disjoint copies of TR_7. We join each vertex $x \in T^i$, $1 \leq i < k$, to two vertices of T^{i+1} by a pair of edges with opposite orientations. Obviously, D_n has $5n - 14$ edges. Let $e_1 \cdots e_{n-1}$ be any binary sequence. Since each T^i realizes every binary sequence B_6, we choose realizations of the binary sequences $e_{7i-6} \cdots e_{7i-1}$ in T^i, $1 \leq i \leq k$. Let these realizations be denoted by $v^i_1, \ldots, v^i_7$. v^{i+1}_1 is chosen as the vertex that beats v^i_7 if $e_{7i} = 0$, or the vertex that loses to v^i_7 if $e_{7i} = 1$. v^1_1 is arbitrarily chosen. Since TR_7 is vertex transitive, such realizations of $e_{7i-6} \cdots e_{7i-1}$ are possible, and the resulting permutation $v^1, \ldots, v^k_7$ is a generalized Hamiltonian path in D_n, realizing $e_1 \cdots e_{n-1}$.

If $n \not\equiv 0$ (mod 7), write $n = 7k + 6m$ $(m \leq 6)$. Let T_i be the tournament described in Fig. 1. T_i admits the automorphism $i \to i+2 \pmod 6$, and is self-complementary ($\{0,2,4\} \to \{1,3,5\}$ under the isomorphism of T_6 and its complement). Below are listed the permutations realizing all 16 binary sequences $1e_2 \cdots e_5$ starting with the vertex 0 (listed in a lexicographic order):

032154	014532	025431	013542
032145	014523	023145	013524
021453	021354	013254	012354
015234	021345	034251	012345

It follows that all these binary sequences can be realized by permutations starting with the vertex 2, and by permutations starting with the vertex 4. Realizing these binary sequences in the complement of T_i, yields realizations in T_6 of all 16 binary sequences $0e_2 \cdots e_5$ by permutations starting with the vertex 1 (3 and 5). To construct D_n, we take k copies of TR_7, $T^1, \ldots, T^k$ and m copies of T_6, $S^1, \ldots, S^m$. We add edges between copies of T^i, $1 \leq i \leq k$, as in the previous case. Each vertex of T^k is joined by two pairs of edges with opposite orientations. One pair is joined to two vertices from $\{0,2,4\}$ of S^1, and the other to two vertices from $\{1,3,5\}$. Each vertex of S^i, $1 \leq i < m$, is joined similarly to four vertices of S^{i+1}. D_n will have $35k + 39m - 24 = 5n + O(1)$ edges. D_n has generalized Hamiltonian sequences realizing any given binary sequence $e_1 \cdots e_{n-1}$; as the proof is similar to the previous case, we omit the details.

Remark. This theorem can be used to prove that almost all tournaments T_n realize all binary sequences. Details will appear elsewhere. Also, by connecting the vertices of T^k (or S^m in the second case) to T^1 in a similar manner, we obtain a digraph with n vertices, $5n + O(1)$ edges, that realizes every binary sequence $e_1 \cdots e_n$ by a generalized Hamiltonian cycle (GHC). GHCs are treated similarly. Of course, a tournament T_n has a GHC realizing the sequence $1 \cdots 1$ iff it is strongly connected. On the other hand, every tournament T_n (with two exceptions) has a GHC realizing the binary sequence $B_n = 11 \cdots 10$ [5]. Thomas-

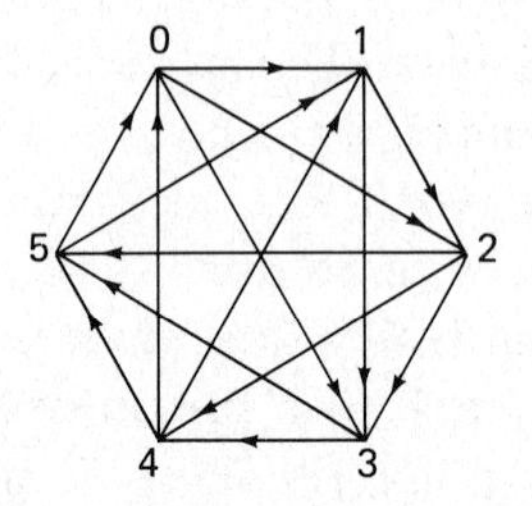

Fig. 1.

sen [11] and Rosenfeld [9] proved that for $n \geq 28$ all tournaments T_{2n} realize the sequence $1010\cdots 10$. Thus, problems of the same nature can be discussed for GHC. Also, it would be interesting to characterize those binary sequences b_n that are always realizable in T_n by a GHC (for $n \geq n_0$).

3. Distribution of realizations of binary sequences

The binary sequence $B_{n-1} = 1\cdots 1$ is obviously realized exactly once in TT_n. On the other hand, it was shown in [9] that every tournament T_n has at least $c \cdot n$ anti-directed Hamiltonian paths. The vertices of a given tournament T_n can be permuted in $n!$ ways, hence for each T_n there is a sequence B_{n-1} which is realized in T_n at least $n!2^{1-n}$ times. Szele [10] has shown, in what is considered to be the first proof by the probabilistic method, that there exists a tournament T_n that has at least $n!\,2^{-n+1}$ Hamiltonian paths. This generalizes immediately to:

Theorem 2. *For each binary sequence $B_{n-1} = e_1 \cdots e_{n-1}$ there is a tournament T_n in which B_{n-1} is realized at least $n!\,2^{-n+1}$ times.*

Proof. There are $2^{\binom{n}{2}}$ distinct tournaments on the labelled vertices $\{1,\ldots,n\}$. Any given permutation $v_1\cdots v_n$ will realize the given binary sequence in exactly $2^{\binom{n}{2}-n+1}$ distinct tournaments. Thus, the total number of realizations of B_{n-1} is $n!2^{\binom{n}{2}-n+1}$. Since the number of tournaments is $2^{\binom{n}{2}}$, at least one of them has $n!2^{-n+1}$ realizations of B_{n-1}.

In the tournament TR_7 (Theorem 1), the binary sequence 111111 is realized 175 times, while the anti-directed path 010101 is realized only 14 times. In [8] it was shown that among all tournaments T_n, the maximal number of realizations of the binary sequence 1010... is achieved in TT_n (the number of up-down permutations). The distribution of the number of realizations of various binary sequences exhibits an intricate behaviour. The following questions seem to be of particular interest.

(1) Determine $A_n = \max_{T_n}\{\max_{B_{n-1}} \#(B_{n-1}; T_n)\}$ (of course $A_n \geq n!2^{-n+1}$).

(2) Characterize the binary sequences B_{n-1} for which this maximum is obtained, and the corresponding T_n's.

(3) For any given sequence B_{n-1} determine $\max_{T_n} \#(b_{n-1}; T_n)$ and $\min_{T_n} \#(B_{n-1}; T_n)$. For $B_{n-1} = 1\cdots 1$ $\min_{T_n} \#(B_{n-1}; T_n) = 1$, while for $B_{n-1} = 1010\cdots \max_{T_n} \#(B_{n-1}; T_n) = R_n$ (R_n is the number of up-down sequences) [9]. It is suspected that $\max_{T_n}(11\cdots 1; T_n)$ is obtained for nearly regular tournaments.

(4) Determine $\max_{B_{n-1}} \#(B_{n-1}; T_n)$ when T_n is a random tournament on n vertices.

4. Digraphs, oriented trees, and complexity of realizations

Similar realizations as well as connections with other properties of graphs can be investigated. The first result in this direction is Gallai's Theorem [3] that every orientation of an n-chromatic graph contains a directed path of length $n-1$ (i.e. realizes the binary sequence $B_{n-1}=1\cdots 1$). A. Gyarfás (private communication) proved that any acyclic orientation of an n-chromatic graph realizes any orientation of every tree with n vertices. We venture:

Problem (5). Any orientation of the edges of an n-chromatic graph, realizes every binary sequence B_{n-1}.

It is easy to see that not every orientation of every tree on n vertices is always realizable in T_n, though it is realizable in TT_n. The following related results and conjectures have been communicated by Brooks Reid.

Conjecture (D. Sumner). Every tournament T_{2n-2} realizes every orientation of a tree with n vertices. Generalizing this conjecture in the spirit of Gallai's Theorem:

Conjecture (C. Thomassen). Every orientation of the edges of a $(2n-2)$-chromatic graph realizes every oriented tree with n vertices.

B. Reid and N. Wormald have obtained some results in this direction. Among others, they proved Sumner's conjecture for near regular tournaments, and for caterpillars with directed spines.

Finally, another direction of investigation was initiated in Hell and Rosenfeld [6]. In this paper the complexity of finding a permutation $v_1\cdots v_n$ realizing a binary sequence B_{n-1} in T_n was analysed in terms of the number of comparisons (v_i, v_j). The complexity of the problem was defined as follows:

input: a tournament T_n and binary sequence B_{n-1};
output: a permutation $v_1\cdots v_n$ realizing B_{n-1};
complexity: smallest number of comparisons (v_i, v_j) required in the worst case to realize B_{n-1}.

Surprisingly, it turned out that distinct binary sequences yield complexities of different order of magnitude. Thus, the binary sequence $1\cdots 1$ has complexity $O(n \log n)$, while the sequence $1010\cdots$ has complexity $O(n)$. Generally, for $0\leq\alpha\leq 1$, the binary sequence $1010.1011\cdots 1$ has complexity $O(n\log^{\alpha} n)$ for an appropriately chosen string of 1's [6]. The complexity of a binary sequence is not known in general.

References

[1] B. Alspach and M. Rosenfeld, Realization of certain generalized paths in tournaments, Discr. Math. (to appear).
[2] R. Forcade, Parity of paths and circuits in tournaments, Discr. Math. 6 (1973) 115–118.
[3] T. Gallai, On directed paths and circuits, in: P. Erdös and G. Katona, Eds., Theory of Graphs (Academic Press, New York, 1968) pp. 115–119.
[4] B. Grünbaum, Antidirected Hamiltonian paths in tournaments, J. Combinat. Theory (B) 11 (1971) 249–257.
[5] F. Harary, Graph Theory (Addison-Wesley, Reading, Mass.) p. 211, problem 16·26.
[6] P. Hell and M. Rosenfeld, The complexity of finding generalized Hamiltonian paths in tournaments, J. Algorithms, to appear.
[7] L. Redéi, Ein Kombinatorische Satz, Acta. Litt. Sci. Szeged 7 (1934) 39–43.
[8] M. Rosenfeld, Antidirected Hamiltonian paths in tournaments, J. Combinat. Theory (B) 12 (1972) 93–99.
[9] M. Rosenfeld, Antidirected Hamiltonian circuits in tournaments, J. Combinat. Theory (B) 16 (1974) 234–242.
[10] T. Szele, Combinatorial investigations concerning directed complete graphs (Hungarian), Mat. Fiz. Lapok 15 (1943) 223–256.
[11] C. Thomassen, Antidirected Hamiltonian circuits and paths in tournaments, Math. Ann. 201 (1973) 231–238.

Received March 1981

Annals of Discrete Mathematics 20 (1984) 271–282
North-Holland

LARGE REGULAR FACTORS IN RANDOM GRAPHS

E. SHAMIR
Mathematics Department, The Hebrew University, Jerusalem, Israel

E. UPFAL
Mathematics Department, The Weizmann Institute of Science, Rehovot, Israel

Does a graph G contain a large regular factor H on 'most' edges? For random graph spaces $G_{n,p}$, $G_{n,N}$ $N = N.w(n)\log n$, $w(n)\to\infty$, we prove the answer is positive with probability $1-o(1)$ even if we take out just a vanishing fraction of the edges. The result holds also for almost regular factors, and it derives a good lower bound on the number of regular graphs in $G_{n,N}$.

1. Introduction

The graph property we investigate here is: 'Given a graph G with n *labelled* vertices, N edges and an integer φ, $\varphi \leqslant \min\deg G$, $n\varphi$ even. *G contains a factor H which is regular of degree φ.*' How large can we take φ?

This property and some variants for almost regular factors are studied here for random graphs. As a result we obtain counting estimates for regular and almost regular graphs with a given degree sequence, and an insight into the structure of the space $G_{n,N}$ of all graphs with n labelled vertices and N edges. This space, $G_{n,N}$, is considered as a probability space with all its graphs being equiprobable. It is used extensively in Erdös–Rényi papers [4,5]. It turns out more convenient to use the spaces $G_{n,p}$: let $0<p<1$, let all $\binom{n}{2}$ edge occurrences be independent trials with probability p for success. Thus, each graph with m specified edges and $\binom{n}{2}-m$ non-edges has probability $p^m(1-p)^{\binom{n}{2}-m}$ in $G_{n,p}$. (Thus, $G_{n,N}$ is in fact a subspace of $G_{n,p}$ obtained by conditioning the number of edges to be N.)

There are many studies of the asymptotic probabilities of graph-properties (usually monotone): if $p=p(n)$, what is $\lim_{n\to\infty}$ Prob $\{G\in G_{n,p}$, G has property $Q\}$? If the limit is 1, we say that Q holds AS (almost surely) for this value of $p=p(n)$.

For our problem we consider a wide range of $p=p(n)$ values; however, not too small:

$$np(n)=\log n.\,w(n),\qquad w(n)\to\infty. \tag{1}$$

Thus, the expected degree $\deg(v)$ of a vertex v in G grows faster than $\log n$. It may go up to $c \cdot n$, in which case an edge probability is $p(n) = c = \text{constant}$.

Condition (1) implies (see the sentence following (18)) that $G \in G_{n,p}$ is AS *almost regular*:

$$(\forall v \in V_G)\left|\frac{\deg(v)}{np(n)} - 1\right| \leq \gamma(n) = w(n)^{-\tau}, \qquad \tau < \tfrac{1}{2}, \tag{2}$$

but this is no longer true for $\tau \geq \frac{1}{2}$.

Our main result is: Given $\varphi(n)$ satisfying

$$1 - \frac{\varphi(n)}{np(n)} \geq w(n)^{-\theta}, \qquad \theta < \tfrac{1}{2}, \tag{3}$$

$G \in G_{n,p}$ contains AS a regular factor of degree $\varphi(n)$. The same holds for a $p(n)n$-almost-regular factor with a given degree sequence. In view of (2), this is the best possible result.

Our proof is based on Tutte's factor theorem [9, 10]. We find it remarkable that this deep theorem of graph theory, which is usually quite cumbersome to apply, turns out to be a convenient tool in the study of factors in random graphs, especially if one looks for (almost) regular factors.

Tutte's theorem is quoted in Section 2, along with our main theorem for regular factors and some lemmas. The main part of the proof is done in Section 3. In Section 4 we bring some extensions and variants, while in Section 5 we apply the main theorem to count the number $R(n, N)$ of regular graphs in $G_{n,N}$. For instance, we have if $N = n(n-1)/4$ (i.e. regular graphs of degree $(n-1)/2$):

$$(\forall \varepsilon > 0)(\exists n_0(\varepsilon))\left[n > n_0 \Rightarrow R\left(n, \frac{n(n-1)}{4}\right) \geq 2^{\binom{n}{2} - n^{3/2+\varepsilon}}\right]. \tag{4}$$

The total number $|G_{n,n(n-1)/4}|$ is about $2^{\binom{n}{2}}$.

We thank Micha Perles and Eytan Koch for asking the right question (about $G_{n,p=\text{constant}}$) which initiated this study, and Joel Spencer for useful conversations.

2. Main theorems

Theorem 1. *Let $p(n)$ and $\varphi(n)$ satisfy $np(n) = 6w(n).\log n$, $w(n) \to \infty$,*

$$(n.p(n) - \varphi(n)) \geq np(n)w(n)^{-\theta} (= 6w(n)^{1-\theta} \log n),$$
$$\theta < \tfrac{1}{2}, \qquad \varphi(n) \geq 4. \tag{5}$$

Then for all n such that $\varphi(n)$, $\frac{1}{2}n\varphi(n)$ are integers, and for $\varepsilon > 0$

$$\text{Prob}\{G \in G_{n,p} \text{ contains a regular factor } H \text{ of degree } \varphi(n)\}$$
$$= 1 - \text{O}[\exp(-w(n)^{1-2\theta-\varepsilon}.\log n], \qquad n \to \infty \tag{6}$$

(The '6' in (5) is just a convenience, but also reminds us how $np(n)$ is replaced.)

The tool we use is Tutte's factor theorem [9,10]:

Theorem 2. *Let $G=(V,E)$ be a graph. Let $0 \leqslant \varphi_v \leqslant \deg(v)$. Let $D=(S,T,U)$ be a decomposition of the set V into 3 disjoint subsets. Let*

$$R(S,T)=\sum_{v\in T}\deg(v)-\lambda(S,T)-\sum_{v\in S}\varphi_v+\sum_{v\in T}\varphi_v, \tag{7}$$

$$\begin{aligned}&\lambda(S,T) \text{ is the number of edges of } G \text{ between } S \text{ and } T,\\ &q(S,T)=\text{the number of `odd' components of } G_U,\end{aligned} \tag{8}$$

where G_U is the induced subgraph on the set of vertices U and a component C of G_U is odd [even] if

$$\sum_{v\in C}[\varphi_v+\lambda(T,v)] \text{ is odd [even]}. \tag{9}$$

Then G contains a factor H with $\deg_H(v)=\varphi_v$ if and only if

$$(\forall S\,\forall T)[q(S,T)\leqslant R(S,T)]. \tag{10}$$

For a random graph G, the number $q(S,T)$ is random; the first two terms on the RHS of (7) are random; the last two terms are non-random and for a regular graph of degree φ:

$$\sum_{v\in S}\varphi_v-\sum_{v\in T}\varphi_v=-\varphi.(t-s), \qquad s=|S|, \qquad t=|T|. \tag{11}$$

For subregular graphs considered in Section 4, the RHS of (10) will be estimated from below by $\varphi\cdot(t-\beta s)$, $0<\beta\leqslant 1$, where $\beta=\beta(n)\to 1$ for almost regular graphs.

Outline of the proof of Theorem 1. We shall establish that (10) holds AS, and estimate the probability it does not hold by summing up the probabilities it does not hold for *some pair* S, T. The general scheme is to prove

$$(\forall S\forall T)[q(S,T)\leqslant s+t+1\leqslant R(S,T)]. \tag{12}$$

The left relation is given in Lemma 2 below, the right relation is treated in the next section. However, in some 'subranges' of S and T we prove (10) directly, e.g. the case $s=t=0$ which follows:

Lemma 1. $\text{Prob}\{q(\emptyset,\emptyset)>R(\emptyset,\emptyset)\}=O(n^{-(1/2)w(n)})$.

Proof. Here $G_U=G$. If G is connected, then the single component is even (cf. (9)) since

$$\sum_{v \in V} \varphi(n) = 0 \ (\mathrm{mod}\ 2),$$

so

$$q(\emptyset,\emptyset) = 0.$$

Also, $R(\emptyset,\emptyset) = 0$. Thus,

$$\{q(\emptyset,\emptyset) > R(\emptyset,\emptyset)\} \subseteq \{G \in G_{n,p} \text{ is not connected}\}.$$

The probability of the RHS is bounded by the probability of having $1 \leqslant k$ vertices not connected to the remaining $n-k$ vertices:

$$\sum_{k=1}^{(1/2)n} \binom{n}{k} \left(1 - \frac{w(n).\log n}{n}\right)^{k(n-k)}$$

$$\leqslant \sum_k n^k \exp\left[-w(n).\log n.\frac{k(n-k)}{n}\right]$$

$$\leqslant \sum_{k=1}^{(1/2)n} n^{-w(n).(1/2)k} = \mathrm{O}(n^{-(1/2)w(n)}), \qquad n \to \infty. \tag{13}$$

Lemma 2. *Let* $s + t \geqslant 1$, $(|S| = s, |T| = t)$,

$$\mathrm{prob}\{(\forall S \forall T) q(S,T) \leqslant s + t + 1\} = 1 - \mathrm{O}(n^{-w(n)/4}). \tag{14}$$

Remark 1. We shall prove a stronger version: After taking out $s+t$ vertices, the induced subgraph on the remaining vertices has AS no more than $s+t+1$ components (odd as well as even). This fact is part of Erdös–Rényi [5] proof of AS existence of a 1-factor in G in the case $p(n) > \log n$. Our proof is simpler, since $p(n)$ grows like $(1/n)w(n).\log n$; also we need the precise probability estimate in (6).

Proof of Lemma 2 (Stronger version). Suppose after taking out $p = s + t$ vertices, the remaining graph decomposes into $p+1$ non-connected parts $A_1, \ldots, A_{p+1}$, $A_i \neq \emptyset$. (Then clearly $|\bigcup_1^{p+1} A_i| \geqslant n/2$.) One of three cases (events) must happen:

$$\mathrm{E}_1: \quad \bigcup_{i=1}^{p} A_i = B, \qquad |B| < n/3,$$

$$\mathrm{E}_2: \quad |A_i| < n/20, \qquad 1 \leqslant i \leqslant p+1,$$

$$\mathrm{E}_3: \quad \text{(the largest one) } |A_{p+1}| \geqslant n/20, \qquad |B| \geqslant n/3.$$

Let E_1 hold. Then B contains at least p vertices, no edge of G connects B with $A_{p+1} = V - (B \cup S \cup T)$. This implies:

E_1': $\exists$ sets of size $r, n-2r$ not connected by an edge, $1 \leqslant r \leqslant n/3$:

$$\text{Prob } E_1 \leqslant \text{prob } E_1' \leqslant \sum_{r=1}^{n/3} \binom{n}{r} \cdot \binom{n-r}{r} \left(1 - \frac{w(n)\log n}{n}\right)^{r(n-2r)}$$

$$\leqslant \sum_r n^{2r} \exp\left[-w(n)\log n \frac{r(n-2r)}{n}\right] = O(n^{-w(n)/4}) \qquad \text{(cf. (13))}.$$

For the other case consider the event (cf. (8)):

$$F_{\alpha\beta} = \{\exists A \exists B[|A| = \alpha n, |B| = \beta n, \lambda(A, B) = 0]\},$$

$$\text{Prob } F_{\alpha\beta} \leqslant \binom{n}{\alpha n}\binom{n}{\beta n}\left(1 - \frac{w(n).\log n}{n}\right)^{\alpha\beta n^2}$$

$$\leqslant \exp(\gamma n - \alpha\beta w(n) n \log n) = O(n^{-n}).$$

Let E_2 hold. We noted that $\Sigma_{i=1}^{p+1}|A_i| \geqslant n/2$. Let j satisfy $\Sigma_{i=1}^{j-1}|A_i| \leqslant n/4$, $\Sigma_1^j|A_i| > n/4$. Let $D = \bigcup_1^{j-1} A_i$, $C = \bigcup_{j+1}^{p+1} A_i$. Since $|A_i| < n/20$, we must have $|D|$ and $|C| \geqslant n/5$, so E_2 implies $F_{\alpha,\beta}$ with $\alpha = \beta = 1/5$.

Let E_3 hold. It clearly implies $F_{\alpha,\beta}$ with $\alpha = 1/20$, $\beta = 1/3$. So Prob E_2 and Prob $E_3 = O(n^{-n})$.

3. Estimates of the right-hand side

$R(S, T) = \Sigma_{v \in T} \deg(v) - \lambda(S, T) - \varphi(n).(t - s)$, the last term $\varphi(n).(t-s)$ is non-random, the rest is random and non-negative (see (16) below). The success of our estimates hinges on using two alternative decompositions of the random part:

$$\sum_{c \in T} \deg(c) - \lambda(S, T)$$

$$= [\lambda(T, V - T) + e(T, T)] + e(T, T) - \lambda(T, S) = X + Y - Z \qquad (15)$$

$$= \lambda(T, V - S - T) + 2e(T, T) = X' + 2Y, \qquad (16)$$

where $Y = e(T, T)$ is the number of edges of G in the subgraph induced by T, while $\lambda(A, B)$ is the number of edges of G between the disjoint sets A, B. Thus, X, X', Y, and Z each consists of independent edge occurrences, i.e. counts the number of successes in a certain Bernoulli sequence trials with probability $p(n)$ for success. It is convenient to write these decompositions formally in terms of the probability distributions

$$B[t.(n - \tfrac{1}{2}(t+1), p(n)] + B[\tfrac{1}{2}t(t-1), p(n)] - B[s.t, p(n)], \qquad (15')$$

$$B[t.(n - s - t), p(n)] + 2B[\tfrac{1}{2}t(t-1), p(n)]. \qquad (16')$$

Here $B(n, p)$ is the binomial distribution. For W which is $B(N, p)$-distributed we shall use two one-sided tail estimates due to Chernoff [2,6, p. 17]:

$$\text{Prob}[W < \lfloor (1-\gamma)Np \rfloor] < \exp(-\tfrac{1}{2}\gamma^2 Np), \qquad 0 \leqslant \gamma \leqslant 1, \tag{17}$$

$$\text{Prob}[W > \lceil (1+\gamma)Np \rceil] < \exp(-\tfrac{1}{3}\gamma^2 Np), \qquad 0 \leqslant \gamma \leqslant 1. \tag{18}$$

For any v, $\deg(v)$ is clearly $B[n-1, p(n)]$, hence (2). (2) does not extend to $\tau \geqslant \frac{1}{2}$ since $\log n . w(n)$ is the variance of $B[n,p]$. We always take $\gamma(n) = w(n)^{-\tau}$, $\theta < \tau < \frac{1}{2}$.

We would like to have (12) hold AS. Our strategy is to divide the whole range of S, T to several subranges determined by (s,t) relations. In each subrange we choose a decomposition (15′) or (16′)) and *replace* a $\pm B(N,p)$ term with a large N by $Np(1 \mp \gamma)$; signs go this way since we want a lower bound. The probability that the bound we obtain after replacement will not hold for some pair (S, T) in the subrange is estimated by summing up the individual tail probabilities given in (17) and (18) over all choices of S, T in the subrange where we made such a replacement of the random terms. Let

$$\theta < \sigma < \sigma + \delta < \tau < \tfrac{1}{2} \quad (\text{so } \sigma + \tau + \delta < 1). \tag{19}$$

The above replacement is done only in subranges where

$$s < 2/3n, \qquad \max(1,s) \leqslant tw(n)^{\delta}. \tag{20}$$

In the subranges where (20) does not hold, the non-random terms will suffice to establish (12).

Now we define the subranges, I, II, III, and IV, and get estimates for the sums of all tail probabilities in each one of them.

I. *s and t large*: *$s \geqslant n/w(n)^{\sigma}$, hence $t \geqslant n/w(n)^{\sigma+\delta}$.*

Here we use decomposition $X + Y - Z$. The tail-sums for X, Y, and Z are:

$$\text{Prob}(\exists T)[X \leqslant t \cdot (n - \tfrac{1}{2}(t+1)) \cdot p(n) \cdot (1 - \gamma(n))]$$

$$\leqslant \sum_t \binom{n}{t} \exp\left[-t \cdot (n - \tfrac{1}{2}(t+1)p) \cdot \frac{\gamma^2}{3}\right]$$

$$\leqslant \sum_t \exp\left[-t \cdot \left(\frac{n}{6} p\gamma^2 - \log n\right)\right] \qquad (t \geqslant nw^{-\sigma-\delta})$$

$$\leqslant n \exp[-w^{-\sigma-\delta} n . \log n (w^{1-2\tau} - 1)]$$

$$\leqslant n^{-nw^{1-2\tau-\sigma-\delta}} = o(1), \qquad n \to \infty;$$

$$\text{Prob}(\exists T)[Y \leqslant \tfrac{1}{2}t(t-1)p(n)(1-\gamma(n))]$$

$$\leqslant n^{-nw^{1-2\tau-2\sigma-2\delta}} = o(1);$$

$$\text{Prob}(\exists S \exists T)[Z \geqslant s.tp(n)(1+\gamma(n))]$$

$$\leqslant \sum_s \sum_t \binom{n}{s}\binom{n}{t} \exp\left(-stp\frac{\gamma^2}{3}\right) \leqslant n^{-nw^{1-2\tau-2\sigma-2\delta}} = o(1).$$

Note that $w(n) \leqslant n/\log n$ so the exponent $n \cdot w^{1-2(\tau+\sigma+\delta)} > n^{\varepsilon'}$, where we can take any $\varepsilon' < 1 - 2\theta$ (see (19)); also the estimates for Y and Z above are similar to that of X above, since $t(t-1)$ and $s.t$ yield $n^2 . w^{-2\sigma-2\delta}$ to the negative exponent of n.

II. *s small, t large but not too close to n*: $s < n/w^{\sigma}$, $n/w^{\sigma+\delta} < t$, $t < n - 2s$.

We use the form (16′), $X' + 2Y$. Here Y is estimated as in I. For X':

$$(*)\ \operatorname{Prob}(\exists S \exists T)[X' \leqq t(n-s-t)\cdot p(n)\cdot(1-\gamma(n))]$$

$$\leqslant \sum_s \sum_t \binom{n}{s}\binom{n-s}{t} \exp\left[-t\cdot(n-s-t)\cdot p(n)\frac{\gamma^2}{3}\right].$$

(if $t < \frac{1}{2}n$)

$$\leqslant \sum_s \sum_t \exp\left[-t\frac{np(n)}{6}\gamma^2 - (s+t)\log n\right]$$

$$\leqslant \sum_s \sum_t \exp[-t\log n(w^{1-2\tau} - 1 - w^{-\delta})] \qquad (t > nw^{-\sigma-\delta})$$

$$\leqslant n^{-nw^{1-2\tau-\sigma-\delta}} = \mathrm{o}(1), \qquad n \to \infty.$$

If, however, $\frac{1}{2}n < t < n - 2s$, then

$$\binom{n}{s}\binom{n}{t} \leqslant \binom{n}{s}\binom{n-s}{n-s-t} \leqslant n^{2(n-s-t)}$$

since $s < n - s - t$, so

$$(*)\ \leqslant \sum_s \sum_{t>(1/2)n} \exp\left[-(n-s-t)\cdot\left(\frac{tp(n)}{6}\gamma^2 - 2\log n\right)\right]$$

$$\leqslant \exp[-\log n(w^{1-2\tau} - 2)] \leqslant n^{-w^{1-2\tau}} = \mathrm{o}(1). \tag{21}$$

III. *s small, t very close to n*:

$$s < \frac{n}{w^{\sigma}}, \qquad t \geqslant n - 2s \geqslant n - \frac{2n}{w^{\sigma}}. \tag{22}$$

Again we use $X' + 2Y$, but replace only Y with tail estimates, as in I.

IV. *s and t are small*:

$$s \leqslant \frac{n}{w(n)^{\sigma}}, \qquad t \leqslant \frac{n}{w(n)^{\sigma+\delta}} \tag{23}$$

We use $X' + 2Y$, replace only X'. Since t is very small, we get instead of (*) in II:

$$\text{Prob}(\exists S \exists T)[X' \leq t(n-s-t)p(n)(1-\gamma(n))]$$
$$\leq n^2 \exp(-w^{1-2\tau} \log n) = n^{-(w^{1-2\tau}-2)} = o(1).$$

We proceed now in proving that Tutte's condition (12) holds AS. Usually we shall go through (12):

(i) (subrange (i)) $s = t = 0$ – done above in Lemma 2.

(ii) $t < sw(n)^{-\delta}$. The random part of R is ≥ 0, while:

$$-\varphi(n)\cdot(t-s) \geq \varphi(n)\cdot s\cdot(1-w(n)^{-\delta})$$
$$\geq 4s(1-w(n)^{-\delta}) \geq 3s \geq s+t+1 \geq q(S,T) \quad \text{AS}.$$

(iii) $s \geq 2n/3$. Here

$$-\varphi(n)\cdot(t-s) \geq \varphi(n)\cdot\frac{n}{3} \geq \frac{4}{3}n \geq q(S,T).$$

In the remaining subranges, (20) holds. Having estimated the sum of tail probabilities, we have justified the replacement of the $\pm B(N,p)$ terms X, X', Y, and $-Z$ by $Np[(1 \mp \gamma) = Np \pm \rho$, $\rho = Np\gamma$. We obtain two forms for $R(S,T)$ (valid in subranges I and II):

$$R(S,T) = (n-s)\cdot tp(n) - \varphi(n)\cdot(t-s) - \rho \text{ terms,} \tag{24}$$

$$R(S,T) \geq (np(n) - \varphi(n))\cdot(t-s) + sp(n)(n-t) - \rho \text{ terms.} \tag{25}$$

The ρ terms are easily checked to add up to

$$\rho = t\cdot(n \pm s)p(n)\gamma(n) \leq 12t\cdot w(n)^{1-\tau}\log n. \tag{26}$$

(iv) $t > s$, $(t-s) \geq \frac{1}{2}t$. Using (25):

$$R(S,T) \geq (np(n) - \varphi(n))(t-s) - \rho$$
$$\geq np(n)w^{-\theta}\tfrac{1}{2}t - \rho \geq 3tw(n)^{1-\theta}\log n - \rho$$
$$\geq t+s+1, \tag{27}$$

since $s \leq w^{\delta}$, $1-\theta-\delta > 1-\tau$, and using our next remark for ρ:

Remark 2. We could absorb $-\rho$ in the other term of (27) because ρ has the estimate (26), and $\tau > \theta + \delta > \theta$. The same holds for terms $\rho' = tw^{1-\sigma-\delta}(\log n)$ and $\rho'' = tw^{1-\tau+\delta}(\log n)$.

(v) $t > s$, $t - s \leq \frac{1}{2}t$. Then $t \leq \frac{1}{3}n$, $n - t \geq n/3$, $s \geq \frac{1}{2}t$. Using (25):

$$R(S,T) \geq sp(n)\frac{n}{3} - \rho \geq tp(n)\frac{n}{6} - \rho \geq tw(n)\cdot\log n - \rho$$
$$\geq t+s+1 \quad \text{(even simpler than (iv))}.$$

(vi) $t \leq s < 2n/3$.

$$R(S,T) \geqslant \frac{n}{3} tp(n) - \rho \geqslant t + s + 1 \quad \text{(as in (v))}.$$

We still have to check subranges III and IV, where we cannot use (24) and (25) since only one random term can be replaced.

(vii) For IV (see (23)). We use the form $X' + 2Y \geqslant X'$ for the random part. X' can be replaced since $(n-s-t)t \geqslant \frac{1}{2}nt$. Thus,

$$\begin{aligned} R(S,T) &\geqslant (n-s-t)tp(n)(1-\gamma) - \varphi(n)(t-s) \\ &\geqslant t[(np(n) - \varphi(n)) - np(n)\gamma - p(n)(s+t)] \\ &\geqslant t[w(n)^{1-\theta}\log n - w(n)^{1-\tau}\log n - 2w(n)^{1-\sigma-\delta}.\log n] \\ &\geqslant tw(n)^{1-\theta} \geqslant s+t+1, \end{aligned}$$

using again Remark 2 above.

(viii) For III (see (22)). We had s small, t very close to n so $Y = e(T,T)$ can be replaced. We use the form $X' + 2Y \geqslant 2Y$ for the random part. We claim that

$$R(S,T) \geqslant t(t-1)p(n)(1-\gamma) - \varphi(n)(t-s) \geqslant n-s-t.$$

The left inequality holds AS by the replacement of Y. It suffices to prove the right inequality since, clearly, $n-s-t \geqslant q(S,T)$. Since $n-s-t \leqslant s$ in this subrange $\varphi(n) \geqslant 4$; it suffices to show

$$t(t-1)p(n)(1-\gamma) - \varphi(n)t \geqslant 0.$$

Dividing by t, using $t \geqslant n-2s$, $s < nw^{-\sigma}$ and dividing again by $np(n)$, we have to show

$$(1-2w^{-\sigma} - n^{-1})(1-w^{-\tau}) \geqslant \varphi.(np)^{-1}$$

or

$$1 - \frac{\varphi(n)}{np(n)} \geqslant \frac{1}{w(n)}\tau + \frac{2}{w(n)}\sigma + \frac{1}{n},$$

or

$$\frac{1}{w(n)^{\theta}} \geqslant \frac{1}{w(n)^{\tau}} + \frac{2}{w(n)^{\sigma}} + \frac{1}{n}.$$

Clearly this holds for large n since, by (19),

$$\tfrac{1}{2} > \tau > \sigma > \theta \quad \text{and} \quad w(n) \leqslant \frac{n}{\log n}.$$

To conclude the proof of Theorem 1 we notice that the sum of all tail probabilities in all subranges is estimated as indicated in (6).

4. Extensions and variants of the main theorem

Let the degree sequence $\varphi_v(n)$ of the required factor H satisfy $\varphi(n) \geq \varphi_v(n) \geq \beta\varphi(n)$, $1 \geq \beta > 0$. The non-random terms in $R(S, T)$ (see (11) and (7)) satisfy:

$$\sum_{v \in S} \varphi_v(n) - \sum_{v \in T} \varphi_v(n) \geq -(t - \beta . s)\varphi(n). \tag{28}$$

It seems that Theorem 1 will hold for such a 'subregular' factor H with the same conditions on φ and integrality. A somewhat different decomposition to subranges is needed, which we did not carry out. One important case we get free from the proof of Theorem 1. Let

$$\left|\frac{\varphi_v(n)}{\varphi(n)} - 1\right| \leq \gamma = w(n)^{-\tau}$$

$$[\textit{then } |\beta(n) - 1| \leq w(n)^{-\tau}, \beta(n) = \beta \text{ in (28)}] \tag{29}$$

(compare with (3)). A graph H with degree sequence $\varphi_v(n)$ satisfying (29) is almost regular (this is an asymptotic property). In this case

$$|(t - s)\varphi(n) - (t - \beta(n)s)\varphi(n)| \leq s . w(n)^{-\tau}\varphi(n)$$

$$\leq sw^{-\tau} np(n) \leq tw^{1-\tau+\delta} \log n. \tag{30}$$

In the case $s \leq tw^{\delta}$, the inequality $q(S, T) \leq R(S, T)$ holds in precisely the same way. In the 'non-random' subranges (ii) and (iii) ($s > tw^{\delta}$, $s \geq \frac{2}{3}n$) it is immediate. In the remaining subranges the difference (30) is absorbed in the other terms, by Remark 2.

Corollary 1. *Theorem 1 holds for an almost regular factor under the suitable integrality conditions. This is important since by (2), a graph in $G_{n,p(n)}$ is almost regular AS.*

If $np(n)$ is decreased to $d \log n$, we have a weak version.

Theorem 3. *If $np(n) = d . \log n$, (6) is replaced by*

$$\varphi(n) \leq \frac{1}{8}\left[\left(1 - \frac{1}{e}\right) . d - 3\right] . \log n, \tag{31}$$

and the other conditions in Theorem 1 remain the same, then

$$\text{Prob}\{G \in G_{n,p} \textit{ contains a } \varphi(n)\textit{-regular factor}\} = 1 - o(1), \qquad n \to \infty.$$

In this case we cannot get factors with all but vanishing fraction of the edges; however, we can get factors below a fixed fraction, given in (31). The proof follows the same lines: first, Lemmas 1 and 2, then the estimates of $R(S,T)$ which is much simpler here. We have to use a variant of the Chernoff estimate (17) for 'far left' tails of $B(N,p)$. Again the proof will work for almost regular factors.

Finally, all the results tansfer to the probability space $G_{n,N}$ upon making the correspondence $N \leftrightarrow \binom{n}{2}p$. For example, instead of (6) we require:

$$N(n) = \tfrac{3}{2}nw(n)\log n, \qquad w(n) \to \infty,$$
$$N(n) - \tfrac{1}{2}n\varphi(n) \geq N(n)w(n)^{-\theta}, \qquad \theta < \tfrac{1}{2}. \tag{32}$$

The proof of the transfer principle (for monotone graph properties, and thresholds) between $G_{n,p}$ and $G_{n,N}$ (with $p(n)$ not too small) is quite simple (cf. [1,2,7]).

5. Counting (almost) regular graphs

Let $N(n) = \frac{1}{2}nw(n)$, $w(n) \to \infty$. (If $w(n) \geq \log^2 n$, the $\log n$ factor which appeared until now can be absorbed.) $G \in G_{n,N(n)}$ has AS the following properties:

(i) $|\deg(v) - w(n)| \leq w(n)^{1-\tau}$, $\tau < \frac{1}{2}$ (G almost regular).

(ii) G contains a $\varphi(n)$-regular factor if

$$|w(n) - \varphi(n)| \geq w(n)^{1-\theta}, \qquad \theta < \tau < \tfrac{1}{2}. \tag{33}$$

(ii) holds also for a prescribed almost regular factor. Moreover, the probability of (i) and (ii) is $\geq 1 - n^{-w^{\varepsilon}}$, for $\varepsilon < \frac{1}{2} - \theta$. It follows that there are about $w(n)^{(1-\tau)n}$ families of almost regular graphs, each family with a given degree sequence. Together they cover $G_{n,N(n)}$ AS. It is reasonable that the families are 'roughly' of the same size. Anyway,

$$\textit{Average size of a family} \approx |G_{n,N(n)}|/w(n)^{(1-\tau)n}. \tag{34}$$

We shall derive from (ii) a somewhat smaller lower board. Let $R(n,N)$ denote the number of regular graphs in $G_{n,N}$. Let

$$N' = N \cdot (1 - w(n)^{-\theta}) = \tfrac{1}{2}n\varphi(n).$$

Starting from all $\varphi(n)$-regular graphs and adding at most $\varphi(n)w(n)^{-\theta} \leq w(n)^{1-\theta}$ edges at each vertex, we cover $G_{n,N}$ AS. Thus, we obtain a lower bound:

$$[1 - n^{-w^{\varepsilon}}] \cdot \frac{\binom{\binom{n}{2}}{N}}{\binom{n}{w^{1-\theta}}^{n}} \leq R(n, N'). \tag{35}$$

The denominator counts all possible ways of adding edges to one regular graph, so it bounds the multiplicity of the source. (Alternatively, we could divide by $\binom{N}{nw^{1-\theta}}$ which will give the same order of magnitude.) It is clear that we can replace N' by N in (35) without noticeable change. The lower bound becomes conspicuous for $w(n)=cn$, i.e. for $R(n, c\binom{n}{2})$. If we take $1-\theta=\frac{1}{2}+\delta$

$$n^{n^{3/2+\delta}} \cdot \binom{\binom{n}{2}}{c\binom{n}{2}} \leqslant R\left(n, c\binom{n}{2}\right).$$

For $c=\frac{1}{2}$ in particular: for $\varepsilon>0$ and $n>n_0(\varepsilon)$:

$$2^{n(n-1)/2-n^{3/2+\varepsilon}} \leqslant R\left(n, \tfrac{1}{2}\binom{n}{2}\right), \tag{36}$$

while $|G_{n,(1/2)\binom{n}{2}}|$ is about $2^{n(n-1)/2}$. By Theorem 3, all these estimates hold for all the almost-regular families of $G_{n,N}$. From the average size estimate (34) for $w(n)=cn$, we may aim at improving and amplifying (36) by replacing $\frac{3}{2}+\varepsilon$ by $1+\varepsilon$ for the lower bound and by $1-\varepsilon$ for the upper bound. (For *small N* see a recent paper of Bollobás [3].)

References

[1] M. Ajtai, J. Komlós and E. Szemerédi, The longest path in a random graph. Combinatorica 1 (1981) 1–12.

[2] D. Angluin and L.G. Valiant, Fast probabilistic algorithms for Hamiltonian circuits and matchings, J. Comput. Syst. Sci. 18 (1979) 155–193.

[3] B. Bollobás, A probabilistic proof of an asymptotic formula for the number of labelled regular graphs, Europ. J. Combinat. 1 (1980) 311–316.

[4] P. Erdös and R. Rényi, On the evolution of random graphs, Publ. Math. Inst. Hung. Acad. Sci. 16 (1960) 17–61.

[5] P. Erdös and A. Rényi, On the existence of a factor of degree one in a connected random graph, Acta Math. Acad. Sci. Hung. 17 (1961) 359–368.

[6] P. Erdös and J. Spencer, Probabilistic Methods in Combinatorics (Academic Press, New York, 1974).

[7] L. Pósa, Hamiltonian circuits in random graphs, Discr. Math. 14 (1976) 359–364.

[8] E. Shamir and E. Upfal, One factor in random graphs based on vertex choice, Discr. Math. 41 (1982) 287–295.

[9] W. Tutte, The factors of graphs, Canad. J. Math. 4 (1952) 314–328.

[10] W. Tutte, The subgraph problem, Ann. Discr. Math. 3 (1978) 289–295.

Received March 1981

Annals of Discrete Mathematics 20 (1984) 283–292
North-Holland

TECHNIQUES FOR INVESTIGATING NEIGHBORLY POLYTOPES

Ido SHEMER
Institute of Mathematics, Hebrew University, Jerusalem, Israel

A $2m$-polytope Q is neighborly if each set of m vertices of Q determines a face of Q. The best known examples of neighborly polytopes are the so-called cyclic polytopes [6, Section 4.7]. P is a subpolytope of Q if vert P is a subset of vert Q. It is shown that the combinatorial structure of a neighborly $2m$-polytope determines the combinatorial structure of every subpolytope. We develop a construction of 'sewing a vertex onto a polytope', which, when applied to a neighborly $2m$-polytope, yields a neighborly $2m$-polytope with one more vertex. Using this construction, we show that the number $g(2m+b,2m)$ of combinatorial types of neighborly $2m$-polytopes with $2m+b$ vertices grows superexponentially as $b \to \infty$ ($m \geq 2$ fixed) and as $m \to \infty$ ($b \geq 4$ fixed).

1. Introduction

A polytope is the convex hull of a finite number of points in some Euclidean space. A polytope P is a d-polytope if P is of dimension $d \geq 0$, and P is k-neighborly if each set of at most k of its vertices determines a proper face of P. A d-polytope is neighborly if it is $[d/2]$-neighborly. We shall deal with neighborly $2m$-polytopes, $m \geq 2$.

There is a well-known family of neighborly polytopes, the so-called cyclic polytopes, whose vertices are on the moment curve $M_d = \{x(t) = (t, t^2, \ldots, t^d) \mid t \in \mathbb{R}\}$.

The facets of a cyclic $2m$-polytope $C(v,2m)$ with $v > 2m$ vertices are characterized by Gale's evenness condition: A set S of $2m$ vertices of $C(v,2m)$ determines a facet of $C(v,2m)$ iff between every two vertices not in S there is an even number of elements of S. (The vertex c is between vertices a and b if c is between vertices a and b on the moment curve.) It is known that there is only one combinatorial type of neighborly $2m$-polytopes with v vertices for $v = 2m+1$, $2m+2$ and $2m+3$ (which is, of course, the combinatorial type of the cyclic polytope). For more details see [6, Section 4.7 and Chapter 7].

Several papers, devoted to enumeration of certain classes of polytopes and triangulated manifolds (e.g. [7, 8, 3, 4, 1]) indicate that there are 'many' neighborly $2m$-polytopes with $v > 2m+3$ vertices.

In this paper we introduce some techniques which we found useful for investigating neighborly polytopes. In Sections 2 and 3 we develop some preliminary concepts, and show that the facial structure of a neighborly $2m$-polytope determines its interior structure. In Sections 4 and 5 we describe the main result, a construction called sewing. In Section 6 we give lower bounds for the number of combinatorial types of neighborly $2m$-polytopes, based on the sewing construction. All the results in this paper appear with detailed proofs in [9].

We denote by $[A_1, A_2, \ldots, A_n]$ the set conv $\bigcup_{i=1}^{n} A_i$, where the A_i's are subsets of $\mathbb{R}^d$ $(n \geq 1)$. If $a \in \mathbb{R}^d$, then $[\ldots, a, \ldots]$ stands for $[\ldots, \{a\}, \ldots]$. All polytopes in this paper are simplicial polytopes but not a simplex, unless otherwise specified. The letter P denotes a d-polytope, and Q denotes a neighborly $2m$-polytope. We denote by vert P the set of vertices of P, and by $\mathscr{F}(P)$ the set of all faces of P (including P itself and the empty set). $\mathscr{B}(P)$ is the set $\mathscr{F}(P) \setminus \{P\}$.

2. Missing faces

A subset S of vert P is a missing face of P if $[S] \notin \mathscr{B}(P)$ but $[T] \in \mathscr{B}(P)$ for every proper subset T of S. Denote by $\mathscr{M}(P)$ the collection of all missing faces of P.

S is a missing k-face if $S \in \mathscr{M}(P)$ and $|S| = k+1$.

We have the following characterization of missing faces:

Proposition 1. *Let S be a subset of* vert P. *Then $S \in \mathscr{M}(P)$ iff the following three conditions hold*:

(1) $P \cap \operatorname{aff} S = [S]$,

(2) $\emptyset \neq [S] \cap [\operatorname{vert} P \setminus S] \subset \operatorname{relint} [S]$, *and*

(3) $|S| = \dim S + 1$.

Let S be a missing face of P. A set B is an opposite set to S with respect to P if $B \subset \operatorname{vert} P \setminus S$, $[B] \cap [S] \neq \emptyset$ and $|B| + |S| \leq \dim P + 2$. (A similar concept appears in [5].)

The next lemma is equivalent to Caratheodory's Theorem.

Lemma 1. *Let A and B be subsets of $\mathbb{R}^d$. If there is a point $z \in [A] \cap [B]$ with a representation $z = h_1 a_1 + \cdots + h_n a_n$, where $h_1 + \cdots + h_n = 1$, $0 < h_i$, $a_i \in A$ $(i = 1, \ldots, n)$, then there are subsets $A' \subset A$, $B' \subset B$ such that $a_1 \in A'$, $[A'] \cap [B'] \neq \emptyset$ and $|A'| + |B'| \leq d+2$.*

From Lemma 1 we can easily deduce:

Lemma 2. *Let S be a missing face of P. If $x \in \operatorname{vert} P \setminus S$, then there is a set B opposite to S with respect to P such that $x \in B$.*

All the definitions and results so far hold for non-simplicial polytopes as well. Simplicial polytopes also have the following useful property:

Lemma 3. *$\mathcal{F}(P)$ is determined by $\mathcal{M}(P)$ as follows. If $T \subset \operatorname{vert} P$, then $[T] \notin \mathcal{F}(P)$ iff T includes an element of $\mathcal{M}(P)$.*

Let Q be a neighborly $2m$-polytope. If $S \in \mathcal{M}(Q)$, then by Lemma 2, $\operatorname{vert} Q$ has a subset B opposite to S (with respect to Q). $[B] \cap [S] \neq \emptyset$, Q is m-neighborly, hence $|B| \geq m+1$ and $|S| \geq m+1$. Since $|B|+|S| \leq 2m+2$, $|S| = |B| = m+1$. Therefore every missing face of Q is a missing m-face and an opposite set to a missing face is a missing face too. $\mathcal{M}(Q)$ consists of precisely those $(m+1)$-subsets of $\operatorname{vert} Q$ which do not determine an m-face of Q, hence

Lemma 4. *$\mathcal{F}(Q)$ is determined by $\operatorname{skel}_m \mathcal{F}(Q)$.*

Let P_1 and P_2 be two polytopes. A 1:1 mapping f of $\operatorname{vert} P_1$ onto $\operatorname{vert} P_2$ induces a mapping $\bar{f}$ from subpolytopes of P_1 to subpolytopes of P_2 ($\bar{f}([T]) = [f(T)]$ for $T \subset \operatorname{vert} P_1$). f is a combinatorial equivalence (or isomorphism) between P_1 and P_2 if $\bar{f}$ maps $\mathcal{F}(P_1)$ onto $\mathcal{F}(P_2)$. We say that P_1 and P_2 are combinatorially equivalent. Usually we do not distinguish between $\bar{f}$ and f.

Theorem 1. *If $f : \operatorname{vert} Q \to \operatorname{vert} Q'$ is a combinatorial equivalence between two neighborly $2m$-polytopes Q and Q', and T is a subset of $\operatorname{vert} Q$, then the restriction $f|_T : T \to f(T)$ is a combinatorial equivalence between $[T]$ and $[f(T)]$.*

We can reformulate the theorem as follows.

Assume $x_i = (x_{i,1}, \ldots, x_{i,2m})$, $x'_i = (x'_{i,1}, \ldots, x'_{i,2m})$, $i = 1, \ldots, v$, are the vertices of Q and Q', respectively. If I is a $(2m+1)$-subset of $\{1, 2, \ldots, v\}$, denote by $\det(x, I)$ [$\det(x', I)$] the determinant of the matrix whose jth row is $(1, x_t)$ [$(1, x'_t)$], where t is the jth element of I. If the mapping $x_i \to x'_i$ is a combinatorial equivalence between Q and Q', then the sign of $\det(x, I) \det(x', I)$ is the same for all $(2m+1)$-subsets I of $\{1, 2, \ldots, v\}$.

If x is a vertex of Q, then Q^x denotes the subpolytope $[\operatorname{vert} Q \setminus \{x\}]$.

In order to prove Theorem 1 it suffices to show that a combinatorial equivalence f between Q and Q' induces a combinatorial equivalence between Q^x and $Q'^{f(x)}$, and this follows from Lemma 4 and the following lemma:

Lemma 5. *Let Q be a neighborly $2m$-polytope in $\mathbb{R}^{2m}$, and let x be a vertex of Q. Assume A is an $(m+1)$-subset of $\operatorname{vert} Q \setminus \{x\}$. Then*

(1) *If* $A \in \mathcal{M}(Q)$ *and if* $(A \setminus \{z\}) \cup \{x\} \in \mathcal{M}(Q)$ *for some* z *in* A, *then* $A \in \mathcal{M}(Q^x)$.

(2) *If* $A \in \mathcal{M}(Q^x)$, *then there is a point* z *in* A *and a set* B *in* $\mathcal{M}(Q^x)$ *such that* B *is opposite to both* A *and* $(A \setminus \{z\}) \cup \{x\}$ *with respect to* Q.

Proof. Assume $A \in \mathcal{M}(Q)$, $x \notin A$, $A \notin \mathcal{M}(Q^x)$. $[A]$ is a face of Q^x. Choose a facet F_1 of Q^x which includes A. $[A]$ is not a face of Q, hence x lies beyond F_1 (with respect to Q^x). Let z be an element of A. Q is m-neighborly, hence $[A \setminus \{z\}]$ is a face of Q. Therefore Q has a facet F_2 which includes $A \setminus \{z\}$, and x lies beneath F_2. $A \setminus \{z\} \subset F_1 \cap F_2$, x lies beyond F_1 and beneath F_2, hence $[A \setminus \{z\}, x] \in \mathcal{F}(Q)$. This proves part (1).

Assume $A \in \mathcal{M}(Q^x)$. Choose a set B in $\mathcal{M}(Q^x)$ opposite to A with respect to Q^x. We claim that $[B] \cap \operatorname{relint}[x, A] \neq \emptyset$. Suppose that $[B] \cap \operatorname{relint}[x, A] = \emptyset$. Take a hyperplane H such that $B \subset H^-$, $\{x\} \cup A \subset H^+$. Choose a point z in $[A] \cap [B]$. By Proposition 1, $z \in \operatorname{relint}[A] \cap \operatorname{relint}[B] \subset H^+ \cap H^- = H$. Hence, $A \cup B$ is in H. We obtain an m-neighborly polytope $[A, B]$, of dimension at most $2m-1$, which has $|A| + |B| = 2m+2$ vertices, and therefore is not a simplex, a contradiction. Therefore $[B]$ intersects $\operatorname{relint}[x, A]$. Each point in $\operatorname{relint}[x, A]$ can be written as $ex + e_1 a_1 + \cdots + e_{m+1} a_{m+1}$, where $A = [a_1, a_2, \ldots, a_{m+1}]$, $e > 0$, $e_j > 0$ $(j = 1, \ldots, m+1)$ and $e + e_1 + \cdots + e_{m+1} = 1$. By Lemma 1 there are sets $B' \subset B$ and $A' \subset A$ such that $[B'] \cap [x, A'] \neq \emptyset$ and $|B'| + |A'| + 1 \leq 2m+2$. Q is m-neighborly and $|B| = m+1$, hence $|A'| \geq m$ and $B' = B$, and therefore $|A'| = m$. This proves part (2), where z is the element of $A \setminus A'$. □

Conclusion. In the notations of Lemma 5, $A \in \mathcal{M}(Q^x)$ iff $A \in \mathcal{M}(Q)$ and there is $D \in \mathcal{M}(Q)$ such that $x \in D$ and $D \setminus \{x\} \subset A$.

We generalize the notion of 'missing face' as follows. Let G be a face of Q. A subset S of $\operatorname{vert} Q \setminus G$ is a missing face of Q relative to G if $[S, G] \notin \mathcal{B}(Q)$, but $[S', G] \in \mathcal{B}(Q)$ for every proper subset S' of S.

Define: $\mathcal{M}(Q/G) = \{S \mid S \text{ is a missing face of } Q \text{ relative to } G\}$.

Note that $\mathcal{M}(Q/\emptyset) = \mathcal{M}(Q)$, and $\mathcal{M}(Q/Q) = \{\emptyset\}$.

3. Universal faces

A face G of Q is a k-universal face of Q if $[G, T] \in \mathcal{B}(Q)$ for every subset T of $\operatorname{vert} Q$ with $|T| \leq k$. (G is a k-universal face of Q iff $k = 0$ and $G \in \mathcal{B}(Q)$ or $k > 0$ and the quotient polytope Q/G is k-neighborly, with $|\operatorname{vert} Q| - |\operatorname{vert} G|$ vertices.)

We say that G is a universal face of Q if G is k-universal with $k = [(2m - |\mathrm{vert}\, G|)/2]$. Equivalently, G is universal if G is a facet of Q, or else if $G \in \mathscr{B}(Q)$ and Q/G is a neighborly polytope with $|\mathrm{vert}\, Q| - |\mathrm{vert}\, G|$ vertices. It is easy to see that G is a k-universal face of Q iff $|S \cap G| \leqslant m - k$ for every $S \in \mathscr{M}(Q)$, or equivalently, if $|S \setminus G| \geqslant k+1$ for every $S \in \mathscr{M}(Q)$. If G is a universal face of Q with $2j$ vertices, then all the members of $\mathscr{M}(Q/G)$ are of cardinality $m - j + 1$. A universal edge of Q is a universal face of dimension 1. From the previous remark it follows that a set $\{a, b\}$ of two vertices of Q determines a universal edge $[a, b]$ iff no element of $\mathscr{M}(Q)$ includes $\{a, b\}$.

We need some facts about the universal faces of a cyclic polytope $K = C(v, 2m)$ $(v > 2m+1, m > 1)$. Suppose the vertices of K in their cyclic order are $a_1, a_2, \ldots, a_v$, a_1. Gale's evenness condition implies that $[a_i, a_{i+1}]$ $(1 \leqslant i < v)$ and $[a_v, a_1]$ are universal edges of K, and if $v \geqslant 2m+3$ then K has no other universal edges. That is, the missing faces of K are exactly those $(m+1)$-subsets of $\mathrm{vert}\, K$ which are isolated in the circuit $a_1, \ldots, a_v, a_1$.

Lemma 6. *Let $H_1, \ldots, H_r$ be pairwise disjoint universal faces of Q. If $|\mathrm{vert}\, H_i| = 2v_i$, $1 \leqslant i \leqslant r$, and $v_1 + v_2 + \cdots + v_r \leqslant m$, then $H = [H_1, H_2, \ldots, H_r]$ is a universal face of Q.*

Proof. It is sufficient to prove this for $r = 2$, since the general case follows by induction on r. Put $H = [H_1, H_2]$. Assume A is a subset of $\mathrm{vert}\, Q \setminus H$ and $[H, A] \notin \mathscr{B}(Q)$. We shall show that $|A| > m - (v_1 + v_2)$. By Lemma 3 there are three pairwise disjoint sets $B \subset A$, $T_1 \subset \mathrm{vert}\, H_1$, and $T_2 \subset \mathrm{vert}\, H_2$, such that $T_1 \cup T_2 \cup B \in \mathscr{M}(Q)$, $|B| + |T_1| + |T_2| = m+1$. $[H_1, T_2, B] \notin \mathscr{B}(Q)$; hence, $|T_2| + |B| \geqslant m - v_1 + 1$. Similarly, $|T_1| + |B| \geqslant m - v_2 + 1$. Thus, $|T_1| + |T_2| + 2|B| \geqslant 2m - v_1 - v_2 + 2$, and therefore $|B| \geqslant m - v_1 - v_2 + 1$, and of course $|A| \geqslant m - v_1 - v_2 + 1$. It follows that if $D \subset \mathrm{vert}\, Q$, $|D| \leqslant m - (v_1 + v_2)$ then $[H, D] \in \mathscr{B}(Q)$. This means that H is a universal face of Q. □

4. Sewing

A tower in Q is a strictly increasing sequence $\mathscr{T} = \{G_j\}_{j=1}^{k}$ of proper faces of Q. For a face G of Q denote by $\mathscr{F}_G$ the set of all facets of Q which include G. For the members G_j of a tower $\mathscr{T}$, denote $\mathscr{F}_{G_j}$ by $\mathscr{F}_j$.

Define: $\mathscr{C} = \mathscr{C}(Q, \mathscr{T}) = \mathscr{F}_1 \setminus (\mathscr{F}_2 \setminus (\cdots \setminus \mathscr{F}_k) \cdots)$. It is easy to see that $\mathscr{C} = (\mathscr{F}_1 \setminus \mathscr{F}_2) \cup (\mathscr{F}_3 \setminus \mathscr{F}_4) \cup \cdots \cup (\mathscr{F}_{k-1} \setminus \mathscr{F}_k)$ if k is even, and $\mathscr{C} = (\mathscr{F}_1 \setminus \mathscr{F}_2) \cup \cdots \cup \mathscr{F}_k$ if k is odd. With the convention that $G_j = Q$ and $\mathscr{F}_j = \emptyset$ for $j > k$ we can write $\mathscr{C} = \bigcup_{i=1}^{\infty} (\mathscr{F}_{2i-1} \setminus \mathscr{F}_{2i})$ and $\mathscr{F}_0 \setminus \mathscr{C} = \bigcup_{i=0}^{\infty} (\mathscr{F}_{2i} \setminus \mathscr{F}_{2i+1})$, where $\mathscr{F}_0$ is the set of all facets of Q.

We say that a point $x \in \mathbb{R}^{2m}$ lies exactly beyond a set $\mathscr{D}$ of facets of Q if x lies beyond every facet of Q that is in $\mathscr{D}$ and beneath every other facet of Q.

Lemma 7. *Let $\mathscr{T}$ be a tower in Q, $\mathscr{C} = \mathscr{C}(Q,\mathscr{T})$. Then there is a point $x \in \mathbb{R}^{2m}$ which lies exactly beyond $\mathscr{C}$.*

Proof. Assume $\mathscr{T} = \{G_j\}_{j=1}^k$, $k \geq 1$. Choose points $p_j \in \operatorname{relint} G_j$ $(1 \leq j \leq k)$ and $p_{k+1} \in \operatorname{int} Q$. Define $x_0 = p_{k+1}$, $x_{j+1} = (1+e)p_{k-j} - ex_j$ $(j = 0,1,\ldots,k-1)$. It follows easily by induction on r, $1 \leq r \leq k$, that if e is positive and sufficiently small, then x_r lies exactly beyond $\mathscr{F}_{k-r+1} \setminus (\cdots(\mathscr{F}_{k-1} \setminus \mathscr{F}_k)\cdots)$. Hence, $x = x_k$ lies exactly beyond $\mathscr{C}$. □

A tower $\mathscr{T} = \{G_j\}_{j=1}^m$ is a universal tower in Q if G_j is a universal face of Q with $2j$ vertices, $1 \leq j \leq m$. We adopt the convention that $G_0 = \emptyset$, and $G_j = Q$ for $j > m$.

From here on we assume that Q is a neighborly $2m$-polytope, $\mathscr{T} = \{G_j\}_{j=1}^m$ is a universal tower in Q, x lies exactly beyond $\mathscr{C} = \mathscr{C}(Q,\mathscr{T})$, and $S_j = \operatorname{vert} G_j \setminus \operatorname{vert} G_{j-1}$ for $j = 1,2,\ldots,m,m+1$. Denote $[Q,x]$ by Q^+.

Theorem 2.

(1) *Q^+ is a simplicial $2m$-polytope and* $\operatorname{vert} Q^+ = \{x\} \cup \operatorname{vert} Q$.

(2) *Q^+ is m-neighborly.*

(3) *If $0 < j \leq m$ is even, then G_j is a universal face of Q^+.*

(4) *If $0 < j \leq m$ is odd, then G_j is not a universal face of Q^+, but if $j < m$, then G_j is still a face of Q^+.*

(5) *If $a \in S_j$ for some $1 \leq j \leq m$, then $[G_{j-1}, a, x]$ is a universal face of Q^+.*

We say that Q^+ is obtained from Q by sewing at x through $\mathscr{T}$.

The last theorem, as well as other results about the structure of $\mathscr{F}(Q^+)$, follows from

Lemma 8.

(1) $\operatorname{vert} Q^+ = \{x\} \cup \operatorname{vert} Q$.

(2) *If $A \subset \{x\} \cup \operatorname{vert} Q$, then $A \in \mathscr{M}(Q^+)$ iff either:*

(a) $A = \bigcup_{r=1}^{j} S_{2r-1} \cup T$ *for some integer $0 \leq j \leq (m+1)/2$ and some set $T \in \mathscr{M}(Q/G_{2j})$, or*

(b) $A = \bigcup_{r=1}^{j} S_{2r} \cup T \cup \{x\}$ *for some integer $0 \leq j \leq m/2$ and some set $T \in \mathscr{M}(Q/G_{2j+1})$.*

Lemma 8 characterizes the missing faces of Q^+ in terms of the missing faces of Q and the blocks S_j $(1 \leq j \leq m+1)$, since one can easily verify that $T \in \mathscr{M}(Q/G_r)$ iff $T = W \setminus G_r$, where $W \in \mathscr{M}(Q)$ and $|T| = m - r + 1$.

In the sequel we shall need the following result:

Lemma 9. *E is a universal edge of Q^+ iff either:*
(1) *$E = [a, x]$ and $a \in S_1$, or*
(2) *E is a universal edge of Q and either*
(a) *$E \cap G_m = \emptyset$, or*
(b) *$E = [a, b]$ with $a \in S_p$ and $b \in S_{p+1}$ for some $1 \leq p \leq m$.*

5. Reconstruction

Assume that Q is a neighborly $2m$-polytope, $\mathscr{T}$ is a universal tower in Q, $\mathscr{C} = \mathscr{C}(Q, \mathscr{T})$, x lies exactly beyond $\mathscr{C}$ and $Q^+ = [Q, x]$. We claim that Q^+ and x determine $\mathscr{T}$. By Theorem 1, $\mathscr{F}(Q^+)$ and x determine $\mathscr{F}(Q)$, and also $\mathscr{C}(Q, \mathscr{T})$. The following algorithm enables us to recover $\mathscr{T}$ from $\mathscr{C}$ (as above, $\mathscr{F}_0$ is the set of facets of Q):

$$G_1 = \cap \mathscr{C},$$

$$\mathscr{F}_1 = \{H \in \mathscr{F}_0 \mid G_1 \subset H\},$$

$$G_2 = \cap (\mathscr{F}_1 \backslash \mathscr{C}),$$

$$\mathscr{F}_2 = \{H \in \mathscr{F}_0 \mid G_2 \subset H\},$$

$$G_3 = \cap (\mathscr{F}_2 \backslash (\mathscr{F}_1 \backslash \mathscr{C})), \quad \text{and so forth.}$$

In Section 4 we saw that if Q^+ is obtained by sewing at x, then certain faces of Q^+ are universal (see Theorem 2(3) and (5)). As a matter of fact, those conditions are also sufficient:

Theorem 3. *Let Q and Q^+ be neighborly $2m$-polytopes. Assume* $\operatorname{vert} Q^+ = \{x\} \cup \operatorname{vert} Q$, *$x \notin Q$, and let $\mathscr{T} = \{G_j\}_{j=1}^m$ be a tower in Q, with $|\operatorname{vert} G_j| = 2j$ $(1 \leq j \leq m)$. If:*
(1) *G_j is a universal face of Q^+ for every even j, $1 \leq j \leq m$, and*
(2) *$[G_j, p, x]$ is a universal face for every even j, $0 \leq j \leq m-1$ and every point $p \in \operatorname{vert} G_{j+1} \backslash \operatorname{vert} G_j$ $(G_0 = \emptyset)$,*
then $\mathscr{T}$ is a universal tower in Q, and Q^+ is obtained from Q by sewing at x through $\mathscr{T}$.

Another useful property of the sewing construction is its 'commutativity'. If $\mathscr{T}$ is a universal tower in Q, denote by $Q(\mathscr{T})$ the class of all polytopes which are obtained from Q by sewing through $\mathscr{T}$. If $\{x_i \mid i \in N\}$ is a (finite) set of points (in $\mathbb{R}^{2m}$), and J is a subset of N, then $x(J)$ denotes the set $\{x_i \mid i \in J\}$. We say that two towers $\mathcal{T}_1$ and $\mathcal{T}_2$ in Q are disjoint if the sets $\cup \mathcal{T}_1$ and $\cup \mathcal{T}_2$ are disjoint (i.e. if $A_1 \in \mathcal{T}_1$, $A_2 \in \mathcal{T}_2$ imply $A_1 \cap A_2 = \emptyset$).

Theorem 4. *For $1 \leqslant i \leqslant p$, let $\mathcal{T}_i$ be a universal tower in Q. If the towers $\mathcal{T}_i$ are pairwise disjoint, then there are points x_i, $1 \leqslant i \leqslant p$, such that $[Q, x(J), x_r] \in [Q, x(J)](\mathcal{T}_r)$ for every subset J of $\{1,\dots,p\}$ and for every r, $r \in \{1,\dots,p\} \setminus J$.*

6. Upper bounds

Denote by $g(v, 2m)$ the number of combinatorial types of neighborly $2m$-polytopes with v vertices. For every $m \geqslant 2$, $g(2m+3, 2m) = 1$, so the first interesting case is $g(2m+4, 2m)$. We give a lower bound by estimating the number of sewn polytopes.

Theorem 5.

$$g(2m+4, 2m) > \frac{(2m+2)!}{3 * 2^{m+3}(m+2)!}.$$

(The right-hand side is asymptotic to $(\sqrt{2}/6)(2m/e)^m$ as $m \to \infty$.)

Proof. Let K be a cyclic polytope of type $C(2m+3, 2m)$. Assume $\operatorname{vert} K = \{a_1, a_2, \dots, a_{2m+3}\}$, and that $a_1, a_2, \dots, a_{2m+3}, a_1$ is the circuit of universal edges of K. $\operatorname{Aut} K$, the group of combinatorial automorphisms of K, is precisely the dihedral group of order $2(2m+3)$ consisting of rotations and reflections of the circuit $a_1, a_2, \dots, a_{2m+3}, a_1$. Consider pairs (Q, z), where Q is a neighborly $2m$-polytope with $2m+4$ vertices and z is a distinguished vertex of Q. We say that two pairs, (Q, z) and (Q', z'), are isomorphic if there is a combinatorial equivalence $f: \operatorname{vert} Q \to \operatorname{vert} Q'$, with $f(z) = z'$. The number of isomorphism types of such pairs is clearly at most $(2m+4)g(2m+4, 2m)$.

Now let us count the number of isomorphism types of pairs (K^+, x), where K^+ is obtained from K by sewing at x through a universal tower $\mathcal{T}$. Consider two such pairs (K_i^+, x_i), with $K_i^+ \in K_i(\mathcal{T}_i)$, $i = 1, 2$. If these pairs are isomorphic by a mapping f of $\operatorname{vert} K_1^+$ onto $\operatorname{vert} K_2^+$ such that $f(x_1) = x_2$, then $h = f|_K$ is an automorphism of K, and h maps $\mathscr{C}(K, \mathcal{T}_1)$ onto $\mathscr{C}(K, \mathcal{T}_2)$. It follows from what precedes Theorem 3 that $h(\mathcal{T}_1) = \mathcal{T}_2$.

The converse is obvious. If $\mathcal{T}_1$ is mapped onto $\mathcal{T}_2$ by an automorphism of K, then the pairs (K_1^+, x_1), (K_2^+, x_2) are isomorphic.

Thus, the number of isomorphism types of pairs (K^+, x) considered above is precisely the number of equivalence classes of universal towers of K under $\operatorname{Aut} K$.

Every universal tower $\{G_j\}_{j=1}^m$ in K can be transformed by a suitable rotation to a tower with $G_1 = [a_1, a_2]$. With G_1 fixed, G_2 can be chosen in $2m+1$ ways, then G_3 in $2m-1$ ways, and so on. (Note that K/G_i is a cyclic polytope of type $C(2(m-i)+3,\ 2(m-i))$, for $0 \leq i < m$.) Therefore the number of universal towers in K with $G_1 = [a_1, a_2]$ is $(2m+1)\ (2m-1)\cdots 7\cdot 5$. There is only one automorphism r of K, except the identity, that maps the edge $[a_1, a_2]$ onto itself. Only one universal tower is fixed by r. It follows that the maximum number of 'non-isomorphic' universal towers in K is $[(2m+1)\ (2m-1)\cdots 5+1]/2$. We conclude that

$$(2m+4)g(2m+4,2m) > \tfrac{1}{2}(2m+1)(2m-1)(2m-3)\cdots 5. \qquad \square$$

The sewing construction can be repeated indefinitely. Indeed, using the notational convention established prior to Theorem 2, we can choose points $t_j \in S_j$, $1 \leq j \leq m$. Define $G'_j = [G_{j-1}, t_j, x]$, $1 \leq j \leq m$. By Theorem 2, $\mathcal{T}' = \{G'_j\}_{j=1}^m$ is a universal tower in Q^+, and we can sew again. Starting with a cyclic $2m$-polytope and sewing it repeatedly as just described, we can prove that $g(v,2m) > c2^v$.

If we sew a large cyclic polytope simultaneously, through several pairwise disjoint universal towers, as in Theorem 4, we can prove:

Theorem 6. $g((2m+1)p+2,2m) \geq \tfrac{1}{2}(pm-p)!$ *for* $p = 2,3,\ldots$

The last estimate of $g(v,2m)$ grows roughly like $[v(m-1)/(2m+1)]!$ as $v \to \infty$.

7. Concluding remarks

The sewing construction yields (Theorem 6) the asymptotic estimate $g(v,2m) > a(m)e^{b(m)v \log v}$, where $b(m) \to 1/2$ as $m \to \infty$.

Using the fact that each vertex x of a neighborly $2m$-polytope Q with v vertices $(v > 2m+3)$ covers exactly $\binom{v-m-2}{m-1}$ facets of Q^*, we can show that $g(v,2m) \leq c(m)e^{d(m)v^2}$, where $c(m)$, $d(m) > 0$ and $d(m) \to 0$ as $m \to \infty$.

There is still a huge gap between the lower bound and the upper bound, even for $m = 2$.

We can combine the sewing construction with the fact that a neighborly $2m$-polytope determines its subpolytopes, and apply to a given polytope successively (in any order) operations of sewing and omitting a vertex. Let us denote by SO(m) the family of neighborly $2m$-polytopes which are obtained in this way from cyclic $2m$-polytopes. We know that for $m = 2$ (dimension 4) any polytope in SO(2) is obtained by repeated sewing from a cyclic polytope. But already for $m = 3$ the situation is different. Fifteen (types of) neighborly 6-polytopes with 10 vertices are obtained by sewing $C(9,6)$ once; 13 more polytopes are obtained by sewing $C(9,6)$ twice and omitting one vertex.

As we noted in Section 6, a sewn polytope can be sewn again. Hence, the family SO(m) has no maximal element. It is most likely that SO(m) does not coincide with the class $N(m)$ of all the neighborly $2m$-polytopes, for any $m \geq 2$. The following question is still open: Is there a maximal neighborly $2m$-polytope? (That is, a neighborly $2m$-polytope which is not a subpolytope of any other neighborly $2m$-polytope.)

An affirmative answer to this question, for any m, would imply that SO(m) is a proper subset of $N(m)$.

Table 1 shows the present state of knowledge as to the number $g(v,2m)$ of combinatorial types of neighborly $2m$-polytopes with v vertices, where by 'totally sewn' we mean 'obtained from a cyclic polytope by repeated sewing'.

Table 1

$2m$	v	$g(v,2m)$	Sources	Remarks
$2m$	$\leq 2m+3$	1	[6, Chapter 6, Section 7.2]	cyclic
2	v	1		
4	8	3	[7]	all sewn
4	9	23	[3,8]	18 sewn
4	10	333–428	[1], recent computation by the author	333 sewn, 287 totally sewn
6	10	28–37	recent computation by the author	15 sewn

Note added in proof. The number of combinatorial types of neighborly 6-polytopes with 10 vertices is 37 (compare Table 1). The geometric realization of 9 of these 37 types is due to J. Bokowski. None of these special 9 types belongs to the family SO(3).

References

[1] A. Altshuler, Neighborly 4-polytopes and neighborly combinatorial 3-manifolds with ten vertices, Canad. J. Math. 29 (1977) 400–420.
[2] A. Altshuler and M.A. Perles, Quotient polytopes of cyclic polytopes, Israel J. Math. 36 (1980) 97–125.
[3] A. Altshuler and L. Steinberg, Neighborly 4-polytopes with 9 vertices, J. Combinat. Theory (A) 15 (1973) 270–287.
[4] A. Altshuler and L. Steinberg, Neighborly combinatorial 3-manifolds with 9 vertices, Discr. Math. 8 (1974) 113–137.
[5] M. Breen, Determining a polytope by Radon partitions, Pacific J. Math. 43 (1972) 27–37.
[6] B. Grünbaum, Convex Polytopes (Interscience, London/New-York/Sydney, 1967).
[7] B. Grünbaum and V.P. Sreedharan, An enumeration of simplicial 4-polytopes with 8 vertices, J. Combinat. Theory 2 (1967) 437–465.
[8] I. Shemer, Neighborly polytopes, M.Sc. thesis (Hebrew), The Hebrew University of Jerusalem (1971).
[9] I. Shemer, Neighborly polytopes, Israel J. Math. 43 (1982) 291–314.

Received 26 April 1981

Annals of Discrete Mathematics 20 (1984) 293–305
North-Holland

EXCHANGE PROPERTIES OF CONVEXITY SPACES

Gerard SIERKSMA*
Econometric Institute, University of Groningen, Groningen, The Netherlands

In several fields of mathematics, exchange properties have played a part, e.g. in matrix and matroid theory. In the theory of convexity, too, fruitful use has been made of exchange properties. In this paper we study the relationship between the number of elements of a set that can be exchanged and the number of elements of the set itself. We also study the exchange number, which is closely related to the Carathéodory number for convexity spaces and convex product spaces. The setting in which these properties are studied is that of a so-called convexity space, also called an algebraic closure system, introduced in 1951 by F.W. Levi, mentioned in 1963 by L. Danzer, B. Grünbaum, V. Klee, and further developed in 1971 by D.C. Kay and E.W. Womble.

1. Introduction

In 1951 Levi [9] introduced the concept of *convexity space* as a pair $(X, \mathscr{C})$ with X a set and $\mathscr{C}$ a collection of subsets of X closed under intersections; the elements of $\mathscr{C}$ are called *convex sets*; the convex hull of a set S in X is then defined as the intersection of all convex sets that contain S and is denoted as $\mathscr{C}(S)$. In his paper Levi also has the following interesting 'exchange' axiom (Axiom C3), namely: The convexity space $(X, \mathscr{C})$ satisfies Axiom C3 iff for each finite set A in X and each p in $\mathscr{C}(A)$ the following holds:

$$\mathscr{C}(A) = \bigcup_{a \in A} \mathscr{C}(p \cup (A \setminus a)).$$

Exchange properties are used later by Reay [10, Lemma 5.9; 11, p. 256], Jamison [6, p. 83], Sierksma [13, 14], Degreef [3], Doignon, Reay and Sierksma [4], and Van der Vel [20, 21]. Convexity spaces are also called *algebraic closure systems* (see Cohn [1]) and have been studied by several more authors, e.g. Eckhoff [5], Kay and Womble [7], Thompson and Hare [19], and Soltan [18]. In Section 2 the 'exchange function' is introduced: it relates the number of elements that can be 'exchanged' to the number of elements of the given set. Also, the exchange function for the convex product space is studied in this section. In

* Part of this paper can also be found in the author's dissertation: Axiomatic convexity theory and the convex product space, University of Groningen, The Netherlands (1976). The author thanks Prof. Dr. J. Ch. Boland for his helpful suggestions.

Section 3 we study the exchange number. For relationships between the exchange number and the classical numbers of Carathéodory, Helly, and Radon see Kay and Womble [7] and Sierksma [14]. When studying dimensions of convexity spaces Van der Vel [21] remarks about the exchange number: '... the exchange number is the only invariant which exactly determines the dimension, and ... plays an essential role in proving the equivalence of Helly's and Carathéodory's theorems'. The exchange number is also used fruitfully to determine the Carathéodory numbers of convex product spaces (see [13]) and to derive a sharper relationship between the numbers of Carathéodory, Helly and Radon (see [14]). All this motivates the more extensive attention to the exchange properties of convexity spaces.

2. The exchange function

For p and A in X and x in $\mathscr{C}(A)$ we define the set

$$\omega(p,x,A)=\{a\in A \mid x\in\mathscr{C}(p\cup(A\setminus a))\},$$

i.e. all elements of A that are *exchangeable* with p in relation to x'. Note that the following properties hold: $p\in\omega(p,x,A)$ if $p\in A$; $\omega(p,x,A)=A$ if $p\in\mathscr{C}(A)$; $A\setminus\{x\}\subset\omega(p,x,A)$ if $x\in A$; $\omega(p,x,A)\subset\omega(p,x,B)$ if $A\subset B$; and $a\in\omega(p,x,A)$ if $a\in\mathscr{C}(A\setminus a)$, i.e. the non-extreme elements (see [16, p. 14]) of A are all exchangeable with p in relation to x. As an example, consider in $\mathbb{R}^2$, with the ordinary convexity conv, the set $A=\{a_1,a_2,a_3\}$ and the points p, x_1, x_2, x_3, as in Fig. 1.

It is then found that $\omega(p,x_1,A)=\{a_3\}$, $\omega(p,x_2,A)=\{a_1,a_2\}$, and that $\omega(p,x_3,A)=A$. Throughout this paper we take in general $X\neq\emptyset$, and define $\mathbb{N}_X=\{1,2,\ldots,|X|\}$ if X is finite and $\mathbb{N}_X=\mathbb{N}$ otherwise. The *exchange function* of a convexity space $(X,\mathscr{C})$ is the mapping $\varepsilon:\mathbb{N}_X\to\mathbb{N}_X$ defined for each $n\in\mathbb{N}_X$ by:

$$\varepsilon(n)=\min\{|\omega(p,x,A)|: p\in X, A\subset X, |A|=n, x\in\mathscr{C}(A)\}.$$

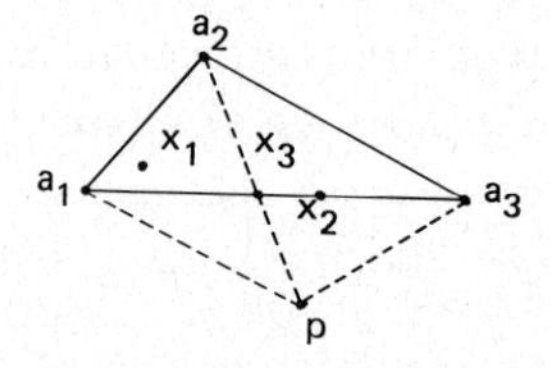

Fig. 1.

So $\varepsilon(n)$ gives the minimum number of elements of a set A with n elements that can be exchanged. Studying the *graph* of the exchange function we first note that the exchange function of $(\mathbb{R}^d, \text{conv})$ satisfies $\varepsilon(n)=0$ if $1 \leqslant n \leqslant d$ and $\varepsilon(n)=n-d$ if $n \geqslant d+1$. It is of interest to know under what conditions the exchange function of an arbitrary convexity space shows the same characteristics as the exchange functions of the ordinary convexity space $(\mathbb{R}^d, \text{conv})$. To that end we call the mapping $f: \mathbb{N}_X \to \mathbb{N}_X$ *n-increasing* iff $f(n)>0$ and $f(k+1)>f(k)$ for each $k \geqslant n$ and $k, k+1 \in \mathbb{N}_X$. In the following theorem we use the 'cone-union condition' (see Sierksma [17] and Van der Vel [21]): a convexity space $(X, \mathscr{C})$ satisfies the *cone-union condition* iff for any element p in X and any collection of convex polytopes (convex hulls of finite sets) $E_1, \dots, E_n$ that cover some convex polytope E, i.e. $E \subset \bigcup_{i=1}^n E_i$, it is true that $\mathscr{C}(p \cup E) \subset \bigcup_{i=1}^n \mathscr{C}(p \cup E_i)$. In [17] it is shown that the cone-union condition is weaker than the well-known 'join-hull commutativity' condition (see, for example, Kay and Womble [7]).

Theorem 1. *Let $(X, \mathscr{C})$ be a convexity space which satisfies the cone-union condition. If $\varepsilon(n)>0$ for some $n \in \mathbb{N}_X$, then ε is n-increasing, i.e. $\varepsilon(n+1) \geqslant \varepsilon(n)+1$.*

Proof. Let there be an element n in $\mathbb{N}_X$ with $\varepsilon(n)>0$. If $n=|X|$ there is nothing to show so we may assume that $n+1 \in \mathbb{N}_X$. Take any p and A in X with $|A|=n+1$ and any $x \in \mathscr{C}(A)$. Furthermore, choose an element a_0 in A and let $A_0 = A \setminus \{a_0\}$. Hence, $|A_0|=n$. As $\varepsilon(n)>0$ we find that $\omega(a_0, y, A_0) \neq \emptyset$ for each $y \in \mathscr{C}(A_0)$. So for each $y \in \mathscr{C}(A_0)$ an element $a \in \omega(a_0, y, A_0)$ exists, hence $y \in \mathscr{C}(a_0 \cup (A_0 \setminus a)) = \mathscr{C}(A \setminus a)$. Therefore $\mathscr{C}(A_0) \subset \bigcup\{\mathscr{C}(A \setminus a) \mid a \in A_0\}$. As $\mathscr{C}(A_0)$ and $\mathscr{C}(A \setminus a)$ are polytopes for each $a \in A_0$ the cone-union condition gives $\mathscr{C}(A) = \mathscr{C}(a_0 \cup A_0) = \mathscr{C}(a_0 \cup \mathscr{C}(A_0)) \subset \bigcup\{\mathscr{C}(a_0 \cup \mathscr{C}(A \setminus a)) \mid a \in A_0\} = \bigcup\{\mathscr{C}(A \setminus a) \mid a \in A_0\} \subset \mathscr{C}(A)$. Hence, $\mathscr{C}(A) = \bigcup\{\mathscr{C}(A \setminus a) \mid a \in A_0\} = \bigcup\{\mathscr{C}(A \setminus a) \mid a \in A\}$. As $x \in \mathscr{C}(A)$, an element $a_1 \in A$ exists such that $x \in \mathscr{C}(A \setminus a_1) \subset \mathscr{C}(p \cup (A \setminus a_1))$. Note that $a_1 \in \omega(p, x, A)$ and that $\omega(p, x, A \setminus \{a_1\}) \subset \omega(p, x, A)$. Consequently, $\omega(p, x, A \setminus \{a_1\}) \cup \{a_1\} \subset \omega(p, x, A)$. Hence, $\varepsilon(n)+1 \leqslant |\omega(p, x, A \setminus \{a_1\})|+1 \leqslant |\omega(p, x, A)|$. By definition, $\varepsilon(n)+1 \leqslant \min\{|\omega(p, x, A)| \mid p \in X, A \subset X, |A|=n+1, x \in \mathscr{C}(A)\} = \varepsilon(n+1)$.

In the following theorem we use the so-called 'continuity property'. A convexity space $(X, \mathscr{C})$ satisfies the *continuity property* iff for each finite set A in X, $B \in \mathscr{C}$, with $A \cap B \neq \emptyset \neq B \setminus A$ it is true that $\mathscr{C}(A) = \bigcap\{\mathscr{C}(x \cup A) \mid x \in B \setminus A\}$. The *shadow cone* generated by p and A in X is defined as:

$$\mathscr{K}(p, A) = \{x \in X \mid p \in \mathscr{C}(x \cup A)\}.$$

Lemma 2. *Let $(X, \mathscr{C})$ be a convexity space. Then the following two assertions are equivalent:*

(i) *$(X, \mathscr{C})$ satisfies the continuity property;*

(ii) *for each p in X and finite A in X with $p \notin \mathscr{C}(A)$ we have $C \cap (X \setminus (A \cup \mathscr{K}(p, A)) \neq \emptyset$ for each $C \in \mathscr{C}$ with $A \cap C \neq \emptyset \neq \mathscr{K}(p, A) \cap C$.*

Note that (ii) is a certain separation property that says that $\mathscr{K}(p, A)$ and A are 'distant' in a way. The proof of this lemma is left to the reader.

Theorem 3. *Let $(X, \mathscr{C})$ be a convexity space which satisfies the continuity property. If $n, n+1 \in \mathbb{N}_X$, then the exchange function satisfies $\varepsilon(n+1) \leq \varepsilon(n)+1$.*

Proof. Suppose there is an element n with $n, n+1 \in \mathbb{N}_X$. Take any p and A in X with $|A| = n$ and $x \in \mathscr{C}(A)$ and with $|\omega(p, x, A)| = \varepsilon(n)$. If $\varepsilon(n) = n$, then $\varepsilon(n+1) \leq n+1 = \varepsilon(n)+1$ and we are done. So we may now assume that $\varepsilon(n) < n$ and therefore that $\omega(p, x, A) \neq A$. Choose an element $a_0 \in A \setminus \omega(p, x, A)$ and let $A_0 = A \setminus \{a_0\}$. Hence, $x \notin \mathscr{C}(p \cup A_0)$. As $x \in \mathscr{C}(A) = \mathscr{C}(a_0 \cup A_0)$ it follows that $a_0 \in \mathscr{K}(x, A_0) \subset \mathscr{K}(x, \{p\} \cup A_0)$. According to Lemma 2 there is an element $q \in \mathscr{C}(p, a_0)$ such that $q \notin \{p\} \cup A_0 \cup \mathscr{K}(x, \{p\} \cup A_0)$. Because $q \notin \mathscr{K}(x, \{p\} \cup A_0)$ we find that $q \neq a_0$ and consequently that $q \in X \setminus A$. It also follows that $x \notin \mathscr{C}(\{p, q\} \cup A_0)$. Now take any $a \in \omega(p, x, A \cup \{q\}) \setminus \{q\}$. Hence, $x \in \mathscr{C}(\{p, q\} \cup (A \setminus a)) = \mathscr{C}(q \cup \mathscr{C}(p \cup (A \setminus a))$. Clearly, $a \neq a_0$. So $a_0 \in A \setminus \{a\}$. From $q \in \mathscr{C}(p, a_0)$ it now follows that $q \in \mathscr{C}(p \cup (A \setminus a))$; hence, $x \in \mathscr{C}(p \cup (A \setminus a))$ or $a \in \omega(p, x, A)$. We conclude that $\omega(p, x, A \cup \{q\}) \subset \omega(p, x, A) \cup \{q\}$. Hence that $\varepsilon(n+1) \leq |\omega(p, x, A \cup \{q\})| \leq \varepsilon(n)+1$.

Corollary 4. *Let $(X, \mathscr{C})$ be a convexity space which satisfies both the cone-union condition and the continuity property. If there is an element $n \in \mathbb{N}_X$ with $\varepsilon(n) > 0$, then*

$$\varepsilon(k+1) = \varepsilon(k)+1$$

for each $k \geq n$ and $k, k+1 \in \mathbb{N}_X$.

Proof. Direct consequence of Theorems 1 and 3.

So convexity spaces that satisfy both the cone-union condition and the continuity property have exchange functions whose graphs have the same characteristics as the graph of the exchange function of the ordinary convexity space. Fig. 2 shows the graph for $(X, \mathscr{C})$ with $|X| = 5$ and $\mathscr{C} = \{\phi, X\}$; Fig. 3 the graph for $(\mathbb{R}^2, \text{conv})$, and Fig. 4 for $(\mathbb{R}^2, \mathscr{C})$ with $\mathscr{C} = \{\mathbb{R}^2, \phi\} \cup \{A \subset \mathbb{R}^2 \mid A \neq \emptyset, A \in \text{conv}, \dim(\text{aff}\, A) \leq 1\}$.

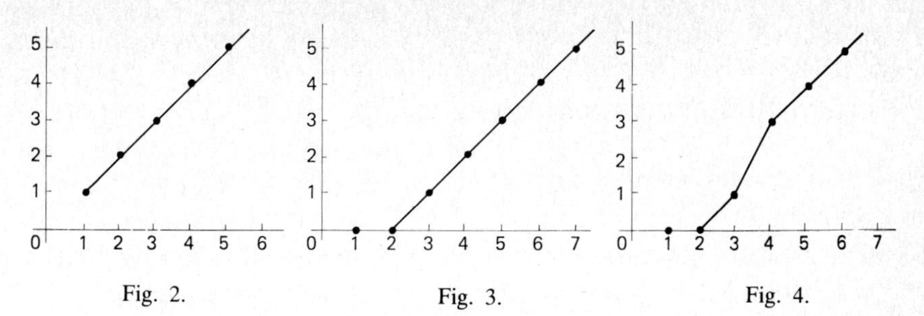

Fig. 2. Fig. 3. Fig. 4.

More peculiar graphs occur when considering more special convexity spaces. In Fig. 5 we have $(X, \mathscr{C}) = (\mathbb{R}^\infty, \text{conv})$; in Fig. 6 we have $|X| = \infty$ and $\mathscr{C} = \{X\} \cup \{A \subset X \mid |A| \leqslant k\}$ with $k \geqslant 2$; and in Fig. 7 the space $(X, \mathscr{C})$ with $X = \mathbb{R}^2$ and $\mathscr{C} = \{\phi, \mathbb{R}^2\} \cup \{A \subset \mathbb{R}^2 \mid n \in \mathbb{N} \Rightarrow M_n \subsetneqq A\}$ with $M_n \subset \mathbb{R}^2$ and $|M_n| = \frac{1}{2}n^2 + \frac{3}{2}n + 1$ and the M_n's are pairwise disjoint.

The convexity spaces of Figs. 2 and 3 satisfy both the cone-union condition and the continuity property. The space of Fig. 4 satisfies only the cone-union condition; note that this graph is 3-increasing. The space of Fig. 6 satisfies both properties; however, the graph is not n-increasing for any n. The convexity space used in Fig. 6 is well known; none of the two properties holds in this case. The space used for the graph of the exchange function for Fig. 7 is quite special; this space has the property that there is never a finite set such that for all larger finite sets at least one of its elements is exchangeable.

The notion of convex product space was introduced by Eckhoff [5] and used later by several authors, namely Jamison [6], Kramer [8], Reay [12], Sierksma [13], Soltan [18], and Thompson and Hare [19]. Let $(X_1, \mathscr{C}_1)$ and $(X_2, \mathscr{C}_2)$ be convexity spaces. Their *convex product space* is the pair $(X_1 \times X_2, \mathscr{C}_1 \oplus \mathscr{C}_2)$, where $X_1 \times X_2$ is the Cartesian product of X_1 and X_2 and $\mathscr{C}_1 \oplus \mathscr{C}_2 = \{A \times B \mid A \in \mathscr{C}_1, B \in \mathscr{C}_2\}$. The $(\mathscr{C}_1 \oplus \mathscr{C}_2)$-hull of a set S in $X_1 \times X_2$ is then equal to $(\mathscr{C}_1 \oplus \mathscr{C}_2)(S) = \mathscr{C}_1(\pi_1 S) \times \mathscr{C}_2(\pi_2 S)$, with π_i the projection of $X_1 \times X_2$ onto X_i, $i = 1, 2$.

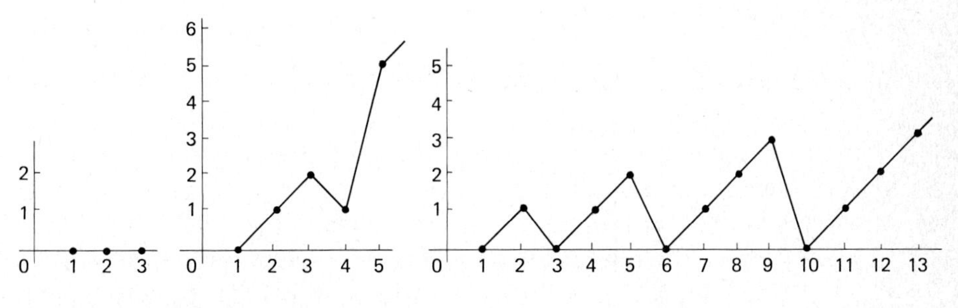

Fig. 5. Fig. 6. Fig. 7.

Theorem 5. *Let $(X_i, \mathscr{C}_i)$ be a convexity space with n_i-increasing exchange functions ε_i, $i = 1, 2$. Then the exchange function ε of $(X_1 \times X_2, \mathscr{C}_1 \oplus \mathscr{C}_2)$ is $(n_1 + n_2 - 1)$-increasing.*

Proof. It is easy to show that for $n \geqslant n_1 + n_2 - 1$ and $n \in \mathbb{N}_{X_1 \times X_2}$ holds $\varepsilon(n) > 0$. We shall prove that $\varepsilon(n+1) \geqslant \varepsilon(n) + 1$ for $n \geqslant n_1 + n_2 - 1$ and $n, n+1 \in \mathbb{N}_{X_1 \times X_2}$. Let there be an element $n \geqslant n_1 + n_2 - 1$ with $n, n+1 \in \mathbb{N}_{X_1 \times X_2}$ and take any $p = (p_1, p_2)$ and E in $X_1 \times X_2$, with $\pi_1 E = E_1$, $\pi_2 E = E_2$ and $|E| = n + 1$. Moreover, let $x = (x_1, x_2) \in (\mathscr{C}_1 \oplus \mathscr{C}_2)(E)$ and *let* $|\omega(p, x, E)| = \varepsilon(n+1)$. *We then distinguish two cases*:

(1) $|E_1| \leqslant n$ and $|E_2| \leqslant n$. Then there are the elements $a_1, a_2, a_3, a_4 \in E$, which are all different except, possibly, a_2 and a_3 and such that $\pi_1(a_1) = \pi_1(a_2)$ and $\pi_2(a_3) = \pi_2(a_4)$.

If $a_2 = a_3$ (see Fig. 8), then $(\mathscr{C}_1 \oplus \mathscr{C}_2)(E) = (\mathscr{C}_1 \oplus \mathscr{C}_2)(E \setminus a_2)$, hence $x \in (\mathscr{C}_1 \oplus \mathscr{C}_2)(E \setminus a_2)$ and we have that $\omega(p, x, E \setminus \{a_2\}) \cup \{a_2\} \subset \omega(p, x, E)$. Hence $\varepsilon(n+1) = |\omega(p, x, E)| \geqslant |\omega(p, x, E \setminus \{a_2\})| + 1 \geqslant \varepsilon(n) + 1$.

If $a_2 \neq a_3$ (see Fig. 9), there are two possibilities again:

(i) $a_1, a_2, a_3, a_4 \notin \omega(p, x, E)$. Clearly $\varepsilon(n+1) > 0$ so $\omega(p, x, E) \neq \emptyset$. Take $a \in \omega(p, x, E)$ and let $\bar{a}_1 = (\pi_1(a), \pi_2(a))$ and $\bar{a}_4 = (\pi_1(a_4), \pi_2(a))$. Furthermore, define $\bar{E} = (E \setminus \{a_1, a_4\}) \cup \{\bar{a}_1, \bar{a}_4\}$, unless $\bar{a}_1 = \bar{a}_4$ then define $\bar{E} = E$ (see Figs. 10 and 11).

Clearly, $\pi_1 \bar{E} = E_1$, $\pi_2 \bar{E} = E_2$ and $|\bar{E}| = |E| = n + 1$. Hence, $x \in (\mathscr{C}_1 \oplus \mathscr{C}_2)(\bar{E})$ and $|\bar{E}| = n + 1$. As $\pi_1(a) = \pi_1(\bar{a}_1)$ and $\pi_2(a) = \pi_2(\bar{a}_4)$ it follows that $|\omega(p, x, \bar{E})| \geqslant \varepsilon(n) + 1$. Moreover, as $a_1, a_4 \notin \omega(p, x, E)$, $a_1, a_4 \notin \omega(p, x, E)$. So $\omega(p, x, \bar{E}) \subset \omega(p, x, E)$, and we find that $\varepsilon(n+1) = |\omega(p, x, E)| \geqslant |\omega(p, x, \bar{E})| \geqslant \varepsilon(n) + 1$.

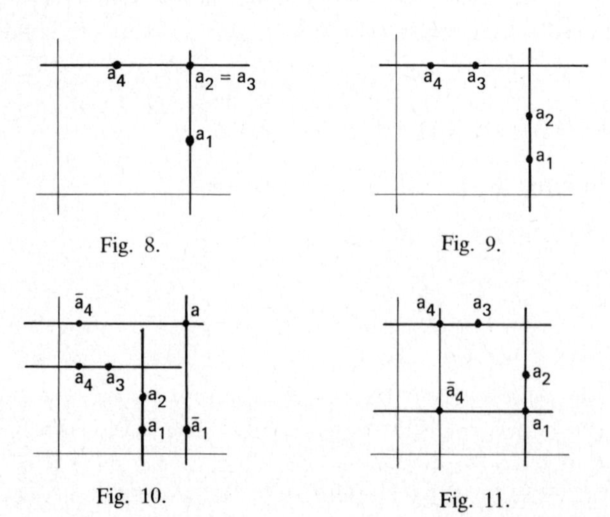

Fig. 8. Fig. 9.

Fig. 10. Fig. 11.

(ii) For some $i \in \{1,2,3,4\}$, say $i=1$, $a_i \in \omega(p, x, E)$. Now let $\bar{a}_4 = (\pi_1(a_4), \pi_2(a_1))$ and $\bar{E} = (E \setminus \{a_4\}) \cup \{\bar{a}_4\}$. Clearly, $\pi_1 \bar{E} = E_1$, $\pi_2 \bar{E} = E_2$ and $|\bar{E}| = |E| = n+1$. Hence, $x \in (\mathscr{C}_1 \oplus \mathscr{C}_2)(\bar{E})$. As $\pi_1(a_2) = \pi_1(a_1)$ and $\pi_2(\bar{a}_4) = \pi_2(a_1)$, we find that $|\omega(p, x, \bar{E})| \geq \varepsilon(n)+1$. As $\omega(p, x, \bar{E}) \setminus \{\bar{a}_4\} \subset \omega(p, x, E) \setminus \{a_4\}$, it follows directly that $\varepsilon(n+1) = |\omega(p, x, E)| = |\omega(p, x, \bar{E})| \geq \varepsilon(n)+1$.

(2) $|E_1| = n+1$ or $|E_2| = n+1$ or both are. Assume that $|E_1| \leq n+1$ and that $|E_2| = n+1$. Let $|E_1| = k$ and σ_1: $E_1 \to E$, σ_2: $E_2 \to E$ be mappings such that $\sigma_1 \pi_1$ and $\sigma_2 \pi_2$ are identity mappings. Note that σ_1 is injective and σ_2 is bijective. We distinguish two cases:

(i) $k = n+1$. Then $|E_1| = |E_2| = |E| = n+1$ and σ_2 is bijective too. Clearly, $H_i = \sigma_i \omega(p_i, x_i, E_i)$, where $H_i = \{z \in E \mid x_i \in \mathscr{C}_i(\pi_i(p \cup E \setminus z))\}$, $i = 1,2$. Since $\omega(p, x, E) = H_1 \cap H_2$ it follows that

$$\begin{aligned}
\varepsilon(n+1) &= |\omega(p, x, E)| = |H_1 \cap H_2| \\
&\geq |\sigma_1 \omega(p_1, x_1, E_1)| + |\sigma_2 \omega(p_2, x_2, E_2)| - (n+1) \\
&= |\omega(p_1, x_1, E_1)| + |\omega(p_2, x_2, E_2)| - n - 1 \\
&\geq [\varepsilon_1(n) + \varepsilon_2(n) - n] + 1.
\end{aligned}$$

(ii) $k \leq n$. Then an element $z \in \sigma_1 E_1$ exists such that $\pi_1(z) \in \pi_1(E \setminus \sigma_1 E_1)$. Obviously, $(E \setminus \sigma_1 E_1) \cup \{z\} \cup \sigma_1 \omega(p_1, x_1, E_1) \subset H_1$, so

$$\begin{aligned}
|H_1| &\leq |E \setminus \sigma_1 E_1| + |\{z\} \cup \sigma_1 \omega(p_1, x_1, E_1)| \\
&\geq n+1-k+\max\{1, |\omega(p_1, x_1, E_1)|\} \\
&\geq n+1-k+\max\{1, \varepsilon_1(k)\}.
\end{aligned}$$

Just as in (i) it follows that $H_2 = \sigma_2 \omega(p_2, x_2, E_2)$. Hence,

$$\begin{aligned}
\varepsilon(n+1) &= |\omega(p, x, E)| = |H_1 \cap H_2| \\
&\geq n+1-k+\max\{1, \varepsilon_1(k)\} + \varepsilon_2(n+1) - (n+1) \\
&\geq [\max\{1, \varepsilon_1(k)\} + \varepsilon_2(n) - k] + 1.
\end{aligned}$$

From the conclusions of (i) and (ii) we derive that

$$\varepsilon(n+1) \geq [\max\{1, \varepsilon_1(k)\} + \varepsilon_2(n) - k] + 1.$$

There exist $q_i \in X_i$, $F_i \subset X_i$ with $|F_1| = k$, $|F_2| = n$ and $y_i \in \mathscr{C}_i(F_i)$ for $i = 1,2$, such that $|\omega(q_1, y_1, F_1)| = \varepsilon_1(k) \geq 0$ and $|\omega(q_2, y_2, F_2)| = \varepsilon_2(n) > 0$. Let $F_1 = \{a_1, a_2, \ldots, a_k\}$ and let the elements $a_1, \ldots, a_k$ be numbered such that $\omega(q_1, y_1, F_1) = \{a_\lambda, \ldots, a_k\}$, where $\lambda = k - \varepsilon_1(k) + 1$, provided $\varepsilon_1(k) > 0$. Furthermore, let $F_2 = \{b_1, b_2, \ldots, b_n\}$ with $\omega(q_2, y_2, F_2) = \{b_1, \ldots, b_\mu\}$, where $\mu = \varepsilon_2(n)$. Define $F = \{(a_1, b_1), \ldots, (a_k, b_k), (a_k, b_{k+1}), \ldots, (a_k, b_n)\}$, $q = (q_1, q_2)$ and $y = (y_1, y_2)$. Then $y \in (\mathscr{C}_1 \oplus \mathscr{C}_2)(F)$ and $\omega(q, y, F) \neq \emptyset$. It is clear that if

$\varepsilon_1(k)>0$, then $\omega(q,y,F)=\{(a_1,b_1),\ldots,(a_k,b_\mu)\}\setminus\{(a_1,b_1),\ldots,(a_{\lambda-1},b_{\lambda-1})\}$, so $|\omega(q,y,F)|=\mu-(\lambda-1)=\mu-\lambda+1=\varepsilon_1(k)+\varepsilon_2(n)-k$.

If $\varepsilon_1(k)=0$, it follows that

$$\omega(q,y,F)=\{(a_1,b_1),\ldots,(a_k,b_k)\}\setminus\{(a_1,b_1),\ldots,(a_{k-1},b_{k-1})\},$$

so $|\omega(q,y,F)|=\mu-(k+1)=1+\varepsilon_2(n)-k$ (Fig. 12). In both cases, $|\omega(q,y,F)|=\max\{1,\varepsilon_1(k)\}+\varepsilon_2(n)-k$. Consequently,

$$\varepsilon(n+1)\geqslant[\max\{1,\varepsilon_1(k)\}+\varepsilon_2(n)-k]+1=|\omega(q,y,F)|+1\geqslant\varepsilon(n)+1.$$

3. The exchange number

In this final section we pay more attention to the well-known exchange number: the relation with the exchange function is described and the exchange number for the convex product space is studied. We first repeat the definition. The *exchange number* e of a convexity space $(X,\mathscr{C})$ is for $X\neq\emptyset$ defined as the smallest positive integer n such that for each p and A in X with $n\leqslant|A|<\infty$ the following holds:

$$\mathscr{C}(A)\subset\bigcup_{a\in A}\mathscr{C}(p\cup(A\setminus a));$$

if $X=\emptyset$ the exchange number is denoted 0. Note that $0\leqslant e\leqslant\infty$ and that if $(X,\mathscr{C})$ is a T_1 space (all singletons are convex; see [7]) and $|X|\geqslant 2$, then $e\geqslant 2$. In Sierksma [14] it is shown that there is the following relationship with the Carathéodory number c, namely $e\leqslant c+1$; this relation is quite similar to the relationship between the numbers of Helly and Radon, namely $h\leqslant r-1$ (Levi's Theorem [9]). More relationships between the various numbers can be found in [14]. Also in [14] can be found various convexity spaces for which the exchange number is calculated. We remark here that for the ordinary convexity space $(\mathbb{R}^d,\text{conv})$ $e=d+1$.

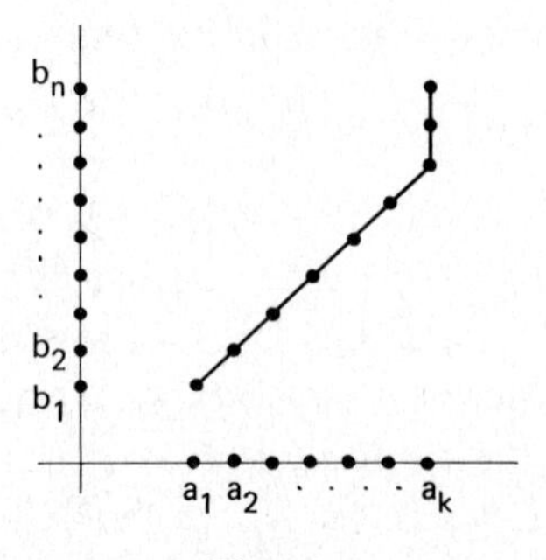

Fig. 12.

The relation with the exchange function is expressed in the following theorems.

Theorem 6. *Let* $(X, \mathscr{C})$ *be a convexity space which satisfies the cone-union condition and let* $k \in \mathbb{N}_X$. *Then the following assertions are equivalent*:

(i) $e \leqslant k$;

(ii) $\mathscr{C}(p \cup A) = \bigcup\{\mathscr{C}(p \cup (A \setminus B)) \mid B \subset A, |B| = |A| - k + 1\}$ *for each* p *and* A *in* X *with* $k \leqslant |A| < \infty$.

Proof.

(i) $\Rightarrow$ (ii). Take any p and A in X with $e \leqslant k \leqslant |A| < \infty$. Clearly, $\bigcup\{\mathscr{C}(p \cup (A \setminus B)) \mid B \subset A, |B| = |A| - k + 1\} \subset \mathscr{C}(p \cup A)$, so we only have to show the opposite inclusion.

Since $\mathscr{C}(p \cup A)$ and $\mathscr{C}(p \cup (A \setminus a))$ are polytopes for each $a \in A$, the cone-union condition implies that $\mathscr{C}(p \cup A) \subset \bigcup\{\mathscr{C}(p \cup \mathscr{C}(p \cup (A \setminus a))) \mid a \in A\} = \bigcup\{\mathscr{C}(p \cup (A \setminus a)) \mid a \in A\}$. Hence, there is an element $a_1 \in A$ such that $x \in \mathscr{C}(p \cup (A \setminus a_1))$; if $|A \setminus \{a_1\}| \geqslant k$, then there is an element $a_2 \in A \setminus \{a_1\}$ such that $x \in \mathscr{C}(p \cup ((A \setminus a_1) \setminus a_2)) = \mathscr{C}(p \cup (A \setminus \{a_1, a_2\}))$; and so on until for $n = |A| - k + 1$, $|A \setminus \{a_1, \ldots, a_n\}| = k - 1$. Let $B = \{a_1, \ldots, a_n\}$, then $x \in \mathscr{C}(p \cup (A \setminus B))$. Therefore $\mathscr{C}(p \cup A) \subset \bigcup\{\mathscr{C}(p \cup (A \setminus B)) \mid B \subset A, |B| = |A| - k + 1\}$.

(ii) $\Rightarrow$ (i). From (ii) it follows that for each p and A in X with $k \leqslant |A| < \infty$,

$$\mathscr{C}(A) \subset \mathscr{C}(p \cup A) = \bigcup\{\mathscr{C}(p \cup (A \setminus B)) \mid B \subset A, |B| = |A| - k + 1\} \subset \bigcup\{\mathscr{C}(p \cup (A \setminus a)) \mid a \in A\}.$$

Hence, $e \leqslant k$.

Theorem 7. *Let* $(X, \mathscr{C})$ *be a convexity space with exchange function* ε *and exchange number* e. *For each* $k \in \mathbb{N}_X$ *the following assertions are equivalent*:

(i) $e \leqslant k$;

(ii) $\varepsilon(n) > 0$ *for each* $n \geqslant k$ *and* $n \in \mathbb{N}_X$.

Proof.

(i) $\Rightarrow$ (ii). Let there be an element $k \in \mathbb{N}_X$ such that $e \leqslant k$. Take any p and A in X with $|A| = n$ and $x \in \mathscr{C}(A)$ and with $n \geqslant k$, $n \in \mathbb{N}_X$. Since $e \leqslant k \leqslant n = |A| < \infty$ it follows that $x \in \mathscr{C}(p \cup (A \setminus a))$ for some $a \in A$, or that $a \in \omega(p, x, A)$ and so $\omega(p, x, A) \neq \emptyset$. Hence, $\varepsilon(n) > 0$.

(ii) $\Rightarrow$ (i). Let there be an element $k \in \mathbb{N}_X$ such that $\varepsilon(n) > 0$ for each $n \geqslant k$ with $n \in \mathbb{N}_X$. Hence, $\omega(p, x, A) \neq \emptyset$ for each p and A in X with $|A| = n$ and each $x \in \mathscr{C}(A)$, $n \geqslant k$, $n \in \mathbb{N}_X$. This implies that for each $p \in X$ and each $A \subset X$ with $k \leqslant |A| < \infty$, $\mathscr{C}(A) \subset \bigcup\{\mathscr{C}(p \cup (A \setminus a)) \mid a \in A\}$. Therefore, $e \leqslant k$.

Corollary 8. *Let $(X, \mathscr{C})$ be a convexity space with exchange function ε. Then the exchange number e of $(X, \mathscr{C})$ satisfies:*

$$e = \inf\{k \in \mathbb{N}_X \mid \varepsilon(n) > 0 \text{ for each } n \geq k \text{ and } n \in \mathbb{N}_X\}.$$

Proof. Direct consequence of Theorem 7.

Corollary 9. *Let $(X, \mathscr{C})$ be a convexity space with k-increasing exchange function ε for some $k \in \mathbb{N}_X$. Then the exchange number e of $(X, \mathscr{C})$ satisfies $e \leq k$.*

Proof. If the exchange function ε is k-increasing for some $k \in \mathbb{N}_X$, then $\varepsilon(n) > 0$ for each $n \geq k$ with $n \in \mathbb{N}_X$. Theorem 7 implies that $e \leq k$.

Theorem 10. *Let $(X, \mathscr{C})$ be a convexity space which satisfies the cone-union condition. For each $k \in \mathbb{N}_X$ the following assertions are equivalent:*

(i) $e \leq k$;

(ii) $\varepsilon(k) > 0$.

Proof. (i) $\Rightarrow$ (ii) is a direct consequence of Theorem 6, and (ii) $\Rightarrow$ (i) follows from Theorem 1 and Corollary 9.

Theorem 11. *Let $(X, \mathscr{C})$ be a convexity space. Then the following assertions are equivalent:*

(i) $\varepsilon(n) = n$ *for each* $n \in \mathbb{N}_X$;

(ii) $e = 1$;

(iii) $\mathscr{C} = \{\emptyset, X\}$.

Proof. Left to the reader.

Theorem 12. *Let $(X, \mathscr{C})$ be a convexity space. Then the following assertions hold:*

(a) $\varepsilon(e-1) = 0$ *if* $e \geq 2$;

(b) $\varepsilon(e) = 1$;

(c) $1 \leq \varepsilon(n) \leq n$ *for each* $n \geq e$ *and* $n \in \mathbb{N}_X$.

Proof. Left to the reader.

Corollary 13. *Let $(X, \mathscr{C})$ be a convexity space with $e \geq 2$. Then the following holds:*

$$e = 1 + \sup\{k \mid \varepsilon(k) = 0\}.$$

Proof. Trivial.

We now turn to the exchange number of convex product spaces. It is well known that if the Carathéodory, Helly, or Radon numbers in the component spaces are finite, then the respective numbers are also *finite* in the convex product. However, this is not the case for the exchange numbers as the following example shows. Let $X_1 = X_2 = \mathbb{R}$ and let $\mathscr{C}$ be defined as in [14, p. 129] and [15, Example 3], i.e. $\mathscr{C} = \{\emptyset, \mathbb{R}\} \cup \{C \subset \mathbb{R} \mid C \subsetneq C_n$ for some $n \geqslant 1\}$ with $C_1 = \{1\}$, $C_2 = \{2, 3\}$, $C_3 = \{4, 5, 6\}$, $C_4 = \{7, 8, 9, 10\}$, etc. In [15] it is shown that the exchange number is 2. It can be shown that the exchange number of $(\mathbb{R} \times \mathbb{R}, \mathscr{C} \oplus \mathscr{C})$ is *in*finite. In what follows, e_1, e_2, and e are the exchange numbers of $(X_1, \mathscr{C}_1)$, $(X_2, \mathscr{C}_2)$ and $(X_1 \times X_2, \mathscr{C}_1 \oplus \mathscr{C}_2)$, respectively.

Theorem 14. *Let* $(X_i, \mathscr{C}_i)$, $X_i \neq \emptyset$ *be a convexity space with* n_i*-increasing exchange functions,* $i = 1, 2$. *Then the following holds:*

$$\max\{e_1, e_2\} \leqslant e \leqslant n_1 + n_2 - 1.$$

If the exchange function is e_i*-increasing* $(i = 1, 2)$, *then*

$$e = e_1 + e_2 - 1.$$

Proof. The first part of the theorem is a direct consequence of Theorem 5. We only show that $e \geqslant e_1 + e_2 - 1$. We may assume that $e_1, e_2 \geqslant 2$. Clearly, there exists a set E_i in X_i with $|E_i| = e_i$, an element $p_i \in E_i$, and an element $x_i \in \mathscr{C}_i(E_i)$ such that $\omega(p_i, x_i, E_i) = \{p_i\}$, $i = 1, 2$. Let $E = [E_1 \times \{p_2\}] \cup [\{p_1\} \times E_2] \setminus \{(p_1, p_2)\}$. It is quite clear that $|E| = e_1 + e_2 - 1$. Note that $x = (x_1, x_2) \in \mathscr{C}_1(E_1) \times \mathscr{C}_2(E_2) = (\mathscr{C}_1 \oplus \mathscr{C}_2)(E)$ (Fig. 13).

We now show that $\omega(p, x, E) = \emptyset$, with $p = (p_1, p_2)$. Suppose, on the contrary, that $a \in \omega(p, x, E)$ for some $a \in E$. We may suppose that $a \in (E_1 \setminus p_1) \times \{p_2\}$. Hence, $a \neq p$ and $x \in (\mathscr{C}_1 \oplus \mathscr{C}_2)(p \cup (E \setminus a))$. So $x_1 \in \mathscr{C}_1(p_1 \cup (E_1 \setminus \pi_1(a)))$ and this implies that $\pi_1(a) \in \omega(p_1, x_1, E_1)$; hence, $p_1 = \pi_1(a)$, which contradicts $p_1 \neq \pi_1(a)$. Consequently, $\omega(p, x, E) = \emptyset$, and $\varepsilon(e_1 + e_2 - 2) = 0$, and therefore we do in fact have that $e \geqslant e_1 + e_2 - 1$.

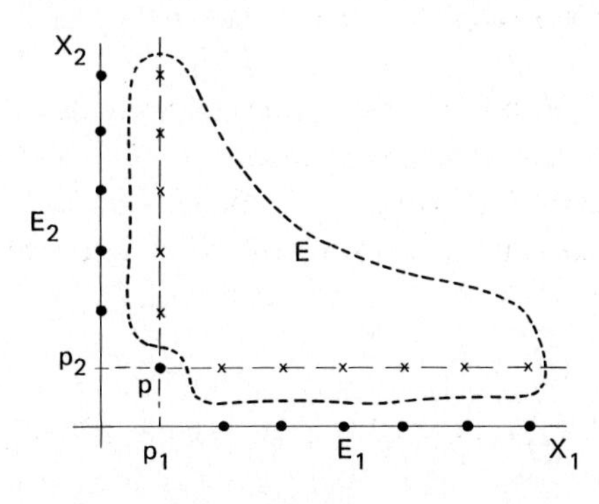

Fig. 13.

Corollary 15. *Let* $(X_1, \mathscr{C}_1)$ *and* $(X_2, \mathscr{C}_2)$, $X_1 \neq \emptyset \neq X_2$, *be convexity spaces both satisfying the cone-union condition. Then the following holds*:

$$e = e_1 + e_2 - 1.$$

Proof. Left to the reader.

Example. Consider the ordinary convexity spaces $(\mathbb{R}^m, \text{conv})$ and $(\mathbb{R}^n, \text{conv})$ where $m, n \in \mathbb{N}$; let e_1 and e_2, respectively, be the exchange numbers. We know that $e_1 = m + 1$ and $e_2 = n + 1$ and that $(\mathbb{R}^m, \text{conv})$ and $(\mathbb{R}^n, \text{conv})$ both satisfy the cone-union condition. Then the exchange number e of $(\mathbb{R}^m \times \mathbb{R}^n, \text{conv} \oplus \text{conv})$ satisfies $e = e_1 + e_2 - 1 = m + n + 1$. As for the exchange function ε of $(\mathbb{R}^m \times \mathbb{R}^n$, $\text{conv} \oplus \text{conv})$ we can add that $\varepsilon(k+1) = \varepsilon(k) + 1$ if $k \geqslant m + n + 1$ and that $\varepsilon(k) = 0$ if $1 \leqslant k \leqslant m + n$.

Example. Let $|X_i| = \infty$, $k_i \in \mathbb{N}$ and $\mathscr{C}_i = \{X_i\} \cup \{A \mid A \subset X_i, |A| \leqslant k_i\}$, $i = 1, 2$. We know that $e_1 = e_2 = 2$ and that ε_i is $(k_i + 1)$-increasing, $i = 1, 2$. The exchange function ε of $(X_1 \times X_2, \mathscr{C}_1 \oplus \mathscr{C}_2)$ is $(k_1 + k_2 - 1)$-increasing. Hence, the exchange number e of $(X_1 \times X_2, \mathscr{C}_1 \oplus \mathscr{C}_2)$ satisfies $2 \leqslant e \leqslant k_1 + k_2 + 1$.

We even have that $e = k_1 + k_2 + 1$. This can be seen as follows. Obviously, we only have to show that $\varepsilon(k_1 + k_2) = 0$. Take some $A_i \subset X_i$ with $|A_i| = k_i$ and some $p_i \in X_i \setminus A_i$, $i \in 1, 2$. Define $E = [A_1 \times \{p_2\}] \cup [\{p_1\} \times A_2]$. Hence, $|E| = k_1 + k_2$. It follows that $\pi_i E = \{p_i\} \cup A$ and that $|\pi_i E| = k_i + 1$, $i = 1, 2$. Therefore $(\mathscr{C}_1 \oplus \mathscr{C}_2)(E) = \mathscr{C}_1(\pi_1 E) \times \mathscr{C}_2(\pi_2 E) = X_1 \times X_2$. Now take $x = (x_1, x_2) \in X_1 \times X_2$ such that $x_i \notin \pi_i E$, $i = 1, 2$. Clearly, $\omega(p, x, E) = \emptyset$, so that $\varepsilon(k_1 + k_2) = 0$.

We have not been able to give an example for which $e < n_1 + n_2 - 1$. This means that we can formulate the following conjecture.

Conjecture. If the exchange functions of $(X_1, \mathscr{C}_1)$ and $(X_2, \mathscr{C}_2)$ are n_1- and n_2-increasing, respectively, then the exchange number e of the convex product space $(X_1 \times X_2, \mathscr{C}_1 \oplus \mathscr{C}_2)$ satisfies $e = n_1 + n_2 - 1$.

We conclude this paper by remarking that the exchange number also has been determined for the so-called *convex sum spaces*; see Degreef [3] and Sierksma [15]. Convex sum spaces deal, roughly speaking, with convexities on unions of sets. In [15] and [16] one can find some open problems concerning sum spaces.

References

[1] P.M. Cohn, Universal Algebra (Harper and Row, New York, 1965).

[2] L. Danzer, B. Grünbaum and V. Klee, Helly's Theorem and its relatives, Proc. of Symposium Pure Math. VII (American Mathematical Society, 1963) pp. 101–180.

[3] E. Degreef, The convex sum space and direct sum space, Simon Stevin, Quart. J. Pure Appl. Math. 56 (1982) 109–119.

[4] J.-P. Doignon, J.R. Reay and G. Sierksma, A Tverberg-type generalization of the Helly Number of a convexity space, J. Geometry 16 (1981) 117–125.

[5] J. Eckhoff, Der Satz von Radon in Konvexen Produktstrukturen I, Monatshefte für Math., Vol. 72 (1968), pp. 303–314.

[6] R.E. Jamison, A general theory of convexity, Doctoral Dissertation, Univ. of Washington, Seattle (1974).

[7] D.C. Kay and E.W. Womble, Axiomatic convexity theory and the relationships between the Carathéodory, Helly, and Radon numbers, Pacific J. Math. 38 (2) (1971) 471–485.

[8] H. Kramer, Supporting spheres for families of sets in product spaces, Revue d'Analyse Num. et de la Théorie de l'Appr., Tone 2 (1973) pp. 49–53.

[9] F.W. Levi, On Helly's Theorem and the axioms of convexity, J. Indian Math. Soc. 15 (1951) 65–76.

[10] J.R. Reay, Generalizations of a theorem of Carathéodory, Amer. Math. Soc. Memoir, No. 54 (1965).

[11] J.R. Reay, Positive bases as a tool in convexity, Proc. of the Colloquium on Convexity, Copenhagen 1965 (1967) pp. 255–260.

[12] J.R. Reay, Carathéodory Theorems in convex product structures, Pacific J. Math. 25 (1) (1970) 227–230.

[13] G. Sierksma, Carathéodory and Helly numbers of convex product structures, Pacific J. Math. 61 (1) (1975) 275–282.

[14] G. Sierksma, Relationships between Carathéodory, Helly, Radon and exchange numbers of convexity spaces, Nieuw Archief voor Wisk. 25 (3) (1977) 115–132.

[15] G. Sierksma, Convexity on unions of sets, Compos. Math. 42 (1981) 391–400.

[16] G. Sierksma, Generalizations of Helly's Theorem: Open problems, Proc. Conf. on Convexity and Related Combinatorial Geometry, Oklahoma, 1980 (Marcel Dekker, Inc., 1981).

[17] G. Sierksma, A cone condition for convexity spaces, Bull. de la Soc. Math. de Belgique XXXIV (1), Ser. B (1982) 41–48.

[18] V.P. Soltan, Some questions in the abstract theory of convexity, Sov. Math. Dokl. 17 (3) (1976) 730–733.

[19] G. Thompson and W.R. Hare, Tverberg-type theorems in convex product structures, Lecture Notes in Mathematics, Vol. 490 (Springer Verlag, 1975) pp. 212–217.

[20] M. van der Vel, Finite dimensional convexity structures I: General results, Report 122, Vrije University, Amsterdam (1980).

[21] M. van der Vel, Finite dimensional convexity structures: The invariants, Report 123, Vrije University, Amsterdam (1980).

Received 24 June 1981

Annals of Discrete Mathematics 20 (1984) 307–310
North-Holland

EXTENDING KOTZIG'S THEOREM: HENSLEY'S LATTICE POLYTOPE THEOREM

Joseph ZAKS
University of Haifa, Israel

1

The *weight* $w(E)$ of an edge E in a graph G is defined as the sum of the valences of the end-points of E. The weight $w(G)$ of the graph G is defined by $w(G) = \min\{w(E) \mid E \in G\}$. Kotzig [4] proved that $w(G) \leqslant 13$ holds for all planar graphs, and Grünbaum and Shephard [2] (see also [1]) proved that $w(G) \leqslant 15$ holds for all toroidal graphs.

Let G_g denote the family of all the graphs which triangulate the orientable 2-manifold S_g of genus g; members of G_g are 1-skeletons of special two-dimensional simplicial complexes. Let MG_g (PG_g) denote the family of all the multigraphs (resp. pseudographs) which dissect S_g into triangles (i.e. regions which have interior homeomorphic to an open disc, and which meet three regions), and have no vertices of valences $\leqslant 2$. Let $w(F)$ be defined for a family F of pseudographs by $w(F) = \max\{w(G) \mid G \in F\}$.

We [6] have the following results.

Theorem 1. *$w(G_g)$ is at most the least odd integer which is greater than $6 + \sqrt{48g + 1}$, for all g, $g \geqslant 1$; the bound is best for all those g, $g \geqslant 2$, making $48g + 1$ a complete square.*

Theorem 2. *$w(MG_g) = 8g + 7$ for all $g \geqslant 1$; the bounds are best for all $g \geqslant 1$.*

Theorem 3. *$w(PG_g) = 24g - 9$ for all $g \geqslant 1$; the bounds are best for all $g \geqslant 1$.*

A pseudograph in PG_g cannot have a large weight if it has too many vertices; more precisely, we have:

Theorem 4. *For every g, $g \geqslant 2$, there is a constant $c = c(g)$ such that if $G \in PG_g$ and $V(G) \geqslant c$, then $w(G) \leqslant 15$.*

Let $v_k(G)$ denote the number of k-valent vertices of G, and let G_g^* be defined by $G_g^* = \{G \mid G \in G_g$ and $v_3(G) = 0\}$. Kotzig [4] proved that $w(G_0^*) = 11$ and Grünbaum and Shephard [2] proved that $w(G_1^*) = 12$. We [6] have the following:

Theorem 5. *For all g, $g \geq 2$, $w(G_g^*)$ is at most the least odd integer which is greater than $5.5 + \sqrt{48g - 17.75}$.*

Extensions of Theorem 1, similar to Theorem 5, are established in [6] for the cases where, in addition, $v_3(G) = v_4(G) = 0$ or $v_3(G) = v_4(G) = v_5(G) = 0$. The bound 15 of Theorem 4 can be reduced to 14 (13, 12) if, in addition, $v_{12}(G)$ ($v_{12}(G) + v_{11}(G)$; $v_{12}(G) + v_{11}(G) + v_{10}(G)$, respectively) is bounded.

We close this section with:

Sketch of the proof of Theorem 1. Suppose $G \in G_g$, n is odd, $n \geq 11$ and n is greater than $6 + \sqrt{48g + 1}$, and $w(G) \geq n$. We will prove that $w(G) = n$ by showing that G has an edge E for which $w(E) = n$.

Let e_{ij} denote the number of edges of G having i-valent and j-valent end-points. $w(G) \geq n$ means that $i + j < n$ implies that $e_{ij} = 0$.

For every k and for every k-valent vertex W of G, at most $[k/2]$ neighbours of W have valence $< n/2$, since otherwise two consecutive (= neighbors) neighbors W' and W'' of W will have valences $< n/2$, implying that $w(W'W'') < n$, contrary to the assumptions on G.

In particular, at most $[k/2]$ neighbors of every k-valent vertex are 3-valent, hence:

$$e_{3,k} \leq \left[\frac{k}{2}\right] v_k \quad \text{holds for all } k.$$

Therefore

$$3v_3 - e_{3,n-3} = \sum_{k \geq n-2} e_{3,k} \leq \sum_{k \geq n-2} \left[\frac{k}{2}\right] v_k.$$

In a similar way, concerning the number of 3-valent or 4-valent vertices which are neighbors of a k-valent vertex, and the number of 3-, 4- or 5-valent vertices which are neighbors of a k-valent vertex, we get:

$$e_{3,k} + e_{4,k} \leq \left[\frac{k}{2}\right] v_k \quad \text{for all } k,$$

implying:

$$3v_3 + 4v_4 - e_{4,n-4} = \sum_{k \geq n-3} (e_{3,k} + e_{4,k}) \leq \sum_{k \geq n-3} \left[\frac{k}{2}\right] v_k$$

and

$$e_{3,k} + e_{4,k} + e_{5,k} \leqslant \left[\frac{k}{2}\right] v_k \quad \text{for all } k.$$

Thus:

$$\begin{aligned} 3v_3 + 4v_4 + 5v_5 - e_{5,n-5} &= \sum_{k \geqslant n-4} (e_{3,k} + e_{4,k} + e_{5,k}) \\ &\leqslant \sum_{k \geqslant n-4} \left[\frac{k}{2}\right] v_k. \end{aligned}$$

By using the well-known equation

$$3v_3 + 2v_4 + v_5 = 12(1-g) + \sum_{k \geqslant 7} (k-6) v_k,$$

we get

$$\begin{aligned} 2e_{5,n-5} + 3e_{4,n-4} + 5e_{3,n-3} \geqslant {} & 120(1-g) + 10 \sum_{k=7}^{n-5} (k-6) v_k \\ & + (9n-95) v_{n-4} + (7.5n - 82.5) v_{n-3} \\ & + 10 \sum_{k \geqslant n-2} \left(k - 6 - \left[\frac{k}{2}\right]\right) v_k. \end{aligned}$$

The rest of the proof is to show, using the assumption on n, that the right-hand side of the inequality is positive; thus $e_{5,n-5} + e_{4,n-4} + e_{3,n-3}$ is positive; hence $w(G) = n$. It follows that $w(G)$ can be at most the least odd integer which is greater than $6 + \sqrt{48g+1}$.

2

A convex polytope P in E^d is called a *lattice polytope* if all of its vertices are lattice points (i.e. points having integral coordinates). Let $G(P)$ ($G^0(P)$) denote the number of lattice points in P (in the interior of P, respectively). Let $g(d,n)$ be defined by $g(d,n) = \sup\{G(P) \mid P$ is a lattice polytope and $G^0(P) = n\}$. Clearly, $g(d,0) = \infty$ for all $d \geqslant 2$, and $g(1,n) = n+2$ for all $n \geqslant 0$.

Scott [5] proved that $g(2,1) = 10$ and $g(2,n) = 3n+6$ for all $n \geqslant 2$. In a joint paper we [7] showed that

$$g(d,n) \geqslant \frac{n+1}{6(d-2)!} 2^{2^{d-a}}$$

for all $d \geqslant 4$, where $a = 0.5856\cdots$; we raised the conjecture that $g(d,n)$ is bounded for all d and n, $n \geqslant 1$.

Recently, Hensley [3] proved that $g(d,n)$ is bounded, i.e. there exist constants $B = B(d,n)$ for all $d \geqslant 3$ and $n \geqslant 1$ such that $g(d,n) \leqslant B(d,n)$ for all $d \geqslant 3$ and $n \geqslant 1$.

It follows that there exist constants $C = C(d,n)$ for all $d \geqslant 3$ and $n \geqslant 1$ such that if P is a lattice polytope in E^d and $G^0(P) = n$, then $\mathrm{Vol}(P) \leqslant C$, where $\mathrm{Vol}(P)$ is the d-volume of P. Thus, if P is a lattice polytope, $G^0(P) \geqslant 1$ and $\mathrm{Vol}(P) > C$, then $G^0(P) > n$. This assertion is closely related to Minkowski's famous theorem, which states that if P is a centrally symmetric (with respect to the origin O) convex body of large volume, and $O \in P$, then P contains many lattice points.

References

[1] B. Grünbaum, New views on some old questions of combinatorial geometry, Colloq. Intn' Teorie Combinatoire, Rome, 1973, Vol. 1 (1976) pp. 451–468.

[2] B. Grünbaum and G. Shephard, Analogues for tilings of Kotzig's Theorem on minimal weights of graphs, in: Theory and Practice of Combinatorics, Annals of Discrete Mathematics, 12 (North-Holland, Amsterdam, 1982) pp. 129–140.

[3] D. Hensley, Lattice vertex polytopes with interior lattice points (to appear). See also Hensley's abstract of his talk at the A.M.S. Meeting in San Francisco, Jan. 1981.

[4] A. Kotzig, Príspevok k téorii eulerovských polyédrov (in Slovak; Russian summary) Mat.-Fyz. Časopis Slovensk Akad. Vied 5 (1955) 101–113.

[5] P.R. Scott, On convex lattice polygons, Bull. Austral. Math. Soc. 15 (1976) 395–399.

[6] J. Zaks, Extending Kotzig's Theorem, Israel J. Math. (to appear).

[7] J. Zaks, J.M. Wills and M.A. Perles, On lattice polytopes having interior lattice points, Elemente der Math. 37 (1982) 44–46.

Received 4 July 1981

Annals of Discrete Mathematics 20 (1984) 311–316
North-Holland

INTERSECTING DIAMETERS IN CONVEX BODIES

Tudor ZAMFIRESCU
University of Dortmund, Dortmund, Fed. Rep. Germany

1. Introduction

We consider here spreads in the sense of Grünbaum (see [3, 4]). Let C be a closed Jordan curve in the plane, D_C the simply connected domain bounded by C, and $x \mapsto -x$ a continuous fixpoint free involution on C. A family $\mathcal{L}$ of Jordan arcs (called *curves*) is said to be a *spread* provided:

(i) for each $x \in C$ there is one curve $L(x) \in \mathcal{L}$ joining x with $-x$ ($L(x) = L(-x)$);

(ii) $\operatorname{inn} \Gamma \subset D_C$ for each $\Gamma \in \mathcal{L}$ ($\operatorname{inn} \Gamma$ means Γ minus its endpoints);

(iii) $\Gamma_1 \cap \Gamma_2$ is a single point for each pair of distinct curves $\Gamma_1, \Gamma_2 \in \mathcal{L}$;

(iv) $L: C \to \mathcal{L}$ is continuous, the topology of $\mathcal{L}$ being that induced by the Hausdorff metric in the space of compact plane sets.

Let cA be the family of all connected conponents of A, and let

$$M_\alpha = \{x \in D_C: \operatorname{card}\{\Gamma \in \mathcal{L}: x \in \Gamma\} \geq \alpha\},$$

$$P_\alpha = \{x \in D_C: \operatorname{card} c\{\Gamma \in \mathcal{L}: x \in \Gamma\} \geq \alpha\},$$

$$T_\alpha = \{x \in D_C: \operatorname{card}\{\Gamma \in \mathcal{L}: x \in \Gamma\} = \alpha\}.$$

Watson [8] proved that there exist spreads for which $M_{\aleph_0} = D_C$, whence $T_1 = \emptyset$. However, in [10] it is proved that, for spreads satisfying certain additional continuity conditions, the boundary of M_2 is a closed Jordan curve different from C. Then $\operatorname{int} T_1 \neq \emptyset$. Such a spread is, for example, that of area bisectors of a planar convex body. Also, it is known that for every *straight* spread (i.e. a spread the curves of which are line-segments) $T_1 \neq \emptyset$ [8, 12]. What, in general, does T_1 look like for straight spreads? Can M_2 be dense in D_C? Can even $M_{\aleph_0}$ be dense in D_C? The last two questions will be answered in this paper.

The word *most* will always be used in the sense of *those in a residual set*, or *all, except those in a set of first Baire category*.

The following result on general spreads will be used below.

Proposition 1. *If $P_{\aleph_0}$ is dense in D_C, then $M_{\aleph_0}$ is residual in D_C.*

Proof. Let $O \subset D_C$ be an open set and $p \in O \cap P_{\aleph_0}$. Since $p \in P_{2n}$ ($n \in \mathbb{N}$), by

Corollary 4 in [11] (see also Theorem 2 of [9]), $p \in \overline{\text{int}\, M_{2n+1}}$, which shows that there is an open set

$$O' \subset M_{2n+1} \cap O \subset M_n \cap O.$$

Thus, $\complement M_n$ is nowhere dense in D_C and

$$\complement M_{\aleph_0} = \complement \bigcap_n M_n = \bigcup_n \complement M_n$$

is of first Baire category. The proof is complete.

If C is a planar, smooth (differentiable), strictly convex curve, then the family of all *diameters* (i.e. chords admitting parallel tangent lines at their endpoints) of C constitutes a spread. In the rest of the paper we shall study (and, without contrary mention, always consider) spreads of diameters. The main attention will be focused on the strange properties that $M_{\aleph_0}$ may have.

Consider such a curve C and let $x, y \in C$, $y \neq \pm x$. Let $d(xy, z)$ denote the distance from $L(x) \cap L(y)$ to z and put

$$\gamma_i^-(x) = \liminf \frac{d(xy, x)}{d(xy, -x)},$$

where y converges to x from the left. $\gamma_s^-(x)$, $\gamma_i^+(x)$, and $\gamma_s^+(x)$ are defined analogously. The following technical lemma, proved elsewhere, will be useful.

Lemma 1 [14]. *Suppose C includes an arc B of a circle and let $x \in C$ be such that $-x \in \text{inn}\, B$. Then*

$$\gamma_i^{\pm}(x) \leq \frac{\rho_i^{\mp}(x)}{\rho^{+}(-x)}; \qquad \gamma_s^{\pm}(x) \geq \frac{\rho_s^{\mp}(x)}{\rho^{+}(-x)}.$$

(For a definition of the left radius of curvature $\rho^-(x)$ and left lower and upper radii of curvature $\rho_i^-(x)$ and $\rho_s^-(x)$ — and analogously of the right ones — see for instance [1].)

We denote by $\mathscr{C}$ the space of all smooth, strictly convex curves in the plane. By results proved independently by Klee [6] and Gruber [2], $\mathscr{C}$ is residual in the Baire space of all convex curves of the plane, the topology being again induced by the Hausdorff metric.

The rest of the paper is organized as follows. In the next section we present an example for which $M_{\aleph_0}$ is residual and null-swept (the definition follows in Section 2). Then we investigate, for most convex curves (and always the spread of diameters), the set $M_{\aleph_0}$ from Baire categories' point of view. Finally, we prove some connectivity properties of $M_{\aleph_0}$ for most convex curves.

2. $M_{\aleph_0}$ can be residual and null-swept

Let μ be the Lebesgue measure in the plane and λ the (one-dimensional) Hausdorff measure on convex curves.

For $C \in \mathscr{C}$, we say that a set $V \subset D_C$ is *null-swept* if there is $H \subset C$ with $\lambda(H) = 0$ such that $V \subset \bigcup_{x \in H} L(x)$.

Following Hammer and Sobczyk [5], a *turning point* is a limit point of $L(x) \cap L(y)$ for $y \to x$. Let U be the set of all turning points in D_C. It is easy to see that U may consist of finitely many points, but may also not be null-swept (see, for instance, the outwardly simple line family constructed at the end of [5]). We have $M_{\aleph_0} \subset U$ [5].

Theorem 1. *There exist curves in $\mathscr{C}$ such that $M_{\aleph_0}$ and U are simultaneously residual and null-swept.*

Proof. By Theorem 2 in [13] and the Theorem in [7], most convex curves $B \in \mathscr{C}$ have the property that:

(i) at each point x of a set $F \subset B$ with $\lambda(B - F) = 0$ of curvature exists and vanishes; and

(ii) at each point y of a dense set $E \subset B$ the lower and upper radii of curvature (from left) satisfy

$$\rho_i^-(y) = 0 \quad \text{and} \quad \rho_s^-(y) = \infty.$$

Take such a curve B, an arc $A \subset B$ at the endpoints of which the tangent lines are parallel, and a semicircle S such that $A \cup S$ is a convex curve C. First, we show that, in D_C, $M_{\aleph_0}$ is residual. Since, for every x in inn $A \cap E$, $\gamma_i(x) = 0$ and $\gamma_s^-(x) = \infty$, it follows from Lemma 1 that all points of $L(x)$ are limit points of $L(y) \cap L(x)$ for $y \to x$ from the left. Thus, inn $L(x) \subset P_{\aleph_0}$, whence $P_{\aleph_0}$ is dense in D_C. By Proposition 1, $M_{\aleph_0}$ is residual in D_C.

Secondly, we prove that

$$U \subset \bigcup_{x \in H} L(x),$$

where $H = A - F$. Let $p \in U$. Clearly, p is a limit point of $L(y) \cap L(x)$ for some point $x \in A$ and $y \to x$ from left or from right, say from left. Then $x \notin F$, since for each $z \in F$ the intersection $L(u) \cap L(z)$ converges to $-z$ when $u \to z$. Hence, $p \in \bigcup_{x \in H} L(x)$ and the theorem is proved.

Hammer and Sobczyk [5] have shown that for every straight spread $\mu(M_{\aleph_0}) = 0$. Also, simple examples show that $M_{\aleph_0}$, even M_4, may be empty, but the preceding example showed that $M_{\aleph_0}$ can also be residual. What does $M_{\aleph_0}$ look like for most convex curves? The answer is given in the next section.

3. In most cases $M_{\aleph_0}$ is residual

Let $\mathscr{C}_n$ be the set of all curves C in $\mathscr{C}$ having an arc A of length n^{-1} such that, for each $x \in \operatorname{inn} A$, there is an endpoint e of $L(x)$ and a component A^* of $A - \{x\}$ such that, for every $y \in A^*$, $d(xy, e) \geq n^{-1}$ $(n \in \mathbb{N})$.

Lemma 2. *$\mathscr{C}_n$ is closed in $\mathscr{C}$.*

Proof. Let $\{C_i\}_{i=1}^{\infty}$ be a sequence of curves in $\mathscr{C}_n$ converging to a curve $C \in \mathscr{C}$. By choosing, if necessary, a subsequence, we arrange that the corresponding sequence of arcs $\{A_i\}_{i=1}^{\infty}$ converges to some arc A of length n^{-1} on C. Let x be a point of inn A. We can choose $x_i \in A_i$ such that $x_i \to x$. The corresponding endpoints e_i coincide with x_i or with $-x_i$ for infinitely many indices i, say with x_i. Then we choose $e = x$. In the same way we may suppose that the sequence of the corresponding components A_i^* converges to some component A^* of $A - \{x\}$. Suppose now there is a point $y \in A^*$ with $d(xy, x) < n^{-1}$. We choose $y_i \in A_i^*$ such that $y_i \to y$. For i large enough, the Hausdorff distance from $L(x_i)$ to $L(x)$ and from $L(y_i)$ to $L(y)$ are so small, that $d(x_i y_i, x_i) < n^{-1}$, which is a contradiction.

Lemma 3. *$\mathscr{C}_n$ is nowhere dense in $\mathscr{C}$.*

Proof. It is well known that an arbitrary curve $C \in \mathscr{C}$ can be approximated as well as we like by a polygon. We can replace the sides of the polygon by arcs of very large circles and the vertices of the polygon by arcs of very small circles, the resulting curve, $C' \in \mathscr{C}$, still remaining near enough to C. On the other hand, C can be approximated well enough by a curve $C'' \in \mathscr{C}$ with $\rho_i^{\pm}(x) = 0$ and $\rho_s^{\pm}(x) = \infty$ on a dense set E of points $x \in C''$ [7]. It is an easy matter now to construct a smooth convex curve C^* such that on one side of a certain diameter of C^*, C^* coincides with half a curve C' and on the other side with half a curve C''. Since

$$\rho_i^{\pm}(x) = 0, \qquad \rho_s^{\pm}(x) = \infty,$$

for every point $x \in C^* \cap E$, by Lemma 1 all points of $L(x)$ are limit points of $L(y) \cap L(x)$ for $y \to x$ from left as well as from right. It follows that $C^* \notin \mathscr{C}_n$. Hence, $\mathscr{C} - \mathscr{C}_n$ is dense in $\mathscr{C}$ and, by Lemma 2, $\mathscr{C}_n$ is nowhere dense in $\mathscr{C}$.

Proposition 2. *For most curves $C \in \mathscr{C}$ the following holds: For every arc A on C and every number $\varepsilon > 0$, there exists a point $x \in \operatorname{inn} A$ such that, for any component A^* of $A - \{x\}$, there are $y, y' \in A^*$ verifying $d(xy, x) < \varepsilon$ and $d(xy', -x) < \varepsilon$.*

Proof. Let $\mathscr{C}^*$ be the set of those curves of $\mathscr{C}$ enjoying the property of the statement. It is easily checked that

$$\mathscr{C} - \mathscr{C}^* = \bigcup_{n=1} \mathscr{C}_n ;$$

hence $\mathscr{C} - \mathscr{C}^*$ is, by Lemma 3, of first Baire category, which proves Proposition 2.

Theorem 2. *For most convex curves C, most points of D_C belong to infinitely many diameters.*

Proof. Let C be a curve enjoying the property of Proposition 2. Put

$$E = \{x \in C\colon \operatorname{inn} L(x) \subset P_{\aleph_0}\}.$$

We show that E is dense on C. Let A_0 be an arc on C; we find $x_1 \in \operatorname{inn} A_0$ and $y_1, y_1' \in \operatorname{inn} A_0$ on the same side of x_1 such that $d(x_1y_1, x_1) < 1$ and $d(x, y_1', -x_1) < 1$. We construct the sequences $\{x_n\}_{n=1}^{\infty}$, $\{y_n\}_{n=1}^{\infty}$, $\{y_n'\}_{n=1}^{\infty}$ inductively as follows. By continuity, there is an arc $A_n \subset A_{n-1}$ containing x_n in its interior, such that for all $x \in A_n$, $d(xy_n, x) < n^{-1}$ and $d(xy_n', -x) < n^{-1}$. Again, we can find a point $x_{n+1} \in A_n$ separating x_n from y_n and y_n', and two points $y_{n+1}, y_{n+1}' \in A_n$ separating x_{n+1} from y_n and y_n' such that $d(x_{n+1}y_{n+1}, x_{n+1}) < (n+1)^{-1}$ and $d(x_{n+1}y_{n+1}', -x_{n+1}) < (n+1)^{-1}$. It is equally guaranteed that $d(x_{n+1}y_m, x_{n+1}) < m^{-1}$ and $d(x_{n+1}y_m', -x_{n+1}) < m^{-1}$ for all $m \leqslant n$.

The sequence $\{x_n\}_{n=1}^{\infty}$ converges to some point $x \in A$ verifying:

$$d(xy_n, x) < n^{-1} \quad \text{and} \quad d(xy_n', -x) < n^{-1},$$

for every $n \in \mathbb{N}$. Also, y_n' lies between y_{n-1} and y_{n+1}, and y_n lies between y_{n-1}' and y_{n+1}'. Thus, it is clear that $\operatorname{inn} L(x) \subset P_{\aleph_0}$ and that E is dense on C. Then $P_{\aleph_0}$ is dense in D_C and the theorem follows from Proposition 1.

4. In most cases $M_{\aleph_0}$ is connected

Simple examples show that, for every cardinal number $\alpha > 3$, M_α may be disconnected, and this seems to be the rule. That — from the point of view of Baire categories — this is not the case, will prove the next result.

Theorem 3. *For most convex curves, the components of T_1 are line-segments (without one or both endpoints), $M_2 - M_{\aleph_0}$ is totally disconnected, and M_α is connected for every $\alpha \leqslant \aleph_0$.*

Proof. Let $p \in T_1$ and $x \in C$ with $p \in L(x)$. Clearly, either inn xp or inn $-xp$ lies in T_1. Say inn $xp \subset T_1$. There is a maximal line-segment S ending in x such that inn $S \subset T_1$. We show that inn S plus, possibly, the endpoint of S different from x is, for most $C \in \mathscr{C}$, the component K of T_1 containing p. Suppose $q \in K - S$. Since, for most convex curves, the set E considered in the proof of Theorem 2 is dense on C, there exists $y \in E$ such that $L(y)$ separates p from q. Since inn $L(y) \subset M_{\aleph_0}$, $q \notin K$, a contradiction.

To see that $M_2 - M_{\aleph_0}$ is totally disconnected, let $s \in M_2 - M_{\aleph_0}$ and let B be a disk around s. We consider x, $y \in C$ such that $L(x) \cap L(y) = \{s\}$. We use again the above set E and see that there are four points, x', x'', y', and y'' in E, such that s lies within a quadrangle $Q \subset B$, with sides on $L(x')$, $L(x'')$, $L(y')$, and $L(y'')$. Since the sides of Q lie in $M_{\aleph_0}$, the component of $M_2 - M_{\aleph_0}$ containing s is a subset of B.

Finally, we show that M_α is connected for every $\alpha \leq \aleph_0$. Let

$$Z = \bigcup_{x \in E} \operatorname{inn} L(x).$$

Clearly, Z is arcwise connected (even an $L_2(\mathscr{L})$-set in the terminology of [9, Section 3]). Since Z is dense in D_C and $Z \subset M_{\aleph_0} \subset M_\alpha$, the assertion follows.

References

[1] H. Busemann, Convex Surfaces (Interscience Publishers, New York, 1958).

[2] P. Gruber, Die meisten konvexen Körper sind glatt, aber nicht zu glatt, Math. Ann. 229 (1977) 259–266.

[3] B. Grünbaum, Continuous families of curves, Can. J. Math. 18 (1966) 529–537.

[4] B. Grünbaum, Arrangements and Spreads, Lectures delivered at a regional conference on Combinatorial Geometry, University of Oklahoma (1971).

[5] P.C. Hammer and A. Sobczyk, Planar line families. II, Proc. Amer. Math. Soc. 4 (1953) 341–349.

[6] V. Klee, Some new results on smoothness and rotundity in normed linear spaces, Math. Ann. 139 (1959) 51–63.

[7] R. Schneider, On the curvatures of convex bodies, Math. Ann. 240 (1979) 177–181.

[8] K.S. Watson, Sylvester's problem for spreads of curves, Can. J. Math. 32 (1980) 219–239.

[9] T. Zamfirescu, On planar continuous families of curves, Can. J. Math. 21 (1969) 513–530.

[10] T. Zamfirescu, On continuous families of curves. VI, Geometriae Dedicata 10 (1981) 205–217.

[11] T. Zamfirescu, Sulle famiglie continue di curve. VII, Rend. Sem. Mat. Univ. Politecn. Torino 36 (1977/78) 183–190.

[12] T. Zamfirescu, Spreads, Abh. Math. Sem. Univ. Hamburg 50 (1980) 238–253.

[13] T. Zamfirescu, The curvature of most convex surfaces vanishes almost everywhere, Math. Z. 174 (1980) 135–139.

[14] T. Zamfirescu, Curvature properties of typical convex surfaces, manuscript.

Received March 1981

Annals of Discrete Mathematics 20 (1984) 317
North-Holland

QUADRATIC FORMS AND GRAPHS

L.J. BEREZINA
University of Haifa, Haifa, Israel

Each simple graph is determined one-to-one by its adjacent matrix $G = \{g_{ij}\}$. We can associate a quadratic form $G(x,x) = X^T \cdot G \cdot X$ with the symmetric matrix G; thus the inertial indices of $G(x,x)$ are arithmetical invariants of the graph. We shall denote the inertial indices p, q, where p is the greater of the two ($p \geq q$). We can choose r linearly independent rows (columns) of G. We will call the corresponding vertices linearly independent vertices. If the row that corresponds to the vertex v_i is a linear combination of the rows which correspond to the vertices, v_k, we will say that the vertex v_i is a linear combination of the vertices v_k. An independent set is a set of vertices which are not connected between themselves.

By using the well-known theory of quadratic forms we can prove:

(1) r linearly independent vertices contain an independent set with at most q vertices.

(2) For the chromatical number $\nu(G)$ we have $\nu(G) \geq r/q$.

(3) A graph G with n-vertices, $rk(G) = r$ and $q = 1$ contains a complete subgraph K_r. The other $n - r$ columns of G repeat some columns that correspond to the vertices of K_r. For such graphs $\nu(G) = r$.

(4) If a graph with n vertices contains an independent set with $n - p$ vertices, then the other p vertices are linearly independent and do not share any linear combinations.

Editor's remark: For references, see the author's papers in J. Geometry 14 (1980) 154–158, and in the Proceedings of the Haifa 1979 Conference on Geometry and Differential Geometry, Springer-Verlag Lecture Notes #792, edited by R. Artzy and I. Vaisman (1980), pp. 20–23.

Annals of Discrete Mathematics 20 (1984) 319–320
North-Holland

APPROXIMATION OF CONVEX BODIES BY POLYTOPES WITH UNIFORMLY BOUNDED VALENCES

Jürgen BOKOWSKI
Ruhr University, Bochum, Fed. Rep. Germany

Peter MANI
University of Bern, Switzerland

Given any convex body K in Euclidean space E^d and any number $\varepsilon > 0$, does there always exist a polytope $P = P(K, \varepsilon)$ in E^d such that the number of vertices of a facet of P and the number of facets meeting in a common vertex are bounded by a constant depending on the dimension d only and such that the Hausdorff-distance $\rho(K, P)$ of K and P is less than ε?

This question was posed by G. Ewald at the Durham Symposium in 1975. For $d = 3$ a more detailed complete answer was given in [1]. For $d = 4$ this problem was solved in [2]. A solution for general d was found by the authors in 1980, independently.

Let us denote by $\mathcal{P}^d$ the class of (convex) polytopes, by $\mathcal{K}^d$ the class of convex bodies in Euclidean d-space E^d, by $B \in \mathcal{K}^d$ the unit ball with center in the origin, and by $+$ the Minkowski sum. Our result and a sketch of a proof can be formulated as follows.

Theorem. *There is a constant c_d such that given any $K \in \mathcal{K}^d$ and any $\varepsilon > 0$ there always exists a polytope $P \in \mathcal{K}^d$ with Hausdorff-distance $\rho(K, P) < \varepsilon$ and*

$$\operatorname{card}\{\textit{facets of } P \textit{ containing a fixed vertex of } P\} \leq c_d,$$

$$\operatorname{card}\{\textit{vertices of } P \textit{ contained in a fixed facet of } P\} \leq c_d.$$

Essential parts in the proof. Let $K \in \mathcal{K}^d$ and $\varepsilon > 0$ be given.

(1) Choose a simplicial $P \in \mathcal{P}^d$ with $\rho(K, P) \leq \varepsilon/3$.

(2) Solve the problem for K being the unit ball B.

(3) Take such an approximation $Q \in \mathcal{P}^d$ of B "good enough", this depends on P.

(4) $P + (\varepsilon/3)Q$ will have the desired properties.

ad 1. This is trivial.

ad 2. This can be done by considering a projection of a lattice on the boundary of B, or independently by an induction argument carrying over a method used in the three-dimensional case.

ad 3. Denote by N (vertex of Q) the cone of outer normals of supporting hyperplanes meeting the correspondent vertex of Q. Then "good enough" means: every such cone contains at most outer normals of faces of P belonging to one common facet of P.

ad 4. Denote by $F(R, v)$ the face of the polytope R belonging to the outer normal v. The equation

$$F\left(P+\frac{\varepsilon}{3} Q, v\right)=F(P, v)+F\left(\frac{\varepsilon}{3} Q, v\right)$$

and properties of P and Q are essential to establish the result.

References

[1] J. Bokowski, Konvexe Körper approximierende Polytopklassen, Elem. Math. 32 (1977) 88–90.
[2] J. Bokowski and Chr. Schulz, Dichte Klassen konvexer Polytope, Math. Z. 160 (1978) 173–182.
[3] D. Larman and C.A. Rogers, Durham Symposium on the relations between infinite- and finite-dimensional convexity, Bull. London Math. Soc. 8 (1976) 1–33.

Received 22 September 1981

Annals of Discrete Mathematics 20 (1984) 321–322
North-Holland

SOME RESULTS ON CONVEXITY IN GRAPHS

Eric DEGREEF
Vrije Universiteit Brussel, Pleinlaan 2, B1050 Brussels, Belgium

A convexity structure $(X, \mathscr{C})$ is a pair $(X, \mathscr{C})$, X being a set and $\mathscr{C} \subset 2^X$ such that (a) $\emptyset, X \in \mathscr{C}$ and (b) $\bigcap \mathscr{F} \in \mathscr{C}$ for each $\mathscr{F} \subset \mathscr{C}$. For each subset S of X we denote by $\mathscr{C}(S)$ the convex hull or $\mathscr{C}$-hull of S, i.e. $\mathscr{C}(S) = \bigcap\{C \in \mathscr{C} \mid S \subset C\}$. A convexity structure $(X, \mathscr{C})$ is called *decomposable in relation to a subset* Y of X, if for each subset S of X, $\mathscr{C}(S) = \mathscr{C}(S \cap Y) \cup \mathscr{C}(S \setminus Y)$. $(X, \mathscr{C})$ is called *totally decomposable* if $(X, \mathscr{C})$ is decomposable in relation to each subset Y of X.

Let $G = (V, E)$ be a finite graph, V the set of the vertices, E of the edges. We define a convexity structure $(V, \mathscr{C})$ in the following way: $\mathscr{C} = \{C \subset V \mid \text{the induced subgraph } \langle C \rangle \text{ is complete}\}$. We then have the following results:

(1) *in the case* $\# V \leq 2$, $(V, \mathscr{C})$ *is totally decomposable.*

(2) *in the case* $\# V \geq 3$, *the following assertions are equivalent*:
  (a) $(V, \mathscr{C})$ *is decomposable in relation to a subset* Y *of* V,
  (b) *for each vertex* y_1 *of* Y *and each vertex* y_2 *of* $V \setminus Y$, y_1 *and* y_2 *are adjacent.*

(3) *in the case* $\# V \geq 3$, $(V, \mathscr{C})$ *is totally decomposable iff the graph* $G = (V, E)$ *is complete.*

Let $(X_i, \mathscr{C}_i)$ be a convexity structure; $i = 1, \dots, n$. We define a structure $\mathscr{C}$ on $\bigcup_{i=1}^n X_i$ in the following way: $\mathscr{C} = \{C \subset \bigcup_{i=1}^n X_i \mid C \cap X_i \in \mathscr{C}_i,\ i = 1, \dots, n\}$. Clearly $(\bigcup_{i=1}^n X_i, \mathscr{C})$ is a convexity structure. We say that it is obtained *by glueing the structures* $(X_i, \mathscr{C}_i)$, $i = 1, \dots, n$. If for each subset S of $\bigcup_{i=1}^n X_i$, the $\mathscr{C}$-hull is given by $\bigcup_{i=1}^n \mathscr{C}_i(S \cap X_i)$, we say that the glueing is *normal.*

Let now $(X_i, \mathscr{C}_i)$, $i = 1, \dots, n$, be a convexity structure, where X_i is a closed segment and $\mathscr{C}_i = \text{conv}|_{X_i}$, conv denoting the classical convexity structure on a real vectorspace. A normal glueing $(\bigcup_{i=1}^n X_i, \mathscr{C})$ where two segments have either an empty intersection, or one endpoint in common, is called *a frame* $(X, \mathscr{C})$.

The endpoints of the segments are precisely the extreme points of the frame. The number of extreme points of a frame will be called *the order of the frame. The graph associated with a frame* $(X, \mathscr{C})$ is a graph where the vertices are the extreme points of $(X, \mathscr{C})$ and the edges are the pairs $\{p, q\}$ of extreme points such that $\mathscr{C}(p, q)$ is a segment. We clearly have the following property:

(4) *each graph without isolated points, is, up to an isomorphism, the graph of a frame.*

We say that *a frame* $(X, \mathscr{C})$ *can be embedded in* $(\mathbb{R}^n, conv)$ if there exists an injection $\phi: X \to \mathbb{R}^n$, such that for each subset S of X,

$$\phi[\mathscr{C}(S)] = \phi(X) \cap \operatorname{conv}[\phi(S)].$$

One can easily verify:

(5) *each frame of order d can be embedded in* $(\mathbb{R}^{d-1}, conv)$.

But is it possible to do better and to characterize which frames are embeddable in which $(\mathbb{R}^n, \operatorname{conv})$? The answer is given by the following results:

(6) *a frame is embeddable in*:
  (a) $(\mathbb{R}, conv)$ *iff the associated graph is isomorphic with* •—•;
  (b) $(\mathbb{R}^2, conv)$ *iff the associated graph is isomorphic with a cycle or with a part of it*;
  (c) $(\mathbb{R}^3, conv)$ *iff the associated graph is isomorphic with a planar one.*

(7) *each frame is embeddable in* $(\mathbb{R}^4, conv)$.

Reference

[1] E. Degreef, Glued convexity spaces, frames and graphs, Compositio Mathematica 47(2) (1982) 217–222.

Annals of Discrete Mathematics 20 (1984) 323
North-Holland

CONVEXITY IN GRAPHS: ACHIEVEMENT AND AVOIDANCE GAMES

Frank HARARY
University of Michigan, Ann Arbor, Michigan, U.S.A.

The *convex hull* of a set S of points of a graph G is the smallest set T containing S such that all the points in a geodesic joining two points of T lie in T. The convex hull T can also be formed by taking all geodesics joining two points of S, and iterating that operation. The number of times this is done to S to get T is $\text{gin}(S)$, the *geodetic iteration number* of S. Then $\text{gin}(G)$ is defined as the maximum of $\text{gin}(S)$ over all sets S of points of G. The smallest number of points in a graph G such that $\text{gin}(G) = n$ was determined by Harary and Nieminen [1].

In an *achievement game*, the first of two players to attain the stated goal wins; in an *avoidance game* he loses. We propose three types of achievement and avoidance games involving the convex hull, the geodetic hull, and geodesics from a point to a set of points; this leads to several unsolved problems.

Wc illustrate the latter games with the cycle C_n. The first player A selects one point u_1 of C_n and colors it green. Then B picks $u_2 \neq u_1$ and colors green u_2 and the points on all $u_1 - u_2$ geodesics. Call the present set of green points S_2. After A chooses $u_3 \notin S_2$, form green S_3 as the union of S_2, u_3, and all points on $u_3 - S_2$ geodesics, and so forth. The games end when all points are green. In the achievement game the last player to move wins; in the avoidance game he loses. Aviezri Fraenkel and I proved that the achievement game is won by A if and only if $n = 2^k - 1$. My student Frederick Teague showed that A wins the avoidance game if and only if $n = 2^k$.

Reference

[1] F. Harary and J. Nieminen, Convexity in graphs, J. Differential Geometry 16 (1981) 185–190.

Annals of Discrete Mathematics 20 (1984) 325–326
North-Holland

ON SCHEDULING THE CONSTRUCTION OF A TREE

Amos ISRAELI
Department of Applied Mathematics, The Weizmann Institute of Science, Rehovot, Israel

Yehoshua PERL
Department of Mathematics and Computer Science, Bar Ilan University, Ramat Gan, Israel

A *schedule of a tree* $T=(V,E)$ of n edges is a one-to-one function $\alpha:\{1,2,\ldots,n\}\rightarrow E$. The *kth stage cost* N_k, $0\leq k\leq n$, of a schedule α of T is the number of internal vertices of the edge subgraph induced by $\{\alpha_1,\alpha_2,\ldots,\alpha_k\}$.

The cost of a schedule α is $c(\alpha)=\sum_{j=1}^{n} N_j$. A schedule of minimum cost is an *optimal schedule.*

We consider the problem of finding an optimal schedule of a tree. This problem has an application in the construction of tree-like communication networks where the internal vertices require expensive transmission equipment (e.g. a directory). The cost of a schedule is linear with the total hiring cost of the transmission equipment during the construction process, assuming each edge is constructed in one unit of time.

In [4] this problem is solved for the case where the edge subgraph induced at each stage of the construction is connected. The general case is investigated in [2], but no complete solution is found. We summarize here the results concerning the general structure of an optimal schedule.

We introduce first several definitions and notations.

Given a schedule α, a vertex u is called a *type*-0, *type*-1, or *type*-2 *vertex* at stage k of the schedule, if it is incident with 0, 1 or at least 2 of the edges $\alpha_1,\ldots,\alpha_k$, respectively.

The *kth stage difference*, $1\leq k\leq n$, is $d_k=N_k-N_{k-1}$.

An edge e is a 0-edge, 1-edge or 2-edge at stage k of a schedule α if its construction at stage $k+1$ adds 0, 1 or 2 to N_k, respectively.

Let α_{i+1} be the first edge adjacent to a previous edge in a schedule α. Then the set of edges $\{\alpha_1,\alpha_2,\ldots,\alpha_i\}$ is a matching of T, called the *initial matching of* α (abbreviated i.m. of α).

A star of vertex u, of order m, at stage k of a schedule α is a subtree containing a vertex u which is not of type 0, called *the center of the star* and m unconstructed edges (u,v_1), $(u,v_2),\ldots,(u,v_m)$ such that v_i, $1\leq i\leq m$, is either of type 0 or of type 2 at stage k.

Note that during the construction of the tree the edges of a star are deleted from the star while they are constructed or while their other end vertex becomes a type-1 vertex. On the other hand, unconstructed edges incident with a center of a star are added to the star while their other end vertex becomes a type-2 vertex.

The *tails subgraph TL* of a tree T, is the subgraph containing all the tails of T.

A schedule is *greedy* if no 2-edge is constructed while a 0-edge or a 1-edge can be constructed.

Let α_k be the first 2-edge of an optimal schedule α with the i.m. $\alpha_1, \ldots, \alpha_i$. We call $\alpha_{i+1}, \ldots, \alpha_{k-1}$ the *star phase* of α, and $\alpha_k, \ldots, \alpha_n$ the *residual subgraph of* α.

The following results are presented in [2].

The i.m. over the tails subgraph is a mximum terminal shifted matching. However, the structure of the i.m. over the other parts of the tree is not entirely clear.

An optimal schedule is starry, i.e. the stars are constructed one at a time, such that all the edges of a star are constructed one by one with no interruption. The first edge of a star is a 1-edge and all the other edges are 0-edges. The order of constructing the different stars in an optimal schedule is still to be determined. Furthermore, an optimal schedule is greedy. Hence, all the stars are constructed consecutively. Then the components of the residual subgraph are constructed one by one in non-increasing order of their size. The construction of each component starts with an arbitrary 2-edge and proceeds in a connected manner with 1-edges.

Thus the construction of an optimal schedule is composed of three phases:

(1) *The initial matching phase*; the stage differences are all equal to zero.

(2) *The stars phase*; the stage differences are either zero or one.

(3) *The residual subgraph phase*; the stage differences are either one or two.

Our conclusion is that releasing the connectivity requirement during the construction yields a difficult problem which is probably NP-complete [1]. We were able to solve only some special cases presented in [3].

References

[1] M.R. Garey and D.S. Johnson, Computers and Intractability: A Guide to the Theory of NP-Completeness (W.H. Freeman and Co., San Francisco, 1979).

[2] A. Israeli and Y. Perl, On scheduling the construction of a tree, Technical Report CS81–02, Weizmann Institute of Science (1981).

[3] A. Israeli and Y. Perl, Optimal scheduling of an external star tree, Technical Report CS81–19, Weizmann Institute of Science (1981).

[4] Y. Perl and Y. Yesha, Mean flow scheduling and optimal construction of a treelike communication network, Networks 11 (1981) 87–92.

Annals of Discrete Mathematics 20 (1984) 327–328
North-Holland

EDGE MAPS AND ISOMORPHISMS OF GRAPHS

A.K. KELMANS
Profsouznaya Str. 130, 117321 Moscow, U.S.S.R.

Let G be an undirected graph and EG denote the edge-set of G. The union of $n \geqslant 1$ chains of G is called an *n-skein* [1] if they have the common end-points (the *terminals* of the skein) and no pair of them has an inner point in common. Let $S_n(G)$ denote the set of n-skeins of G. A 1–1 map $e : EG \to EF$ is called an *n-skein isomorphism of G onto F* if $A \in S_n(G) \Leftrightarrow e(A) \in S_n(F)$. In [1] the following result is obtained.

Theorem 1. *Suppose that G and F are $(n+1)$-vertex connected, $n \geqslant 2$. Then G and F are isomorphic if and only if they are n-skein isomorphic, and any n-skein isomorphism of G onto F is induced by an isomorphism of G onto F.*

The same assertion for n-skeins in n-connected graphs is wrong, as Whitney [2] showed for $n = 2$. Also, it is easy to see [1] that every permutation of the edge-set of K_4 is a 3-skein automorphism of K_4, but clearly not every such map is induced by an automorphism of K_4. In [1] we read: "... we were not able to find more complicated examples in the case $n = 3$, nor could we find for $n \geqslant 4$ an example of an n-skein isomorphism of an n-vertex connected graph which was not induced by an isomorphism". Here we see that no other example of this kind exists for 3-connected graphs.

Theorem 2. *Let G and F be graphs where G is 3-connected and has at least 5 vertices and F has no isolated vertices; let $e: EG \to EF$ be a 1–1 map such that $A \in S_3(G) \Leftrightarrow e(A) \in S_3(F)$. Then there exists an isomorphism of G onto F inducing e, and this isomorphism is unique.*

Let $S_n^r(G)$ denote the set of n-skeins of G with the distance $\leqslant r$ between the terminals. Then the following strengthened version of Theorem 1 is also true.

Theorem 3. *Let G and F be graphs where G is $(n+1)$-vertex connected and F has no isolated vertices; let $e: EG \to EF$ be a 1–1 map such that $A \in S_n^2(G) \Leftrightarrow e(A) \in S_n^2(F)$. Then an isomorphism of G onto F exists and is the only one which induces e.*

Note added in proof. Three different proofs of Theorem 2 were given in [3–5]. A.K. Kelmans [6] found an example of a 4-skein automorphism of a 4-connected graph which was not induced by an automorphism of the graph and proved that this example is unique among the 4-skein isomorphisms of 4-connected graphs. The results of the note have been reported to the Moscow Discrete Mathematics Seminar (Institute of Control Sciences, Moscow, October 1979), to the All-Union Graph Theory Seminar (Odessa, U.S.S.R., September 1980) and to the All-Union Conference on Statistic and Discrete Analyses of Non-Number Information (Alma-Ata, U.S.S.R., September 1981) [7].

References

[1] R. Halin and H.A. Jung, J. London Math. Soc. 42 (1967) 254–256.
[2] H. Whitney, Amer. J. Math. 55 (1933) 245–254.
[3] R.L. Heminger, H.A. Jung and A.K. Kelmans, On 3-skein isomorphisms of graphs, Combinatorica 4 (1982) 373–376.
[4] A.K. Kelmans, On 3-skeins in a 3-connected graph, submitted to Combinatorica in July 1981.
[5] A.K. Kelmans, A short proof and a strengthening of the Whitney 2-isomorphism theorem on graphs, in print.
[6] A.K. Kelmans, On homeomorphic embeddings of graphs with given properties, in print.
[7] A.K. Kelmans, 3-skeins in a 3-connected graph, in: First All-Union Conference on Statistic and Discrete Analyses of Non-Number Information, Moscow–Alma-Ata, 1981 (in Russian).

Annals of Discrete Mathematics 20 (1984) 329
North-Holland

A MULTIFLOW PROBLEM WITHOUT THE MAX-FLOW MIN-CUT PROPERTY

M.V. LOMONOSOV

Chertanovskaya 34–1–286, 143525 Moscow, U.S.S.R.

Let V be a finite set of vertices, $\binom{V}{2}$ denote the set of 2-subsets $[x, y]$ of V, and $\mathscr{E}$ denote the Euclidean space of functions $\binom{V}{2} \to \mathbb{R}$. Let $\mathscr{U}$ denote the set of metrics on V (the "metric cone" in $\mathscr{E}$). For $X, Y \subset V$, $X \cap Y = \emptyset$, put $[X, Y] = \{[x, y] \in \binom{V}{2};\ x \in X,\ y \in Y\}$; for $f \in \mathscr{E}$ and $E \subseteq \binom{V}{2}$ put $f(E) = \Sigma\{f(e); e \in E\}$. A pair $(c, d) \in \mathscr{E}_+ \times \mathscr{E}_+$ is *feasible* if there exists a multiflow $F = \{f_e\, ; e \in \binom{V}{2}\}$ (i.e. a set of two-terminal flows $f_e \colon V \times V \to \mathbb{R}_+$) such that: (a) $e =$ [source, sink] of f_e and the magnitude of f_e equals $d(e)$ for each e, and (b) $\Sigma\{f_e(x, y) + f_e(y, x); e \in E\} \leqslant c[x, y]$ for each $[x, y]$. The following four facts are known:

(1) (c, d) is feasible iff $c - d$ belongs to the dual of $\mathscr{U}$ (Lomonosov).

(2) Given $c \in \mathscr{E}_+$ and $E \subseteq \binom{V}{2}$ put $M(E, c) = \max\{d(E); (c, d) \text{ is feasible}\}$; there exists $m \in \mathscr{U}$ satisfying $m(e) = 1$, $e \in E$, such that $M(E, c) = mc$.

(3) The graph $S = (T, E)$, where T is the set of vertices covered by E, is called the *scheme* of the problem (2); the following features of S are equivalent: (i) whatever are V $(\supseteq T)$ and $c \in \mathscr{E}_+$, m may be chosen as a positive linear combination of "cuts" (i.e. indicators of subsets $[X, \bar{X}] \subset \binom{V}{2}$), and (ii) the intersections graph of the maximal independent sets of S has no triangles (Karzanov, Lomonosov, Pevzner). In fact for this class of problems we have $M(E, c) = \Sigma\{\alpha(X)\hat{c}(X); X \subset T\}$, where $\alpha \colon 2^T \to \mathbb{R}_+$ and $\mathscr{E}(X) = \min\{c[Y, \bar{Y}];\ Y \subset V,\ Y \cap T = X\}$.

(4) $M(\binom{T}{2}, c) = \frac{1}{2}\Sigma\{\hat{c}(t); t \in T\}$ (Cherkassaki) and $\min\{d(e);\ (c, d) \text{ is feasible},\ d(\binom{V}{2}) = \max\} = r(e, c) = \frac{1}{2}(\hat{c}(x) + \hat{c}(y) - \hat{c}(x, y))$, where $[x, y] = e$ (Lomonosov).

Here we are concerned with the simplest scheme violating 3(ii): $E = \binom{A}{2} \cup \binom{B}{2}$ for some A, B, $A \cup B = T$. The problem (2) with such S generalizes both the "free problem" (4) and the two-flow problem (T. Hu).

Theorem. $M(E, c) = \min(M\binom{T'}{2} - \frac{1}{2}\Sigma\{r(e, c');\ e \in [A' - B',\ B' - A']\})$ *over all contractions* $g \colon V \to V'$ *satisfying* $g(x) \neq g(y)$ *for* $x, y \in E$; *here* $T' = g(T)$, $A' = g(A)$, $B' = g(B)$ *and* $c'[x', y'] = c[g^{-1}(x'), g^{-1}(y')]$. *Moreover, if* c *is integer-valued, then the problem has a solution* $F = \{f_e\}$ *whose values are multiples of* $\frac{1}{4}$.

Annals of Discrete Mathematics 20 (1984) 331
North-Holland

POLYHEDRAL MANIFOLDS WITH FEW VERTICES

Ch. SCHULZ*

Fernuniversität Hagen, D-5800 Hagen, Fed. Rep. Germany

A (polyhedral) realization of the closed orientable surface $\mathcal{M}_g$ of genus g is a cell complex $M_g \subset E^d$, whose underlying point set is homeomorphic to $\mathcal{M}_g$. So, loosely speaking, M_g is a model of $\mathcal{M}_g$, made up of plane convex polygons.

The proof of the Heawood map color theorem by Ringel and Youngs gives rise to a series of $M_g \subset E^5$ with the number of vertices only growing by the order of $\sqrt{g}$ (which is, for simple combinatorial reasons, the best possible). The minimal number of vertices of the so far known realizations in E^3 depends linearly on the genus, and it looks somehow plausible that one cannot do better, since this would imply the existence of polyhedral manifolds in ordinary 3-space with the number of handles exceeding the number of vertices. However we have:

Theorem 1. *There are $M_g \subset E^3$ for all g, such that the number of vertices of M_g grows by the order of $g/\log g$.*

The proof is based on the construction of $M_g \subset E^3$ (made up of quadrilaterals), which are isomorphic to the well-known 2-manifolds in the boundary complex of the d-cube containing its whole 1-skeleton, originally discovered by Coxeter. So, these M_g are equivelar (see the abstract of J.M. Wills):

Corollary. *There exist manifolds $\{4, q; g\}$ in E^3 for all q.*

By variations of our construction we can show that for each fixed q there are $\{4, q; g\}$ for infinitely many g. Showing the existence of the duals of the M_g in E^3, McMullen proved:

Theorem 2. *There exist manifolds $\{p, 4; g\}$ in E^3 for all p.*

* Joint paper with P. McMullen and J.M. Wills.

Annals of Discrete Mathematics 20 (1984) 333–334
North-Holland

REGULAR POLYHEDRAL MANIFOLDS

J.M. WILLS*
University of Siegen, Siegen, Fed. Rep. Germany

A polyhedral 2-manifold is a geometric model of an abstract 2-manifold, made up of convex polygons. We consider particularly closed oriented polyhedral 2-manifolds in the ordinary Euclidean E^3.

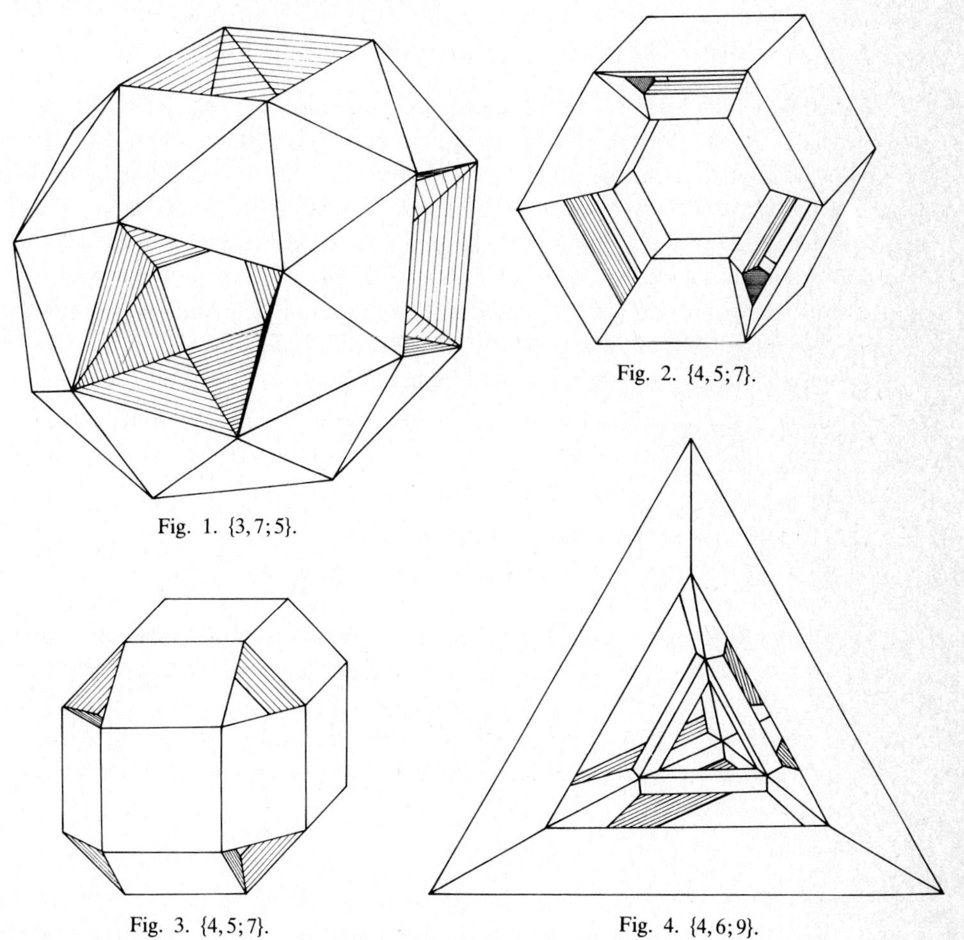

Fig. 1. {3, 7; 5}.

Fig. 2. {4, 5; 7}.

Fig. 3. {4, 5; 7}.

Fig. 4. {4, 6; 9}.

* Joint paper with P. McMullen and Ch. Schulz.

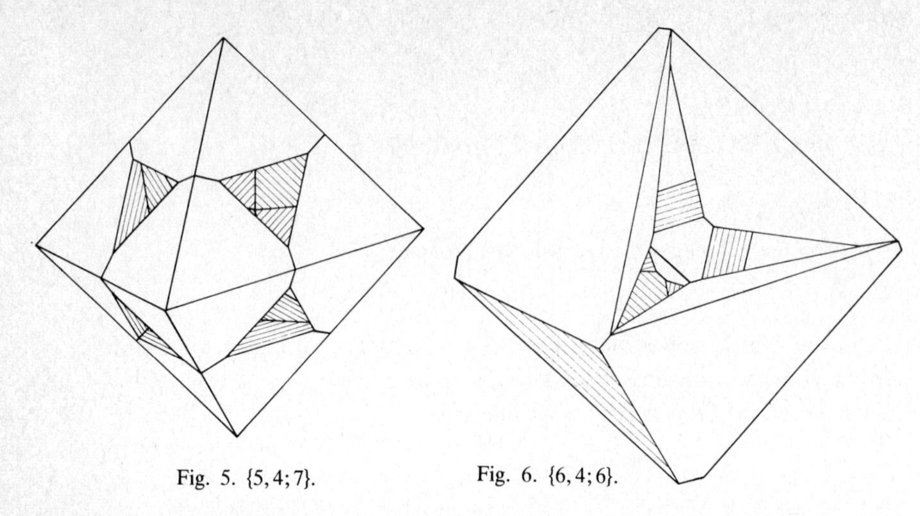

Fig. 5. $\{5,4;7\}$.

Fig. 6. $\{6,4;6\}$.

We call a polyhedral 2-manifold regular or (more precisely) equivelar (i.e. with equal flags) if each of its facets has the same number p of edges, and if each of its vertices belongs to the same number q of edges. The class of such polyhedra we denote by $M_{p,q}$ and for a particular manifold of genus g we write $\{p,q;g\}$. For $g=0$ there are the 5 Platonic solids, for $g=1$ there are infinitely many $\{4,4;1\}$ and $\{3,6;1\}$ and no other regular tori. For $g \geqslant 2$ regular manifolds behave quite different from $g=0$ and $g=1$. Especially they have unique f-vectors and simple minimality properties. In our paper [1] we look for those classes $M_{p,q}$ for which all (resp. all but finitely many) $\{p,q;g\}$ exist. The main result is

Theorem. *There exist*:

(a) $\{3,7;g\}$ *for* $g \geqslant 2$,

(b) $\{3,8;g\}$ *for* $g \geqslant 4$,

(c) $\{4,5;g\}$ *and* $\{5,4;g\}$ *for* $g=5,7$ *and* $g \geqslant 9$,

(d) $\{3,9;g\}$; $\{4,6;g\}$ *and* $\{6,4;g\}$ *for* $g=6,9,10$ *and* $g \geqslant 12$.

In what follows we show simple examples of the seven classes mentioned except the cases $M_{3,8}$ and $M_{3,9}$, which can be obtained by local subdivisions from $M_{6,4}$ and $M_{4,6}$. Especially we show two examples of $\{4,5;7\}$ with (of course) the same f-vector (48, 120, 60) but different symmetry group. In Figs. 1, 3 and 5 $S_2 \times S_4$ (S_4: octahedral group, S_2: a suitable inversion) acts transitively on the 48 vertices (resp. faces).

Reference

[1] P. McMullen, Ch. Schulz and J.M. Wills, Equivelar polyhedral manifolds in E^3, Israel J. Math. 41 (1982) 331–346.

Received 26 June 1981

Annals of Discrete Mathematics 20 (1984) 335–339
North-Holland

OPEN PROBLEMS

Editor: J. ZAKS
University of Haifa, Israel

1 (P. Duchet). Prove the following generalization of Fáry's Theorem: "Every graph which is representable on surface S (in $\mathbb{R}^3$) is representable in such a way that every line (corresponding to edges) is a geodesic."

2 (R. Häggkvist). Let G be a graph on $4n$ vertices each of degree at least $2n + o(n)$. Prove, or disprove, that G contains a set of n vertex-disjoint four-cycles.

3 (A. Kotzig). Let n and r be integers such that $n \geq 5$; $2 \leq r \leq (n+1)/2$ and let us define the graph $P_n(r)$ as follows: $V(P_n(r)) = \{v_1, v_2, \ldots, v_n\}$ and $E(P_n(r)) = \{(v_i, v_{i+1}), (v_i, v_{i+r}) \mid i = 1, \ldots, n\}$, where the indices should be taken modulo n. Clearly $P_n(r)$ is a 4-regular and edge-4-connected graph. It has been shown that for every admissible value of n, the graph $G = P_n(2)$ has the following property (we shall call it the "H-property"): *At least one factor in an arbitrary 2-factorization of G is a Hamiltonian circuit of G.* It is known (and the result will be published) that with a very small number of exceptions the following assertion is true: "If $G = P_n(r)$, where $2 \neq r \neq (n+1)/2$, then G has not the H-property." Clearly, $P_{2k+1}(k)$ is isomorphic with $P_{2k+1}(2)$ and therefore $P_{2k+1}(k)$ has the H-property.

Problem. Do there exist other (= non-isomorphic with any $P_n(r)$) 4-regular edge-4-connected graphs with the H-property?

4 (A. Kotzig). Let n be an arbitrary integer > 2. Let $D^{(n)} = \{D_1^{(n)}, D_2^{(n)}, \ldots, D_{r(n)}^{(n)}\}$ be the set of all decompositions of K_{2n+1} into n Hamiltonian circuits. Let $G(D^{(n)})$ be the graph constructed as follows: $D^{(n)}$ is the vertex-set of $G(D^{(n)})$ and the vertices $D_i^{(n)}$ and $D_j^{(n)}$ of $D^{(n)}$ $(i \neq j)$ are joined by an edge in $G(D^{(n)})$ if and only if $D_i^{(n)}$ and $D_j^{(n)}$ have exactly $n-2$ Hamiltonian circuits in common.

Problem. What is the smallest n for which the graph $G(D^{(n)})$ is disconnected? Does there exists such an n?

5 (A. Kotzig). Let n be an integer > 4 and let $\mathscr{G}_n$ be the set of all the directed graphs G with n vertices and with the property that for every vertex of G the following holds: the indegree equals the outdegree and both equal two. Let $h(G)$ denote the number of distinct decompositions of G into two directed Hamiltonian cycles, and define the value of $h(n)$ by: $h(n) = \max\{h(G) \mid G \in \mathscr{G}_n\}$. One can easily find that $h(n) \leqslant 2^{-1+(n/2)}$ and for $n = 4k$ graphs $G \in \mathscr{G}_n$ were found satisfying $h(G) \geqslant 2^{-2+(n/2)}$.

Problem. Can $h(n)$ be greater than 2^{r-2}, where $r = 2[n/4]$?

6 (D. Larman). Let S be a family of subsets of n distinct objects, such that any two subsets overlap in at least k objects. Can S be divided into n sets $S_1, \ldots, S_n$ such that any two sets in S_i overlap in at least $k+1$ objects? In particular, prove or disprove it under the extra assumption that each member of S has the same cardinality ($\geqslant k+1$).

7 (D. Larman). Let P be a convex polygon in E^2 with diameter 1 and 2^n equal sides. For each side (a, b) of P there exists a vertex c of P with $\|a-c\| = \|b-c\| = 1$. Does P exist for $n \geqslant 3$? (P does not exist when $n = 2$.)

8 (J. Srivastava). Let m and N be positive integers, and let $K(N \times m)$ be an arbitrary (0, 1) matrix. Let w_m be the set of integers $\{1, 2, \ldots, m\}$ and let Ω_m be the class of all the 2^m distinct subsets of w_m, so that Ω_m includes the empty set (denoted by μ). Thus $\Omega_4 = \{\mu, 1, 2, 3, 4;\ 12, 13, 14, 23, 24, 34;\ 123, 124, 234;\ 1234\}$. Let the ith ($i = 1, \ldots, m$) column of K be denoted by $\mathbf{k}_i$, and let K be assumed to be over $GF(2)$. Let $A\,(N \times 2^m)$ be the (0, 1) matrix whose columns correspond to the elements of Ω_m, and which is obtained as follows. The column of A corresponding to μ has 0 everywhere. Also, if $(i_1, \ldots, i_u)$, $1 \leqslant u \leqslant m$, is any element of Ω_m, then the column of A corresponding to $(i_1, \ldots, i_u)$ is $(\mathbf{k}_{i_1} + \mathbf{k}_{i_2} + \cdots + \mathbf{k}_{i_u})$. Let A^* be the matrix over the real field obtained from A by replacing 0 by 1 and 1 by (-1). Now, if t is a positive integer, then any matrix B is said to have the property P_t, if every set of t columns of B is linearly independent. Thus, for certain choices of K, the matrix A^* will have property P_t. There are two questions:

(1) What is the minimum value of N for which there exists a matrix $K(N \times m)$, such that $A^*(N \times 2^m)$ has P_t.

(2) Obtain a K with minimal N such that A^* has P_t. ("Any person who obtains a complete solution to the above problem (for general m and t) before 19 March 1991 will receive an award of $1000 from J.S., within one month after presentation of the solution to J.S.")

9 (F. Hering). On the complete bipartite graph $K_{n,m}$ with $n+m$ vertices a two-colouring π of the edges is *nested*, if the vertices of any two disjoined edges carrying the same colour are joined by at least one additional edge of the same colour. Let $a_{r,m,n}(\pi)$ denote the number of paths on $r+1$ vertices, such that the colours of the r edges of that path alternate. Then $a_{r,m,n}$ is the maximum of $a_{r,m,n}(\pi)$ over all nested colourings π. The problem is to determine $a_{r,m,n}$. This problem has been posed by the author in the language of (0, 1)-sequences in *Ann. Math. Monthly* 81 (1974) 883. A partial solution for small r and small $m-r$ exists by H. Passing; Über ein kombinatorisches Problem bei Dualzahlen, Dissertation, Bonn (1973) [in German].

10 (G. Kalai). Let S be a finite set of points in $\mathbb{R}^d$. For $k \geq 1$ define:

$$f(k, S) = \max \left\{ \dim \bigcap_{i=1}^{k} \operatorname{conv} S_i : (S_1, \ldots, S_k) \text{ a partition of } S \right\} .$$

Conjecture. $\sum_{k=1}^{|S|} f(k, S) \geq 0$.

11 (S. Schreiber). Let $C_1 = (v_0, v_1, \ldots, v_{p-1})$ be a Hamiltonian cycle of the complete graph K_p, where $p = 2m+1$ is an odd prime. For every k, $2 \leq k \leq m$, the edges connecting pairs of vertices at distance k in C_1 form another cycle, to be denoted by C_k. If one partitions the vertices into two nonempty sets W and B, and colours them *white* and *black*, respectively, and in addition each edge between two vertices of the same colour is coloured *red* and each edge between vertices of unlike colours is coloured *blue*, it is trivial to show that for every k, $1 \leq k \leq m$, each cycle C_k contains an even number $2b_k$ of blue edges.

Call a sequence $b = (b_1, b_2, \ldots, b_m)$ of $m = (p-1)/2$ integers *nice* if it can be obtained from a vertex-and-edge colouring of K_p as described above; clearly, the same sequence will result if one:

(1) interchanges B and W,

(2) permutes the vertices cyclically,

(3) reverses their cyclical order.

Queries:

(α) Given a sequence of m positive integers, can one decide whether it is nice?

(β) Given a nice sequence, can one recover the partition W, B from it (up to the modifications (1)–(3) above)?

Known necessary conditions:

(a) $1 \leqslant b_i \leqslant m$ for $1 \leqslant i \leqslant m$;

(b) $\sum_{i=1}^{m} b_i = (p^2 - (|W| - |B|)^2)/8$;

(c) for all $\omega \neq 1$, $\omega^p = 1$, and $1 \leqslant k \leqslant m$, set $c_k(\omega) = \omega^k + \omega^{-k}$; then $\sum_{k=1}^{m} b_k c_k(\omega) \leqslant 0$.

Thus, for given $||W| - |B||$, the lattice point b lies on the hyperplane (b) and inside the polytope given by the inequalities (a) and (c).

12 (Y. Kupitz). The d-cube Q^d has 2^d vertices and no triangular 2-face. Does every d-polytope ($d \geqslant 2$) with fewer than 2^d vertices have a triangular 2-face? (True for $d = 2, 3$.)

13 (Y. Kupitz).

Conjecture. If V is any finite subset of $\mathbb{R}^2$, not necessarily in general position, then there is a line l that contains at least two points of V, such that

$$|\#(V \cap l^+) - \#(V \cap l^-)| \leqslant 1,$$

where l^+ and l^- are the two open halfplanes determined by l.

14 (M.C. Golumbic and A.I. Reisner). What is the maximum number N of circles having diameter 1 which may be packed into the circle which circumscribes a 4×4 square. We have a configuration for which $N = 24$ and conjecture that this is the optimum.

This problem was suggested to the submitters by a discussion in tractate Sukkah of the Talmud dealing with the proper dimensions of a sukkah and in particular of a round sukkah. A sukkah is the temporary booth or shelter with a roof of branches and leaves which is built to celebrate the Jewish holiday of Sukkot. Here a person is assumed to require the area of a unit circle, and our circumscribing sukkah will accommodate 24 people. (See also W.M. Feldman, *Rabbinical Mathematics and Astronomy*, Sepher-Hermon Press, 1978, p. 28.)

15 (P. Erdös). Let n_r be the smallest integer for which $n_r \rightarrow (3)_r$. In other words, n_r is the smallest integer for which if one colors the edges of the complete graph $K(n_r)$ by r colors there always is a monochromatic triangle. Is it then true that if we color the edges of $K(n_r)$ by $r+1$ colors, there either is a monochromatic triangle or a vertex which is incident to edges of all the $r+1$ colors?

This is very easy for $r = 2$ and with a little trouble can be proved for $r = 3$, but is open for $r \geq 4$. One of the difficulties is that n_r is not known for $r > 3$; in fact, it is not even known if $n_r^{1/r}$ tends to infinity.

Originally I conjectured that if $\mathcal{G}$ is any graph for which $\mathcal{G} \rightarrow (3)_r$, then if we color the edges of $\mathcal{G}$ by $r+1$ colors, there either is a monochromatic triangle or a vertex with edges of all the colors incident to it. This conjecture was very soon disproved (independently) by F. Chung and Enamoto for $r = 2$.

16 (P. Erdös). Let $\mathcal{G}(n)$ be a Hamiltonian graph of n vertices, each vertex has degree $> 2n/5$, and further $\mathcal{G}(n)$ is not bipartite. Schmeichel conjectured that then $\mathcal{G}(n)$ is pancyclic.

I proved that there is an absolute constant c so that for every $3 \leq k < cn$ our $\mathcal{G}(n)$ has a circuit C_k. The real difficulty seems to be to prove that $\mathcal{G}(n)$ contains a C_{n-1}. This led me to ask the following question. Let $n = 2m + 1$. What is the smallest $f(n)$ for which every Hamiltonian $\mathcal{G}(n)$ each vertex of which has degree $\geq f(n)$ has a circuit C_{n-1}? If Schmeichel's conjecture holds, $f(n) \geq (2n/5) + 1$ suffices, but it is not impossible that $f(n)$ can be much smaller.

In trying to prove Schmeichel's conjecture I proved that if $\mathcal{G}(n)$ contains a C_{2k+1} and each vertex of $\mathcal{G}(n)$ has degree greater than $[n/2(2k+1)] + c$, then $\mathcal{G}(n)$ also contains a C_{2k+3}. Apart from the value of c, this result is the best possible.

Remark. Problem 5 appears in a revised version (8 August 1981). R. Häggkvist remarked (21 March 1981) that the original version of the problem can be corrected to its present form; he proved in his remark that the maximum number of Hamilton decompositions is at most $2^{n/2-1}$, and gave an example of a directed graph on $4p$ vertices having 2^{2p-2} distinct Hamilton decompositions. Kotzig's revision was independent of Häggkvist's remark.